Methods in Molecular Biology™

Series Editor
John M. Walker
School of Life Sciences
University of Hertfordshire
Hatfield, Hertfordshire, AL10 9AB, UK

For further volumes:
http://www.springer.com/series/7651

Protein Microarrays

Methods and Protocols

Edited by

Ulrike Korf

Division of Molecular Genome Analysis, German Cancer Research Center (DKFZ), Heidelberg, Germany

Editor
Ulrike Korf
Division of Molecular Genome Analysis
German Cancer Research Center (DKFZ)
Heidelberg, Germany
u.korf@dkfz-heidelberg.de

ISSN 1064-3745 e-ISSN 1940-6029
ISBN 978-1-61779-285-4 e-ISBN 978-1-61779-286-1
DOI 10.1007/978-1-61779-286-1
Springer New York Dordrecht Heidelberg London

Library of Congress Control Number: 2011934254

Printed on acid-free paper

Humana Press is part of Springer Science+Business Media (www.springer.com)

Preface

Proteins are involved in almost any aspect of cellular function. The cellular proteome is subjected to a steady flow of dynamic changes, and therefore is a very suitable readout for the functional properties of a cell or an organism. Proteins, for example, build the cellular architecture, and are essential components of membranous compartments confining a cell, as well as subcellular organelles. Networks of tightly regulated enzymes are in command of the energy supply, and provide molecular building blocks, such as carbohydrates, lipids, and nucleic acids. Other proteins are involved in replication and transcriptional processes, and assist in the translation of new proteins. Proteins in extracellular fluids maintain the communication between cells of a tissue as well as within an organism and may serve as disease biomarkers. The number of different proteins encoded by the genome is increased by at least an order of magnitude, due to the introduction of posttranslational modifications, such as glycosylation, lipid-modifications, acetylation, and by protein phosphorylation which is the best-studied mode of cellular regulation.

Understanding protein function and the regulation of signaling networks requires large-scale efforts which enable the dynamic analysis of numerous samples in parallel. Progress in functional proteomics has been limited for a long time, partially because of limitations in assay sensitivity and sample capacity. Protein microarrays have the ability to overcome these limitations so that a highly parallel analysis of hundreds of proteins in thousands of samples is attainable. Advancements in the field of robotics and signal detection have facilitated an increase in sensitivity and sample capacity and, therefore, have contributed to the evolution of an increasing number of robust protein microarray applications. Thus, due to the robustness and flexibility of this experimental platform, diverse applications can now be implemented in principles of different types of biochemical assays.

This volume presents an up-to-date collection of robust strategies in the field of protein microarrays, and summarizes recent advantages in the field of printing technologies, the development of suitable surface materials, as well as detection and quantification technologies. Parallel to the advancement of wet-lab techniques, new software tools were developed for data analysis in order to deal with large data sets generated by protein microarray applications.

Thanks to all article authors for taking the time to prepare a chapter for this book, the series editor for shaping the idea for this volume, people at Springer for their uncomplicated and helpful advice, and special thanks to my family for their patience and cooperation while I edited the articles in this book to their completion.

I am confident that this book will stimulate the application and further advancement of this powerful technology in labs worldwide. I am very much looking forward to the future of protein microarray-based applications.

Heidelberg, Germany *Ulrike Korf*

Contents

Part II Antibody Microarrays

Part III Protein Microarrays

Part IV Sample Immobilization Strategies

Contributors

MOHAMED S.S. ALHAMDANI • *Functional Genome Analysis, German Cancer Research Center (DKFZ), Heidelberg, Germany*
KRISTI AMBROZ • *Director of Biotechnology Reagent Operations and Technical Support, LI-COR, Lincoln, NE, USA*
JOHN AUSTIN • *Aushon BioSystems Inc., Concord, MA, USA*
ANDREA BAUER • *Functional Genome Analysis, German Cancer Research Center (DKFZ), Heidelberg, Germany*
MIKAEL BAUER • *Department of Biophysical Chemistry, Lund University, Lund, Sweden*
KARL-FRIEDRICH BECKER • *Institut für Pathologie, Technische Universität München, Munich, Germany*
CLAUDIO BELLUCO • *CRO-IRCCS, National Cancer Institute, Aviano, Italy*
DANIELA BERG • *Institut für Pathologie, Technische Universität München, Munich, Germany*
JONATHAN M. BLACKBURN • *Division of Medical Biochemistry & Institute for Infectious Disease & Molecular Medicine, University of Cape Town, Cape Town, South Africa*
CARL A.K. BORREBAECK • *Department of Immunotechnology, Lund University, Lund, Sweden; CREATE Health, Lund University, Lund, Sweden*
JAN C. BRASE • *Division of Molecular Genome Analysis, German Cancer Research Center (DKFZ), Heidelberg, Germany*
DOLORES J. CAHILL • *Translational Science, School of Medicine and Medical Sciences, UCD Conway Institute, University College Dublin, Dublin, Ireland*
KEVIN R. COOMBES • *Departments of Bioinformatics and Computational Biology, The University of Texas M.D. Anderson Cancer Center, Houston, TX, USA*
DANIEL D'DORAZIO • *Center for Proteomic Chemistry, Novartis Pharma AG, Basel, Switzerland*
HOLGER ERFLE • *BioQuant, University of Heidelberg, Heidelberg, Germany*
ANASTASIA ESKOVA • *BioQuant, University of Heidelberg, Heidelberg, Germany*
VIRGINIA ESPINA • *Center for Applied Proteomics and Molecular Medicine, George Mason University, Manassas, VA, USA*
DORIANO FABBRO • *Center for Proteomic Chemistry, Novartis Pharma AG, Basel, Switzerland*
KURT FELLENBERG • *Chair of Proteomics and Bioanalytics, Technical University Munich, Freising, Germany*
FRANK GÖTSCHEL • *Division of Molecular Genome Analysis, German Cancer Research Center (DKFZ), Heidelberg, Germany*
BRIAN B. HAAB • *Van Andel Research Institute, Grand Rapids, MI, USA*
DOROTHEA HAASEN • *Center for Proteomic Chemistry, Novartis Pharma AG, Basel, Switzerland*
DUNCAN HALL • *Arrayjet Ltd., MIC, Roslin, UK*
MINGYUE HE • *The Babraham Institute, Cambridge, UK*

Frauke Henjes • *Division of Molecular Genome Analysis, German Cancer Research Center (DKFZ), Heidelberg, Germany*
Jörg D. Hoheisel • *Functional Genome Analysis, German Cancer Research Center (DKFZ), Heidelberg, Germany*
Antonia H. Holway • *Associate Director, Translational Research, Lahey Clinic, Burlington, MA, USA*
Anette Jacob • *Functional Genome Analysis, German Cancer Research Center (DKFZ), Heidelberg, Germany*
Anika Jöcker • *Division of Molecular Genome Analysis, German Cancer Research Center (DKFZ), Heidelberg, Germany*
Ulrike Korf • *Division of Molecular Genome Analysis, German Cancer Research Center (DKFZ), Heidelberg, Germany*
Steven M. Kornblau • *Departments of Stem Cell Transplantation and Cellular Therapy, The University of Texas M.D. Anderson Cancer Center, Houston, TX, USA*
Marisa Chong Kwan • *Arrayjet Ltd., MIC, Roslin, UK*
Sara Linse • *Department of Biophysical Chemistry, Lund University, Lund, Sweden*
Lance A. Liotta • *Center for Applied Proteomics and Molecular Medicine, George Mason University, Manassas, VA, USA*
Katharina Malinowsky • *Institut für Pathologie, Technische Universität München, Munich, Germany*
Heiko Mannsperger • *Division of Molecular Genome Analysis, German Cancer Research Center (DKFZ), Heidelberg, Germany*
Georg Martiny-Baron • *Center for Proteomic Chemistry, Novartis Pharma AG, Basel, Switzerland*
Iain McWilliam • *Arrayjet Ltd., MIC, Roslin, UK*
Claudius Mueller • *Center for Applied Proteomics and Molecular Medicine, George Mason University, Manassas, VA, USA*
Peter Nilsson • *SciLifeLab Stockholm, KTH – Royal Institute of Technology, Tomtebodav, Sweden*
Satoshi S. Nishizuka • *Molecular Therapeutics Laboratory, Department of Surgery, Iwate Medical University School of Medicine, Uchimura, Japan*
John K. O'Brien • *Wellcome Trust Genome Campus, Cambridge, UK*
David J. O'Connell • *Conway Institute of Biomolecular & Biomedical Research, University College Dublin, Dublin, Ireland*
Sara L. O'Kane • *Conway Institute of Biomolecular & Biomedical Research, University College Dublin, Dublin, Ireland*
Emanuel F. Petricoin III • *Center for Applied Proteomics and Molecular Medicine, George Mason University, Manassas, VA, USA*
Mariaelena Pierobon • *Center for Applied Proteomics and Molecular Medicine, George Mason University, Manassas, VA, USA*
Bilge Reischauer • *Institut für Pathologie, Technische Universität München, Munich, Germany*
Jürgen Reymann • *BioQuant, University of Heidelberg, Heidelberg, Germany*
Özgür Sahin • *Division of Molecular Genome Analysis, German Cancer Research Center (DKFZ), Heidelberg, Germany*

MARTA SANCHEZ-CARBAYO • *Tumor Markers Group, Spanish National Cancer Research Center, Madrid, Spain*
URSULA SAUER • *Health & Environment Department, Biosensor Technologies, AIT Austrian Institute of Technology GmbH, Seibersdorf, Austria*
CHRISTINA SCHOTT • *Institut für Pathologie, Technische Universität München, Munich, Germany*
CHRISTOPH SCHRÖDER • *Functional Genome Analysis, German Cancer Research Center (DKFZ), Heidelberg, Germany*
JOCHEN M. SCHWENK • *SciLifeLab Stockholm, KTH – Royal Institute of Technology, Tomtebodav, Sweden*
AUBREY SHOKO • *Centre for Proteomic & Genomic Research, University of Cape Town, Cape Town, South Africa*
JOHANNA SONNTAG • *Division of Molecular Genome Analysis, German Cancer Research Center (DKFZ), Heidelberg, Germany*
VYTAUTE STARKUVIENE • *BioQuant, University of Heidelberg, Heidelberg, Germany*
ODA STOEVESANDT • *Protein Technology Group, Babraham Bioscience Technologies Ltd, Cambridge, UK*
HOLGER SÜLTMANN • *Division of Molecular Genome Analysis, German Cancer Research Center (DKFZ), Heidelberg, Germany*
MICHAEL J. TAUSSIG • *Protein Technology Group, Babraham Bioscience Technologies Ltd, Cambridge, UK*
STEFAN UHLMANN • *Division of Molecular Genome Analysis, German Cancer Research Center (DKFZ), Heidelberg, Germany*
JOHANNES VOSHOL • *Center for Proteomic Chemistry, Novartis Pharma AG, Basel, Switzerland*
CHRISTER WINGREN • *Department of Immunotechnology, Lund University, Lund, Sweden; CREATE Health, Lund University, Lund, Sweden*
CLAUDIA WOLFF • *Institut für Pathologie, Technische Universität München, Munich, Germany*
TINGTING YUE • *Van Andel Research Institute, Grand Rapids, MI, USA*

Part I

Reverse Phase Protein Arrays

Chapter 1

Reverse Phase Protein Microarrays for Clinical Applications

Mariaelena Pierobon, Claudio Belluco, Lance A. Liotta, and Emanuel F. Petricoin III

Abstract

Phosphorylated proteins represent one of the most important constituents of the proteome and are under intense analysis by the biotechnology and pharmaceutical industry because of their central role for cellular signal transduction. Indeed, alterations in cellular signaling and control mechanisms that modulate signal transduction, functionally underpin most human cancers today. Beyond their central role as the causative components of tumorigenesis, these proteins have become an important research focus for discovery of predictive and prognostic biomarkers. Consequently, these pathway constituents comprise a powerful biomarker subclass whereby the same analyte that provides prediction and/or prognosis is also the drug target itself: a theranostic marker. Reverse phase protein microarrays have been developed to generate a functional patient-specific circuit "map" of the cell signaling networks based directly on cellular analysis of a biopsy specimen. This patient-specific circuit diagram provides key information that identifies critical nodes within aberrantly activated signaling that may serve as drug targets for individualized or combinatorial therapy. The protein arrays provide a portrait of the activated signaling network by the quantitative analysis of the phosphorylated, or activated, state of cell signaling proteins. Based on the growing realization that each patient's tumor is different at the molecular level, the ability to measure and profile the ongoing phosphoprotein biomarker repertoire provides a new opportunity to personalize therapy based on the patient-specific alterations.

Key words: Proteomics, Biomarkers, Cell signaling, Phosphoproteins, Oncology, Personalized therapy

1. Introduction

The era of personalized therapy for cancer treatment has begun in earnest with new FDA approved molecularly targeted therapeutics coming on-line on a yearly basis. In the near future, the treating oncologist will have a large armamentarium of precise therapeutics to select from. Indeed, since drugs such as imatinib mesylate and traztuzumab have had a dramatic impact on GIST, CML, and c-erbB2+ breast cancers, respectively, the emphasis for patient therapy

Ulrike Korf (ed.), *Protein Microarrays: Methods and Protocols*, Methods in Molecular Biology, vol. 785,
DOI 10.1007/978-1-61779-286-1_1,

decisions will shift from the therapy itself to the biomarkers that are used to stratify and personalize the therapy. These biomarkers will serve as "gatekeepers" for therapeutic decision-making processes as a companion diagnostic and provide the physician with critical missing information on helping to guide which targeted therapies to consider. Consequently, the discovery of biomarkers that provide predictive and prognostic ability for patient stratification/therapy selection, that is the companion diagnostics of the future, is taking on an increasingly intense focus in all areas of translational research. Because of the central, causal role that alterations in cell signaling and aberrant cell signaling have in tumorigenesis (1–8), phosphoprotein pathway biomarkers may be among the most important class of biomarkers for prediction, prognosis, and patient-tailored therapy (4, 8–10). The hope that gene expression analysis will provide a direct route to unraveling and elucidating ongoing protein signaling events and provide an effective molecular surrogate for protein pathway biomarkers has largely dissipated as recent studies have revealed little correlation between gene expression and protein expression (11, 12). Moreover, protein expression levels themselves are not able to predict the phosphorylation levels of signaling activation, which points to the need for technologies that can directly assess and measure the activation state of the cellular "circuitry" and generate the pathway biomarker information that is critically needed.

2. Cell Signaling Activation Alterations in Human Cancer

Post-translational protein modifications (PTM), mainly phosphorylation, are now known to control the kinase-driven signaling networks that are abarrently activated in human cancers (13–27). The vast majority of protein phosphorylation occurs on serine and threonine residues with the remainder (approximately 10%) occurring on tyrosine. Many growth factor receptor (e.g., vascular endothelial growth factor receptor (VEGFR), epidermal growth factor receptor (EGFR), c-erbB2)-mediated signaling are based on receptors that are themselves kinase enzymes, and mainly utilize tyrosine phosphorylation-based PTM. Upon ligand binding, the receptors dimerize, self-phosphorylate, which then form structural alterations and new binding sites for downstream protein kinase interactions (13–27). Downstream signaling cascades are comprised of enzymatic networks of kinases and phosphatases and their substrates, linking together based on defined phosphorylation events that then provide the necessary substrates for structural interactions such as through SH2 and SH3 domains (13–27). How the cell orchestrates coordinate control of these signaling networks is also under intense investigation, and new approaches using mathematical modeling of the networks are now being explored in order to both reconstruct signaling networks de novo and/or exploit the pathway architecture to identify optimal therapeutic strategies (28–35). While cancer, at a functional level, is a disease of the signaling pathway network, the complexity of the human "kinome," comprised of less than a thousand proteins (36) is of relatively low-dimensional space compared to the genome or the entire proteome. Recent extensive genomic analysis of individual human tumor specimens has revealed a complex heterogeneous portrait of hundreds of independent somatic genetic mutations (5–7). Which of these specific mutations represent the tipping points for transition into different stages of tumorigenesis and metastasis remains unknown. While the mutational portraits of cancer appear complex and highly heterogeneous, the cells containing mutations that ultimately and functionally provide a survival advantage are selected out. This functional selection is manifest in cell signaling pathway changes that are responsible for altered cell growth, death, motility, differentiation, and metabolism. As complex as signaling networks may be in the myriad of possible connections and permutations of protein–protein linkages, cell signaling ultimately must abide by chemistry and physical heuristics. Based on this, one would predict that disparate tumor types, defined in the past by location and histology, would share common signaling alteration "themes" regardless of the apparent differences at the somatic mutational backdrop within each patient. Indeed, this appears to be the case as a growing cadre of data points to an entirely new categorization of human cancer, based on functional protein pathway activation themes, and not on mutational status, location, tumor grading, and gene expression. An example of this is the ubiquitous nature of AKT/mTOR pathway derangements, growth factor receptor-mediated signal pathway activation, and ras–raf–ERK network activation in a large number of human cancers, regardless of location and organ microenvironment (37–42).

3. Reverse Phase Protein Microarrays: Enabling Technology for Patient-Tailored Therapeutics

Protein microarrays represent a technology platform that could address the limitations of previous platforms through the analysis and quantitative measurement of many phosphoprotein biomarkers at once from a clinical biopsy specimen. In particular, the reverse phase protein microarray (RPMA) is proving to be a powerful enabling technology for the analysis of clinical material for pathway phosphoprotein biomarker profiling (43–52) (Fig. 1).

In contrast to a forward phase format (e.g., antibody array) where the analyte detecting molecule is immobilized, with the RPMA format, cellular lysates from individual test samples are printed directly and immobilized on the array surface such that a finished array could be comprised of lysates from cells from different

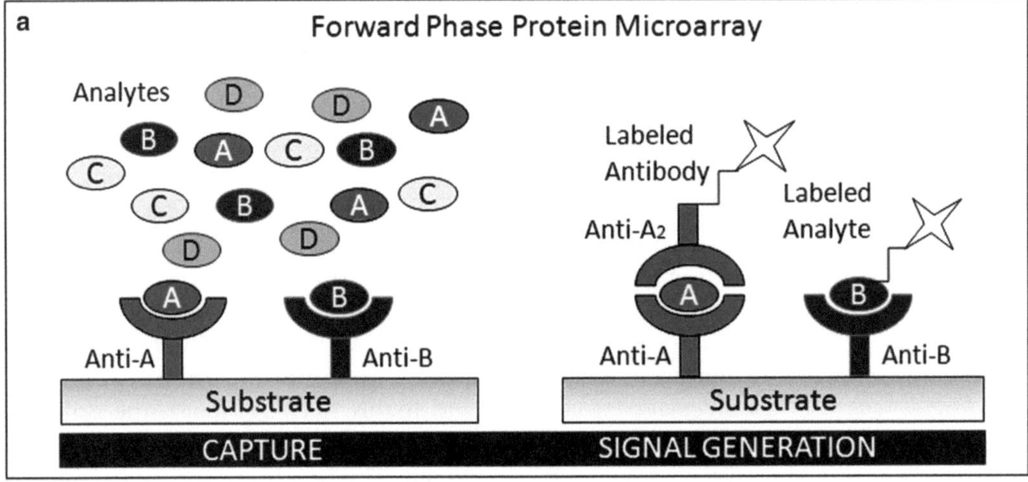

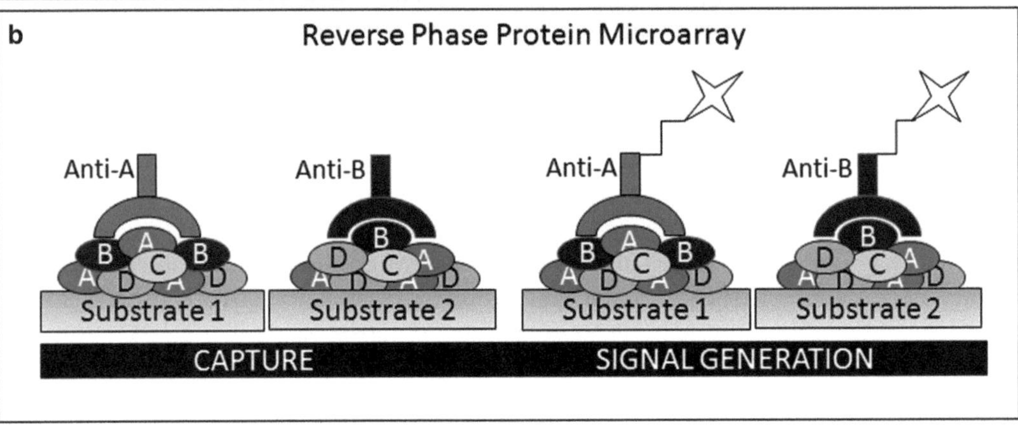

Fig. 1. Comparison between forward phase microarray (**a**) and reverse phase protein microarray (RPMA) (**b**). While the forward phase microarray format is based on immobilization of analyte capture reagents (e.g., antibodies) into a solid support, the RPMA is characterized by immobilization of analytes into the substrate, allowing direct comparison of hundreds of samples. Instead of a sandwich assay-based approach that requires two well-performing analyte capture reagents, the reverse phase array requires only one well-performing analyte detection reagent.

patient biopsy samples or cellular lysates. Depending on the size of the pin used to print the lysates, which normally vary between 80 and 400 μm, it is possible to print a few hundred to several thousand spots on each slide. Since each printing deposits as little as 1–5 nl, it is possible to as many as 100 slides from a lysate of 1,000 microdissected cells. Each slide is then incubated with one specific primary antibody, and a single analyte endpoint is measured and directly compared across multiple samples on each slide (Fig. 2). Direct quantitative measurements can be achieved by printing on each array high and low controls, and a series of calibrators (prepared in dilution series) that serve as an internal standard curve. While the RPMA format was initially designed for colorimetric detection, recently, adaptation to fluorescent detection (53) has increased its capacity by obviating printing in dilution curves

Reverse Phase Protein Microarray Layout

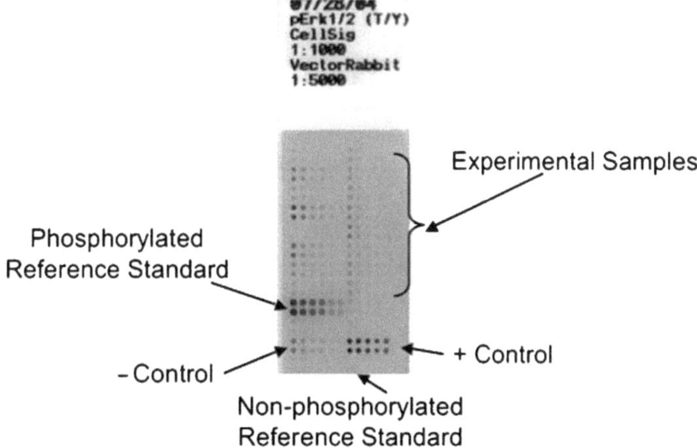

Fig. 2. An example of a typical RPMA layout. Denatured cellular lysates, either from cell lines or whole tissue, or from laser capture microdissected material, is spotted directly onto a nitrocellulose-coated slide, and multiple samples are simultaneously probed with the same antibody. Each sample may be printed in a step-wise dilution curve (shown) or as a single replicate spot (not shown) with colorimetric or florescent detection, respectively. Similar to an ELISA or immunoassay, high and low controls and calibrators are printed on every slide with the RPMA format to ensure inter- and intra-assay reproducibility, process QA/QC, and fidelity of data generated.

(necessary for colorimetric detection) and by increasing the within-spot dynamic range about 300 times. The RPMA format is capable of extremely sensitive analyte detection, for example, with reported levels of a few hundred molecules of EGFR per spot, and a CV of less than 10% (51). The sensitivity of detection for the RPMAs is such that low abundance phosphorylated pathway biomarkers can be measured from a spotted lysate representing less than 10 cell equivalents (51), which is critical if the starting input material is only a few hundred cells from a needle biopsy or fine needle aspirate specimen. Since the RPMA technology requires only one specific antibody for each analyte (e.g., phospho-specific antibodies), the ability to perform quantitative broad profiling measurements of multiplexed phosphoprotein pathway biomarkers concomitantly is currently unmatched. The platform is dependent on the availability of high quality, specific antibodies, particularly those recognizing PTM or active states of proteins. Antibody availability is a major limiting factor for the successful implementation of any immunoassay-type platforms.

Because human tissues are composed of interacting cell populations, such as stromal, epithelial, and immune cells, RPMAs provide an opportunity for pathway marker studies in each cellular compartment within the context of the tumor microenvironment (44, 54). The use of laser capture microdissection (LCM) (55) combined with RPMA enables the facile detailed analysis of discreet cell populations within a clinical biopsy specimen and provides cell signaling analysis and phosphoprotein pathway marker profiling (43–52). Indeed, recent analysis whereby pathway profiling was performed comparing patient-matched undissected and LCM procured tumor epithelium revealed significant and numerous differences in pathway activation portraits between the two (47, 56). Despite employing case studies where the tumor epithelium comprised over 75% of the cellular content, lysates from the undissected whole tissue lysates were not able to accurately recapitulate the pathway conclusions obtained from the LCM pure tumor epithelium (47, 56).

Key technological components of the RPMA offer unique advantages over other array-based platforms such as tissue arrays (57) or antibody (forward phase) arrays (58). The RPMA can employ denatured lysates, so that antigen retrieval, a significant limitation for tissue arrays, antibody arrays, and immunohistochemistry technologies, is not problematic. RPMAs only require a single class of antibody per analyte protein and do not require direct tagging of the protein as readout for the assay. Other technologies, such as suspension bead array platforms, have significant limitations in the portfolio of analytes that can be measured, even in multiplex, because of the requirement of a two-site assay. The ability to generate quantitative data from minute quantities of cellular input without a two-site assay also enables a marked improvement in reproducibility, sensitivity, and robustness of the assay over other techniques.

4. Use of Reverse Phase Arrays for Signal Pathway Profiling of Human Cancer

Recent case studies demonstrate the ability of RPMA for the analysis of surgically obtained tissues and the potential for aiding in therapeutic decision-making by providing information about the activity of signaling proteins. The first published demonstration of RPMA signal pathway profiling by Paweletz et al. revealed that members of the PI3 kinase/pro-survival protein pathways are activated at the invasion front during prostate cancer progression (43). In another study, Zha et al. examined the differences in pro-survival signaling between Bcl-$2^{+/-}$ lymphomas (59). Comparison of various pro-survival proteins in Bcl-2^{+} and Bcl-2^{-} follicular lymphoma subtypes by RPMAs suggested that there are pro-survival signals independent of Bcl-2 (59). Evidence for signaling changes in colonic tumor cells undergoing epithelial mesenchymal transition (EMT) was found (50) whereby LCM procured tumor, normal epithelium, and matched stromal cells next to each compartment were compared using RPMA analysis. VanMeter et al. (46) used RPMA to analyze the signaling events from NSCLC tumor specimens from patients with and without EGFR mutations and found that a specific EGFR phosphorylation profile perfectly correlated with the presence or absence of the mutation revealing that the signaling events ultimately manifest any underpinning activating mutation.

RPMA profiling has indentified prognostic signatures in human cancer that correlate with outcome and response to therapy. A signature composed of members of the AKT-mTOR pathway was found coordinately activated in children with rhabdomyosarcoma who did not respond to therapy and progressed rapidly compared to children whose tumors were relatively quiescent within the pathway (52). A protein pathway activation signature composed of COX2-EGFR signaling networks was found differentially activated in the primary tumors of patients with colorectal cancer that appeared with synchronous liver metastasis and died rapidly vs. those patients who presented with primary cancer only (44). Such pathway marker sets are attractive therapeutic targets and go beyond prognosis alone. Indeed, shutting off those activated pathways could be a rational approach to delay or eliminate cancer recurrence in these indications.

RPMAs have also been used to compare cell signaling portraits in patient-matched primary and metastatic cancer lesions (9). Because the tissue microecology of the metastatic lesion is inherently different from the environment within the primary tumor, cell signaling events may be significantly altered depending on the site of metastasis. Since the signaling changes in the metastasis would be the most appropriate for the selection of targeted therapy due to the fact that metastasis most often determines mortality, it might be critical to develop a profile of metastatic cells themselves. In a view of the future, a patient who presents with advanced stage disease and multiple metastatic sites could be treated with a selected combination of different targeted therapies, tailored to the different signaling changes. Preliminary published data support this concept. A small case study set of three laser capture microdissected, patient-matched primary colorectal tumor cells and the corresponding cells from the hepatic metastasis (obtained simultaneously at surgery) were analyzed for the status of multiple phosphoprotein endpoints involved in mitogenesis and survival including growth factor receptors, signal transducing proteins, and nuclear transcription factors (9). Unsupervised hierarchical clustering of the data suggested that cell signaling in metastatic hepatic lesions differed significantly from the matched primary lesions, yet, appeared very similar to each other (9). Significant changes in cell signaling pathways depending on the underpinning microenvironment were also observed in a similarly constructed study of six primary ovarian tumors and patient-matched omental metastases taken simultaneously at surgery (54). In this study, signaling within the metastatic lesion was dramatically changed compared with their matched primary counterparts, with phosphorylation of c-Kit dramatically elevated in five of the six metastatic tumors compared to the primary lesions. The clinical implications that the metastatic cell signaling is so dissimilar to the primary tumor are important, if validated in further studies. Patient-tailored therapy that is designed to mitigate the metastatic process could have significant implications at the clinic. RPMAs are also well suited to the analysis of clinical trial material in that they can provide signaling network information that complements standard histological analysis of patient specimens collected before, during, and after treatment. This technology is being applied to several ongoing clinical trials in a variety of cancers.

5. Use of Reverse Phase Protein Arrays: A View to the Future

Molecular profiling of the ongoing signaling cascades produced within and as a consequence of the tumor microenvironment holds great promise in effective selection of therapeutic targets as well as patient stratification. As our understanding of human diseases such as cancer expands, we are now beginning to understand the true patient-specific nature of cancer at the molecular level (4–7). Protein-based analysis where phosphorylation-driven information can be gleaned is particularly useful in this area since these endpoints are the direct drug targets themselves. Knowledge of the activation state of these networks will provide the data needed for a rationally based formulation of targeted therapies, perhaps in combination with each other. The promise of proteomic-based profiling, that is critically distinct from gene transcript profiling, is that the resulting prognostic signatures are derived from drug targets (e.g., activated kinases) not genes, so the pathway analysis provides a direction to therapy. In effect, the phosphoproteomic pathway analysis becomes both a diagnostic/prognostic signature as well as a guide to therapeutic intervention.

References

1. Faivre S, Djelloul S, Raymond E. (2006) New paradigms in anticancer therapy: targeting multiple signaling pathways with kinase inhibitors. 1: Semin Oncol. 33(4):407–20.
2. Huang PH, Mukasa A, Bonavia R, Flynn RA, Brewer ZE, Cavenee WK, et al. (2007) Quantitative analysis of EGFRvIII cellular signaling networks reveals a combinatorial therapeutic strategy for glioblastoma. Proc Natl Acad Sci USA. 31;104(31):12867–72.
3. Engelman JA, Zejnullahu K, Mitsudomi T, Song Y, Hyland C, Park JO, et al. (2007) MET amplification leads to gefitinib resistance in lung cancer by activating ERBB3 signaling. Science. 18;316(5827):1039–43.
4. Sawyers CL. (2008) The cancer biomarker problem. Nature. 3;452(7187):548.
5. Parsons DW, Jones S, Zhang X, Lin JC, Leary RJ, Angenendt P, et al. (2008) An Integrated Genomic Analysis of Human Glioblastoma Multiforme. Science 26;321(5897):1807–12.
6. Jones S, Zhang X, Parsons DW, Lin JC, Leary RJ, Angenendt P, et al. (2008) Core signaling pathways in human pancreatic cancers revealed

by global genomic analyses. Science. Sep. 26;321(5897):1801–6.

7. Wood LD, Parsons DW, Jones S, Lin J, Sjöblom T, Leary RJ, et al. (2007) The genomic landscapes of human breast and colorectal cancers. Science. 16;318(5853):1108–13.
8. Liotta LA, Kohn EC, and Petricoin EF. (2001) Clinical proteomics: personalized molecular medicine. JAMA. 286(18):2211–4.
9. Petricoin EF 3rd, Bichsel VE, Calvert VS, Espina V, Winters M, Young L. et al. (2005) Mapping molecular networks using proteomics: a vision for patient-tailored combination therapy. J Clin Oncol. 23:3614–21.
10. Wulfkuhle JD, Edmiston KH, Liotta LA, Petricoin EF. (2006) Technology Insight: pharmacoproteomics for cancer-promises of patient-tailored medicine using protein microarrays. Nat Clin Pract Oncol. 3(5):256–68.
11. Anderson L, Seilhamer J.(1997) A comparison of selected mRNA and protein abundances in human liver. Electrophoresis 18(3–4):533–7.
12. Gygi SP, Rochon Y, Franza BR, Aebersold R. (1999) Correlation between protein and mRNA abundance in yeast. Mol Cell Biol. 19(3): 1720–30.
13. Irish JM, Hovland R, Krutzik PO, Perez OD, Bruserud Ø, Gjertsen BT, et al. (2004) Single cell profiling of potentiated phospho-protein networks in cancer cells. Cell. 23;118(2):217–28.
14. Irish JM, Anensen N, Hovland R, Skavland J, Børresen-Dale AL, Bruserud O, et al. (2007) Flt3 Y591 duplication and Bcl-2 overexpression are detected in acute myeloid leukemia cells with high levels of phosphorylated wild-type p53. Blood. 15;109(6):2589–96.
15. Stern DF. (2005) Phosphoproteomics for oncology discovery and treatment. Expert Opin Ther Targets. 9(4):851–60.
16. Moran MF, Tong J, Taylor P, Ewing RM. (2006) Emerging applications for phosphoproteomics in cancer molecular therapeutics. 1: Biochim Biophys Acta. Dec 1766(2):230–41.
17. Hunter, T. (2000) Signaling-2000 and beyond. Cell 100, 113–127.
18. Figlin RA. (2008) Mechanisms of Disease: survival benefit of temsirolimus validates a role for mTOR in the management of advanced RCC. Nat Clin Pract Oncol. 5(10):601–9.
19. Jin Q, Esteva FJ. (2008) Cross-talk between the ErbB/HER family and the type I insulin-like growth factor receptor signaling pathway in breast cancer. J Mammary Gland Biol Neoplasia. 13(4):485–98.
20. Guha U, Chaerkady R, Marimuthu A, Patterson AS, Kashyap MK, Harsha HC, et al. (2008) Comparisons of tyrosine phosphorylated proteins in cells expressing lung cancer-specific alleles of EGFR and KRAS. Proc Natl Acad Sci USA. 105(37):14112–7.
21. Cui Q, Ma Y, Jaramillo M, Bari H, Awan A, Yang S, et al. (2007) A map of human cancer signaling. Mol Syst Biol. 3:152.
22. Haura EB, Zheng Z, Song L, Cantor A, Bepler G. (2005) Activated epidermal growth factor receptor-Stat-3 signaling promotes tumor survival in vivo in non-small cell lung cancer. Clin Cancer Res. 11(23):8288–94.
23. Zandi R, Larsen AB, Andersen P, Stockhausen MT, Poulsen HS. (2007) Mechanisms for oncogenic activation of the epidermal growth factor receptor. Cell Signal. 19(10):2013–23.
24. Swanton C, Futreal A, Eisen T. (2006) Her2-targeted therapies in non-small cell lung cancer. Clin Cancer Res. 12(14 Pt 2):4377s–4383s.
25. Casalini P, Iorio MV, Galmozzi E, Ménard S. (2004) Role of HER receptors family in development and differentiation. J Cell Physiol. 200(3):343–50.
26. Wiley HS. (2003) Trafficking of the ErbB receptors and its influence on signaling. Exp Cell Res. 284(1):78–88.
27. Arteaga CL.(2002) Epidermal growth factor receptor dependence in human tumors: more than just expression? Oncologist. 7 Suppl 4:31–9.
28. Smock RG, Gierasch LM. (2009) Sending signals dynamically. Science. 324(5924):198–203.
29. Ventura AC, Jackson TL, Merajver SD. (2009) On the role of cell signaling models in cancer research. Cancer Res. 69(2):400–2.
30. Araujo RP, Liotta LA, Petricoin EF. (2007) Proteins, drug targets and the mechanisms they control: the simple truth about complex networks. Nat Rev Drug Discov. 6(11):871–80.
31. Geho DH, Petricoin EF, Liotta LA, Araujo RP. (2005) Modeling of protein signaling networks in clinical proteomics. Cold Spring Harb Symp Quant Biol. 70:517–24.
32. Iadevaia S, Lu Y, Morales FC, Mills GB, Ram PT. (2010) Identification of optimal drug combinations targeting cellular networks: integrating phospho-proteomics and computational network analysis. Cancer Res. Jul 19. [Epub ahead of print].
33. Araujo RP, Liotta LA. (2006) A control theoretic paradigm for cell signaling networks: a simple complexity for a sensitive robustness. Curr Opin Chem Biol. 10(1):81–7.
34. Araujo RP, Petricoin EF, Liotta LA. (2005) A mathematical model of combination therapy using the EGFR signaling network. Biosystems. 80(1):57–69.
35. Napoletani D, Sauer T, Struppa DC, Petricoin E, Liotta L. (2008) Augmented sparse reconstruction of protein signaling networks. J Theor Biol. 255(1):40–52.

36. Johnson SA, Hunter T. (2005) Kinomics: methods for deciphering the kinome. Nat Methods. 2(1):17–25.
37. Hennessy BT, Smith DL, Ram PT, Lu Y, Mills GB. (2005) Exploiting the PI3K/AKT pathway for cancer drug discovery. Nat Rev Drug Discov. 4(12):988–1004.
38. O'Reilly KE, Rojo F, She QB, Solit D, Mills GB, Smith D, et al. (2006) mTOR inhibition induces upstream receptor tyrosine kinase signaling and activates Akt. Cancer Res. 66(3):1500–8.
39. Grünwald V, Soltau J, Ivanyi P, Rentschler J, Reuter C, Drevs J. (2009) Molecular targeted therapies for solid tumors: management of side effects. Onkologie. 32(3):129–38.
40. Huang Z, Brdlik C, Jin P, Shepard HM. (2009) A pan-HER approach for cancer therapy: background, current status and future development. Expert Opin Biol Ther. 9(1):97–110.
41. Ramos JW. (2008) The regulation of extracellular signal-regulated kinase (ERK) in mammalian cells. Int J Biochem Cell Biol. 40(12):2707–19.
42. McCubrey JA, Steelman LS, Chappell WH, Abrams SL, Wong EW, Chang F, et al. (2007) Roles of the Raf/MEK/ERK pathway in cell growth, malignant transformation and drug resistance. Biochim Biophys Acta. 1773(8): 1263–84.
43. Paweletz CP, Charboneau L, Roth MJ, Bichsel VE, Simone NL, Chen T, et al. (2001) Reverse phase proteomic microarrays which capture disease progression show activation of pro-survival pathways at the cancer invasion front. Oncogene. 12;20(16):1981–9.
44. Pierobon M, Calvert V, Belluco C, Garaci E, Deng J, Lise M, et al. (2009) Multiplexed Cell Signaling Analysis of Metastatic and Nonmetastatic Colorectal Cancer Reveals COX2-EGFR Signaling Activation as a Potential Prognostic Pathway Biomarker. Clin Colorectal Cancer. 8(2):110–7.
45. Gulmann C, Sheehan KM, Conroy RM, Wulfkuhle JD, Espina V, Mullarkey MJ, et al. (2009) Quantitative cell signalling analysis reveals down-regulation of MAPK pathway activation in colorectal cancer. J Pathol. 218(4):514–9.
46. Vanmeter AJ, Rodriguez AS, Bowman ED, Harris CC, Deng J, Calvert VS, et al. (2008) LCM and protein microarray analysis of human NSCLC: Differential EGFR phosphorylation events associated with mutated EGFR compared to wild type. Mol Cell Proteomics. 7(10):1902–24.
47. Wulfkuhle JD, Speer R, Pierobon M, Laird J, Espina V, Deng J, et al. (2008) Multiplexed Cell Signaling Analysis of Human Breast Cancer: Applications for Personalized Therapy. J of Prot Res. 7(4):1508–17.
48. Sanchez-Carbayo M, Socci ND, Richstone L, Corton M, Behrendt N, Wulkfuhle J, et al. (2007) Genomic and Proteomic Profiles Reveal the Association of Gelsolin to TP53 Status and Bladder Cancer Progression. Am J Pathol. 171(5):1650–8.
49. Zhou, J, Wulfkuhle J, Zhang H, Gu P, Yang Y, Deng J, et al. (2007) Activation of the PTEN/mTOR/STAT3 pathway in breast cancer stem-like cells is required for viability and maintenance. PNAS. 104(41):16158–63.
50. Sheehan KM, Gulmann, C, Eichler GS, Weinstein, J, Barrett HL, Kay EW, et al. (2007) Signal Pathway Profiling of Epithelial and Stromal Compartments of Colonic Carcinoma Reveal Epithelial-Mesenchymal Transition Oncogene. 27(3):323–31.
51. Rapkiewicz A, Espina V, Zujewski JA, Lebowitz PF, Filie A, Wulfkuhle J, et al. (2007) The needle in the haystack: Application of breast fine-needle aspirate samples to quantitative protein microarray technology. Cancer. 111(3):173–84.
52. Petricoin EF, Espina V, Araujo RP, Midura B, Yeung C, Wan X, et al. (2007) Phosphoprotein Signal Pathway Mapping: Akt/mTOR Pathway Activation Association with Childhood Rhabdomyosarcoma Survival. Cancer Research. 67(7):3431–4.
53. Calvert VS, Tang Y, Boveia V, Wulfkuhle J, Schutz-Geschwender Olive DM, et al. (2004) Development of Multiplexed Protein Profiling and Detection Using Near Infrared Detection of Reverse-Phase Protein Microarrays. Clinical Proteomics. 1(1):81–90.
54. Sheehan KM, Calvert VS, Kay EW, Lu Y, Fishman D, Espina V, et al. (2005) Use of reverse-phase protein microarrays and reference standard development for molecular network analysis of metastatic ovarian carcinoma. Mol Cell Proteomics. 4, 346–55.
55. Emmert-Buck MR, Bonner RF, Smith PD, Chuaqui RF, Zhuang Z, Goldstein SR, et al. (1996) Laser capture microdissection. Science. 274(5289):998–1001.
56. Silvestri A, Colombatti A, Calvert VS, Deng J, Mammano E, Belluco C, et al. (2010) Protein pathway biomarker analysis of human cancer reveals requirement for upfront cellular-enrichment processing. Lab Invest. 90(5):787–96.
57. Avninder S, Ylaya K, Hewitt SM. (2008) Tissue microarray: a simple technology that has revolutionized research in pathology. J Postgrad Med. 54(2):158–62.
58. Haab BB. (2005) Antibody arrays in cancer research. Mol Cell Proteomics. 4(4):377–83.
59. Zha H, Raffeld M, Charboneau L, Pittaluga S, Kwak LW, Petricoin E 3rd, Liotta LA et al. (2004) Similarities of prosurvival signals in Bcl-2-positive and Bcl-2-negative follicular lymphomas identified by reverse phase protein microarray. Lab Invest. 84, 235–44.

Chapter 2

Impact of Blocking and Detection Chemistries on Antibody Performance for Reverse Phase Protein Arrays

Kristi Ambroz

Abstract

Careful selection of well-qualified antibodies is critical for accurate data collection from reverse phase protein arrays (RPPA). The most common way to qualify antibodies for RPPA analysis is by Western blotting because the detection mechanism is based on the same immunodetection principles. Western blots of tissue or cell lysates that result in single bands and low cross-reactivity indicate appropriate antibodies for RPPA detection. Western blot conditions used to validate antibodies for RPPA experiments, including blocking and detection reagents, have significant effects on aspects of antibody performance such as cross-reactivity against other proteins in the sample. We have found that there can be a dramatic impact on antibody behavior with changes in blocking reagent and detection method, and offer an alternative method that allows detection reagents and conditions to be held constant in both antibody validation and RPPA experiments.

Key words: Reverse phase protein array, Blocking buffer, Antibody validation, Detection chemistry, Near-infrared fluorescence

1. Introduction

Reverse phase protein array (RPPA) analysis is a high-throughput technique that has been used to characterize cancer signaling pathways (1–4) and identify characteristic changes which may define a set of diagnostic and prognostic biomarkers (5). Lysates of whole cells, microdissected tissues, or other patient samples are applied to nitrocellulose-coated glass slides followed by probing with one or two analyte-specific antibodies that can be detected by colorimetric, chemiluminescent, amplified fluorescent, or near infrared methods (6, 7). The most critical aspect of RPPA success is validation and selection of appropriate antibodies for detection.

Ulrike Korf (ed.), *Protein Microarrays: Methods and Protocols*, Methods in Molecular Biology, vol. 785, DOI 10.1007/978-1-61779-286-1_2, © Springer Science+Business Media, LLC 2011

Antibodies must be highly specific as demonstrated by a single band on a Western blot (8). Western blot chemistries consist of not only antibodies, but also blocking agent for decreasing background, and several different signal generating approaches. Since RPPA detection is based on the same immunodetection principles as Western blot detection, such changes can also significantly affect antibody reactivity in RPPA and therefore, impact the quantification and analysis of the experiment. The use of blocking conditions for RPPA analysis that differs from those used for the initial Western blot antibody validation has been shown to significantly alter the data obtained from the RPPA experiment (9, 10).

Near infrared detection for antibody validation using Western blots and RPPA slides offers a sensitive, quantitative, and accurate way to identify protein changes in RPPA. The method below is designed to optimize antibody performance by testing multiple blocking buffers. The RPPA is then detected using the optimized blocking buffer with the exact same antibodies and near infrared detection procedure, thereby eliminating any bias caused by changing detection chemistries.

2. Materials

2.1. SDS-Polyacrylamide Gel Electrophoresis for Blocker Optimization

1. Tissue Lysate: Mouse and rat thymus, liver, and brain tissue available from BIOMOL International L. P. (Plymouth Meeting, PA). Store at –80°C (see Note 1).
2. 4–20% Tris–Glycine Novex™ Gel, 15-well (Invitrogen, Carlsbad, CA). Store at 4°C (see Note 2).
3. Running buffer: 25 mM Tris, 192 mM glycine. Store at room temperature.
4. Protein loading buffer (2×): 62.5 mM Tris–HCl, pH 6.80, 25% (v/v) glycerol, 2% (v/v) SDS, 1% (w/v) Orange G, 5% (v/v) β-mercaptoethanol. Store at room temperature.
5. Prestained molecular weight markers: Two-Color Protein Markers (LI-COR, Lincoln, NE). Store in aliquots at –20°C.

2.2. Western Blotting for Blocker Optimization

1. Odyssey® nitrocellulose membrane from LI-COR and 3 MM Chr chromatography paper from Whatman, Maidstone, UK (see Note 3).
2. Transfer buffer: 25 mM Tris, 192 mM glycine, 0.1% SDS, 20% (v/v) methanol. Store at –20°C.
3. 1× Phosphate-buffered saline (PBS): 137 mM NaCl, 2.7 mM KCl, 4.3 mM Na_2HPO_4, 1.47 mM KH_2PO_4, pH 7.4. Store at room temperature (see Note 4).

4. 1× PBS Tween®-20 (PBST): 137 mM NaCl, 2.7 mM KCl, 4.3 mM Na_2HPO_4, 1.47 mM KH_2PO_4, pH 7.4, 0.1% (v/v) Tween-20. Store at room temperature.
5. Blocking buffers: Odyssey blocking buffer (LI-COR, Lincoln, NE), 5% (w/v) nonfat dry milk in PBS, 5% (w/v) bovine serum albumin (BSA) in PBS. Store at 4°C (see Note 5).
6. Primary and secondary antibody diluents: Odyssey blocking buffer, 0.2% (v/v) Tween-20; 5% (w/v) nonfat dry milk in PBS, 0.2% (v/v) Tween-20; 5% (w/v) BSA in PBS, 0.2% (v/v) Tween-20. Store at 4°C.
7. Primary antibody: Rabbit anti-ERK 1 (K-23) (Santa Cruz Biotechnology, Inc., Santa Cruz, CA). Store at 4°C (see Note 6).
8. Secondary antibody: IRDye® 800CW Goat anti-rabbit (LI-COR). Store at 4°C (see Note 7).

2.3. Reverse Phase Protein Array Detection Using Optimized Detection Chemistry

1. Panorama™ Mouse/Rat Tissue Extract Protein Array (Sigma, St. Louis, MO). Store at room temperature (see Note 8).
2. 1× PBS: 137 mM NaCl, 2.7 mM KCl, 4.3 mM Na_2HPO_4, 1.47 mM KH_2PO_4, pH 7.4. Store at room temperature (see Note 4).
3. 1× PBS Tween-20 (PBST): 137 mM NaCl, 2.7 mM KCl, 4.3 mM Na_2HPO_4, 1.47 mM KH_2PO_4, pH 7.4, 0.1% (v/v) Tween-20. Store at room temperature.
4. Blocking buffers: Odyssey blocking buffer (LI-COR), 5% (w/v) nonfat dry milk in PBS, 5% (w/v) BSA in PBS. Store at 4°C (see Note 9).
5. Primary and secondary antibody diluents: Odyssey blocking buffer, 0.2% (v/v) Tween-20; 5% (w/v) nonfat dry milk in PBS, 0.2% (v/v) Tween-20; 5% (w/v) BSA in PBS, 0.2% (v/v) Tween-20. Store at 4°C (see Note 10).
6. Primary antibody: Rabbit anti-ERK 1 (K-23) (Santa Cruz Biotechnology, Inc., Santa Cruz, CA). Store at 4°C (see Note 6).
7. Secondary antibody: IRDye 800CW Goat anti-rabbit (LI-COR). Store at 4°C (see Note 7).

3. Methods

3.1. SDS-PAGE for Blocker Optimization

1. Prepare samples by placing 20 μL (100 μg) each of mouse brain, rat brain, mouse liver, rat liver, mouse thymus, and rat thymus tissue extracts into different 0.5 mL microcentrifuge tubes and label with contents (see Note 11). Add 20 μL of 2× protein loading buffer to each extract sample. Mix by gently pipetting up and down. Cap tubes and place at 100°C for 5 min.

Remove from heat and place directly on ice until ready to load gel. Centrifuge briefly to collect sample to bottom of tube.

2. These instructions utilize the XCell SureLock™ Mini-Cell Electrophoresis Apparatus (Invitrogen) for electrophoresis. Remove tape strip from bottom of two 4–20% Tris–glycine Novex gels. Assemble according to XCell SureLock Mini-Cell Instruction Manual. Remove gel combs. Fill box with 1× running buffer. Using pipette gently rinse out the wells with buffer in buffer tank.
3. For two gels, load lanes 1, 8, and 15 with 5 μL of Two-Color Protein Markers. In lanes 2–7, place 10 μL tissue extract samples in above order and repeat in lanes 9–14.
4. Fully assemble XCell SureLock Mini-Cell Electrophoresis Apparatus and plug into power supply. Run the gel at a fixed voltage of 125 V for 100 min.

3.2. Western Blotting for Blocker Optimization

1. Following electrophoresis, transfer the samples to supported nitrocellulose membrane. It is important that the membrane only be handled by the edges with clean forceps. Take great care to never touch the membrane with bare or gloved hands (see Note 12).
2. This procedure describes the use of Bio-Rad Mini Trans-Blot® Electrophoretic Transfer Cell tank system (see Note 13). While gel is running, fill Bio-Ice cooling unit with distilled water and place in −20°C. Cut two pieces of the nitrocellulose membrane to 7×8 cm size with a paper cutter designated for membrane cutting (i.e., does not get used for general purpose). Place two pieces of cut nitrocellulose membrane into a Rubbermaid container (710 mL rectangle). Place four pieces of Whatman paper 3 MM, 7×8 cm into the same plastic container (710 mL rectangle). Cover the nitrocellulose and Whatman paper with transfer buffer. Place four fiber pads into a different plastic container (710 mL rectangle). Cover the fiber pads with transfer buffer. Let fiber pads, nitrocellulose, and Whatman paper soak while gel is running.
3. After electrophoresis open the gel cassette using the gel tension wedge. Trim the bottom of the gel off just above the loading dye. Pour 200 mL of 1× transfer buffer into a plastic container (710 mL rectangle). Carefully place gels in 1× transfer buffer. Gently shake on platform shaker for 5 min.
4. Prepare gel sandwich and assemble transfer cell according to the Mini Trans-Blot Electrophoretic Transfer Cell Instruction Manual, Bio-Rad. Plug into power supply and run at a fixed voltage of 100 V for 65 min.
5. Disassemble transfer cells and remove blots from transfer unit. Place in between two sheets of Whatman paper. Let air dry

overnight. Blots can be stored dry at 4°C for up to 3 months before being processed.

6. Cut both membranes down the middle of the protein marker in lane 8, being careful not to touch the membrane. Label with pencil appropriately.
7. Place membranes into three different Western blot incubation boxes. There will be one extra membrane that can be used as a backup or for a fourth blocking condition. In box 1 add 10 mL of Odyssey blocking buffer; in box 2 add 10 mL of 5% (w/v) nonfat dry milk in PBS; in box 3 add 5% (w/v) BSA in PBS and block the membranes for 1 h with gentle shaking.
8. Dilute ERK 1 primary antibody 1:1,000 in 10 mL of diluent as follows: For box 1, dilute in Odyssey blocking buffer diluent; for box 2, dilute in 5% (w/v) nonfat dry milk diluent; for box 3, dilute in 5% (w/v) BSA diluent (see Note 6).
9. For all blots, decant off blocking buffer and add the diluted ERK antibody. Incubate blots overnight at 4°C with gentle shaking (see Note 14).
10. Decant off primary antibody solution. Rinse membrane with 1× PBST. Cover blot with 10 mL of 1× PBST. Shake vigorously on platform shaker at room temperature for 5 min. Decant off wash solution. Repeat three additional times.
11. Dilute IRDye® 800CW Goat anti-rabbit antibody 1:10,000 in 10 mL of diluent as follows: For box 1, dilute in Odyssey blocking buffer diluent; for box 2, dilute in 5% (w/v) nonfat dry milk diluent; for box 3, dilute in 5% (w/v) BSA diluent (see Note 7).
12. Add the diluted secondary antibody to the appropriate boxes. Incubate blots for 1 h at room temperature with gentle shaking. Protect membranes from light during incubation by covering with foil or a cardboard box.
13. Decant off secondary antibody solution. Rinse membrane with 1× PBST. Continue to protect the membranes from light during washes. Cover blot with 10 mL of 1× PBST. Shake vigorously on platform shaker at room temperature for 5 min. Decant off wash solution. Repeat three additional times.
14. Rinse membranes with 1× PBS to remove residual Tween-20. The membranes are now ready to be imaged.
15. Scan wet blots on Odyssey Infrared Imaging System by placing them face down on the glass surface and image with the following settings: resolution = 169 μm, quality = medium, focus offset = 0.0 mm, intensity = 5(800) (see Note 15).
16. Select a "blocking buffer" condition from the three that gives the least amount of nonspecific banding and the highest signal intensity to move forward with the RPPA processing (see Note 16). For this target, Odyssey blocking buffer is the buffer of choice. An example of the results produced is shown in Fig. 1.

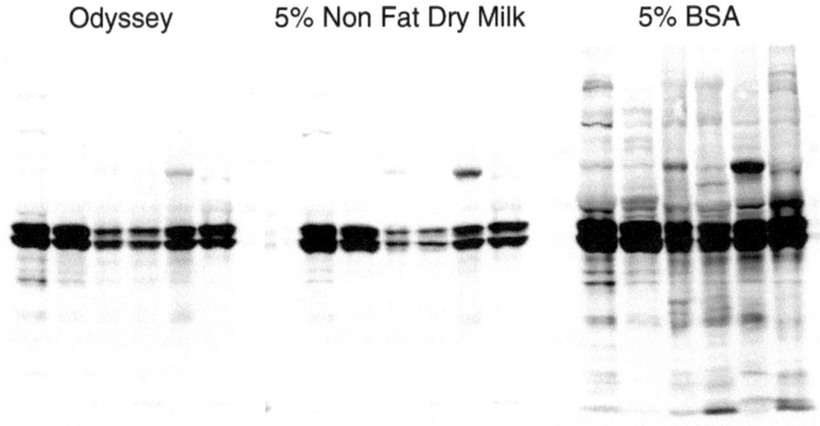

Fig. 1. ERK 1 antibody performance in Odyssey blocking buffer, 5% nonfat dry milk, and 5% BSA. Tissue lysates are as follows: *lane 1* – mouse brain, *lane 2* – rat brain, *lane 3* – mouse liver, *lane 4* – rat liver, *lane 5* – mouse thymus, *lane 6* – rat thymus. Odyssey blocking buffer was chosen for detection of the RPPA. Blots were imaged on Odyssey Infrared Imaging System at the following settings: resolution = 169 μm, quality = medium, focus offset = 0.0 mm, intensity = 5(800).

3.3. Reverse Phase Protein Array Detection Using Optimized Detection Chemistry

1. Label Panorama™ Mouse/Rat Tissue Extract Protein Array with target using a pencil.
2. Place slide in a small incubation box. Incubate the slide for 10 min in pre-incubation buffer. Aspirate the pre-incubation buffer from the box. Add enough Odyssey blocking buffer to the incubation box to completely submerge the slide. Incubate with gentle rocking for 40 min at room temperature.
3. Dilute ERK 1 primary antibody 1:1,000 in 4 mL of Odyssey blocking buffer diluent (see Note 17).
4. Aspirate Odyssey blocking buffer out of the incubation box. Add diluted primary antibody to the slide. Incubate slide overnight at 4°C with gentle shaking.
5. Aspirate primary antibody solution. Cover slide with 1× PBST. Shake vigorously on platform shaker at room temperature for 5 min. Decant off wash solution. Repeat three additional times.
6. Dilute IRDye 800CW Goat anti-rabbit antibody 1:10,000 in 4 mL of Odyssey blocking buffer diluent (see Note 17).
7. Add diluted secondary antibody to the slide. Incubate slide for 1 h at room temperature with gentle shaking. Protect slide from light during incubation by covering with foil or a cardboard box.
8. Aspirate secondary antibody solution. Cover slide with 1× PBST. Continue to protect the membranes from light during washes. Shake vigorously on platform shaker at room temperature for 5 min. Pour off wash solution. Repeat three additional times. Rinse slide with 1× PBS to remove residual Tween-20.

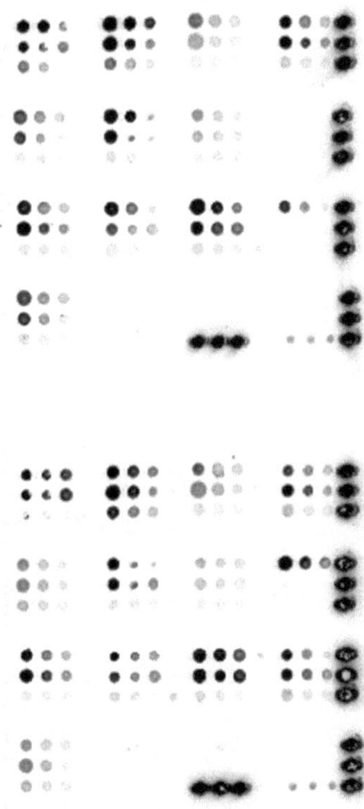

Fig. 2. ERK1 antibody performance using Odyssey blocking buffer on a Sigma Panorama mouse and rat tissue RPPA. Array was imaged on Odyssey Infrared Imaging System at the following settings: resolution = 42 μm, quality = medium, focus offset = 0.0 mm, intensity = 5(800).

9. Using a slide carrier, centrifuge the slide dry to eliminate as much liquid as possible. Allow slide to air dry in the dark for 30 min before imaging.
10. Scan dry slide on Odyssey Infrared Imaging System by placing it protein side down on the glass surface and image with the following settings: resolution = 42 μm, quality = medium, focus offset = 0.0 mm, intensity = 5(800) (see Note 15). An example of the results produced is shown in Fig. 2.

4. Notes

1. Mouse and rat tissue were chosen for this example. When validating antibodies for RPPA, a representative sample from the array should be used for Western blot validation.
2. This protocol should be adapted for an SDS-polyacrylamide gel electrophoresis (SDS-PAGE) electrophoresis system that is

optimal for both protein sample as well as the size of the target protein. Buffer composition and percent acrylamide can be altered.

3. Not all nitrocellulose is optimal for use on the Odyssey Infrared Imaging System. Some nitrocellulose has more background fluorescence than others. Use caution if using another brand of nitrocellulose.
4. If the primary antibody being evaluated performs optimally in Tris-buffered saline (TBS) replace all PBS buffers with TBS.
5. Three blocking buffers were chosen for this procedure; however, it can be modified for other blocker choices. The addition of 0.2% Tween-20 to the blocking buffer is critical for the primary and secondary antibody diluents to reduce background. It is important to note that some blockers already have detergent in them and additional Tween-20 may not be optimal.
6. Any primary antibody can be substituted. Dilutions of primary antibody may need to be optimized. Vendor recommended dilutions for Western blot applications are generally the best place to begin.
7. The choice of secondary antibody will vary depending on the host species of the primary antibody being evaluated. IRDye 800CW conjugated secondary antibodies are optimal as there is very little autofluorescence of the membrane surfaces and biomolecules in the 800 nm range of the spectrum. Secondary antibody dilutions may need to be optimized. Typical dilution recommendations are 1:5,000–1:25,000.
8. Panorama Mouse/Rat tissue extract arrays were used to optimize this procedure. Any RPPA on nitrocellulose-coated glass slides can be substituted. RPPA with spot sizes greater than 200 μm in diameter will result in the best quantification results with infrared detection on the Odyssey Infrared Imaging System.
9. The blocking buffer that is chosen from the Western blot antibody optimization should be the blocking buffer used for the RPPA.
10. The primary and secondary antibody diluents for use with RPPA will correspond to the blocking treatment that was chosen in Western blot antibody optimization.
11. This procedure utilized 25 μg of tissue lysate in each well of the gel. Depending on the sample type and target of interest 5–25 μg of sample may be optimal.
12. When detecting membranes in the near infrared it is important not to contaminate the membrane. The most common contaminants are blue pen, Coomassie stain, poorly cleaned incubation containers, and fingerprints.

13. This procedure can be adapted to most wet transfer units by following manufacturer's recommendations.
14. Primary antibody incubation may need to be optimized. Typical recommendations are 1–4 h at room temperature or overnight at 4°C with gentle shaking.
15. The scan intensity may need to be optimized depending on the sensitivity needs of the blot. If the image contains saturation the scan intensity will need to be reduced. Weak band signal could be improved by increasing the scan intensity. It is important to recognize that background may increase as well.
16. Quantification of the blots can be done using the Odyssey application software.
17. Volume will vary depending on the size of container used for incubations. The volume in this procedure is for use with the incubation plates that come with the Panorama arrays.

References

1. Grubb, R.L., Calvert, V.S., Wulkuhle, J.D., Paweletz, C.P., Linehan, W.M. et al. (2003) Signal pathway profiling of prostate cancer using reverse phase protein arrays *Proteomics* **3**, 2142–2146.
2. Nishizuka, S., Carboneau, L., Young, L., Major, S., Reinhold, W.C. et al. (2003) Proteomic profiling of the NCI-60 cancer cell lines using new high-density reverse-phase lysate microarrays *Proc. Natl. Acad. Sci. USA.* **100**, 14229–14234.
3. Sheehan, K.M., Calvert, V.S., Kay, E.W., Lu, Y., Fishman, D., Espina, V. et al. (2005) Use of reverse phase protein microarrays and reference standard development for molecular network analysis of metastatic ovarian carcinoma *Mol. Cell. Proteomics* **4**,346-355.
4. Korf, U., Löboke, C., Sahin, Ö., Haller, F., Sültmann, H., and Poustka, A. (2009) Reverse-phase protein arrays for application-oriented cancer research *Proteomics* **3**, 1140–1150.
5. Wulfkuhle, J.D., Aquino, J.A., Calvert, V.S., Fishman, D.A., Coukos, G. et al. (2003) Signal pathway profiling of ovarian cancer from human tissue specimens using reverse-phase protein microarrays *Proteomics* **3**, 2085–2090.
6. Paweletz, C.P., Charboneau, L., Bichsel, V.E., Simone, N.L., Chen, T. et al. (2001) Reverse phase protein microarrays which capture disease progression show activation of pro-survival pathways at the cancer invasion front *Oncogene* **20**,1981–1989.
7. Calvert, V.S., Tang, Y., Boveia, V., Wulfkuhle, J., Schutz-Geschwender, A. et al. (2004) Development of multiplexed protein profiling and detection using near infrared detection of reverse-phase protein microarrays *Clin Proteomics* **1**, 81–89.
8. Towbin, H., Staehelin, T., and Gordon, J. (1979) Electrophoretic transfer of proteins from polyacrylamide gels to nitrocellulose sheets: procedure and some applications. *Biotechnology* **24**, 145–149.
9. Aoki, H., Iwaldo, E., Eller, M., Kondo, Y. et al. (2007) Telomere 3'overhanging-specific DNA oligonucleotides induced autophagy in malignant glioma cells. *FASEB J.* **21**, 2918–2930.
10. Ambroz, K.L.H., Zhang, Y., Schutz-Geschwender, A., Olive, D.M. (2008) Blocking detection chemistries affect antibody performance on reverse phase protein arrays *Proteomics* **8**, 2379–2383.

Chapter 3

Phosphoprotein Stability in Clinical Tissue and Its Relevance for Reverse Phase Protein Microarray Technology

Virginia Espina, Claudius Mueller, and Lance A. Liotta

Abstract

Phosphorylated proteins reflect the activity of specific cell signaling nodes in biological kinase protein networks. Cell signaling pathways can be either activated or deactivated depending on the phosphorylation state of the constituent proteins. The state of these kinase pathways reflects the in vivo activity of the cells and tissue at any given point in time. As such, cell signaling pathway information can be extrapolated to infer which phosphorylated proteins/pathways are driving an individual tumor's growth. Reverse phase protein microarrays (RPMAs) are a sensitive and precise platform that can be applied to the quantitative measurement of hundreds of phosphorylated signal proteins from a small sample of tissue.

Pre-analytical variability originating from tissue procurement and preservation may cause significant variability and bias in downstream molecular analysis. Depending on the ex vivo delay time in tissue processing, and the manner of tissue handling, protein biomarkers such as signal pathway phosphoproteins will be elevated or suppressed in a manner that does not represent the biomarker levels at the time of excision. Consequently, assessment of the state of these kinase networks requires stabilization, or preservation, of the phosphoproteins immediately post-tissue procurement. We have employed RPMA analysis of phosphoproteins to study the factors influencing stability of phosphoproteins in tissue following procurement. Based on this analysis we have established tissue procurement guidelines for clinical research with an emphasis on quantifying phosphoproteins by RPMA.

Key words: Cell signaling, Kinase, Phopshoprotein, Pre-analytical variablity, Reverse phase protein microarray, Stability

1. Introduction

The instant a tissue biopsy is removed from a patient, the cells within the tissue react and adapt to the absence of vascular perfusion, ischemia, hypoxia, acidosis, accumulation of cellular waste, absence of electrolytes, and temperature changes (1). It would be expected that a large surge of stress-related, hypoxia-related, and wound repair-related protein signal pathway proteins and

Ulrike Korf (ed.), *Protein Microarrays: Methods and Protocols*, Methods in Molecular Biology, vol. 785,
DOI 10.1007/978-1-61779-286-1_3,

transcription factors will be induced in the tissue immediately following procurement (2, 3). Investigators in the past have worried about the effects of vascular clamping and anesthesia, prior to excision, on the fidelity of molecular data in tissues. A much more significant and underappreciated issue is the fact that excised tissue is alive and reacting to ex vivo stress (1). The promise of tissue protein biomarkers to provide revolutionary diagnostic and therapeutic information will never be realized unless the problem of tissue protein biomarker instability is recognized, studied, and solved. There is a critical need to develop standardized protocols and novel technologies that can be used in the routine clinical setting for seamless collection and immediate preservation of tissue biomarker proteins, particularly those that have been postranslationaly modified such as phosphoproteins. This need extends beyond the large research hospital environment to the private practice, where most patients receive therapy. The fidelity of the data obtained from a diagnostic assay applied to tissue must be monitored and verified, otherwise a clinical decision can be based on incorrect molecular data. To date, clinical preservation practices routinely rely on protocols that are decades old, such as formalin fixation, and are designed to preserve specimens for histologic examination, not molecular analysis.

1.1. Tissue Processing Delays in Clinical Tissue Procurement

Two categories of variable time periods that define biomarker stability during human tissue procurement are the (a) postexcision delay time and (b) processing delay time. The postexcision delay time is the variable timeframe between specimen excision and the point at which the specimen is placed in a stabilized state, e.g., immersed in fixative or snap-frozen in liquid nitrogen. During the postexcision timeframe the tissue may reside at room temperature, or it may be refrigerated, either in a closed or open container. The second variable time period is the processing delay time. Common variables associated with processing delay time are the permeation rate of the fixative through the tissue and length of time to freeze the specimen.

In addition to the uncertainty about the length of these two time intervals, a host of known and unknown variables can influence the stability of tissue molecules during these time periods prior to measurement. These include (1) patient hypoxia, (2) tissue ischemia, (3) presence of imaging dyes and contrast media, (4) temperature fluctuations prior to fixation or freezing, (5) preservative chemistry and rate of tissue penetration, (6) size of the tissue specimen, (7) extent of handling, cutting, and crushing of the tissue, (8) fixation and staining prior to microdissection, (9) tissue hydration and dehydration, and (10) the introduction of phosphatases or proteinases from the environment at any time (1).

1.2. Formalin Fixation May Be Unsuitable for Quantitative Protein Biomarker Analysis in Tissue

Proteins can be extracted with variable yield from formalin-fixed tissue (4). The yield depends on the time, chemistry of formalin fixation, and the tissue geometry and density. Formalin penetrates tissue at a variable rate, reported to be within the range of millimeter/hour (5–7). During this time the portion of the living tissue deeper than several millimeters would be expected to undergo significant fluctuations with regard to phosphoprotein analytes. When one considers the volume of a typical 16-gauge core needle biopsy (7 mm × 1.6 mm (diameter); volume is approximately 14.1 mm^3) the cellular molecules in the depth of the tissue will have significantly degraded by the time formalin permeates the tissue (5, 8). Moreover, penetration rate is not synonymous with fixation. In aqueous solutions formaldehyde becomes hydrated, forming methylene glycol (5, 7). Methylene glycol penetrates the tissue, yet it is the small percentage of carbonyl formaldehyde component that covalently cross-links with proteins and nucleic acids and causes tissue fixation (5, 7). Formalin cross-linking, the formation of methylene bridges between amide groups of protein, blocks analyte epitopes as well as decreases the yield of proteins extracted from the tissue. Typically, the dimensions of the tissue and the depth of the block from which samples are prepared are unknown variables. Consequently, formalin fixation would be expected to cause significant variability in protein and phosphoprotein stability for molecular diagnostics (5, 9, 10).

1.3. Phosphoprotein Stability Is a Balance Between Kinase and Phosphatase Activity

Kinases phosphorylate a substrate amino acid and phosphatases remove the phosphate group from the amino acid (Fig. 1). At any point in time within the tissue cellular microenvironment, the phosphorylated state of a protein is a function of the local stoichiometry of associated kinases and phosphatases specific for the phosphorylated residue. Thus, in the absence of kinase activity, proteins may be dephosphorylated by phosphatases, reducing the level of a phosphoprotein analyte causing a false-negative result. This can be prevented by a variety of chemical- and protein-based phosphatase inhibitors (11, 12). However if the kinase remains active, then the addition of a phosphatase inhibitor alone will result in an augmentation of the phospho-epitope, generating a false-positive result. Optimally, a stabilizing chemistry should arrest both sides of the kinase/phosphatase balance in order to prevent positive or negative fluctuations in phosphorylation events as the excised tissue reacts to the ex vivo conditions (1).

During the ex vivo time period, because the tissue cells are alive and reactive, phosphorylation of certain kinase substrates may transiently increase due to the persistence of functional signaling, activation by hypoxia, or some other stress-response signal (1, 13–15). While these reactive changes would be expected to increase protein phosphorylation, the availability of ubiquitous cellular phosphatases would be expected to ultimately decrease

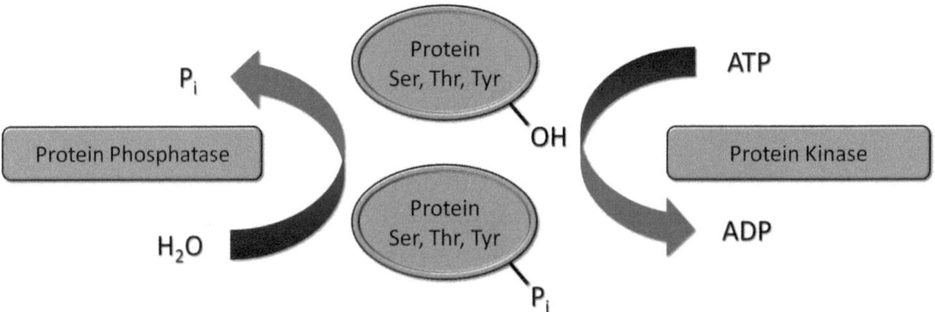

Fig. 1. Stoichiometry between protein kinases and protein phosphatases. Phosphorylated cell signaling cascades are regulated by a series of kinases and phosphatases which act in concert to activate/deactivate a protein. Phosphatase inhibitors, in the absence of kinase inhibitors, will cause a false elevation of phosphorylated proteins if the kinase remains functional and active.

phosphorylation sites, given enough time (1–3). These imbalances will significantly distort the molecular signature of the tissue compared to the state of the markers in vivo. This physiologic fact must be taken into consideration for tissue protein biomarker analysis in the hospital or clinic, where the living, reacting tissue may remain in the collection container for hours (Fig. 2).

Application of RPMA phosphoprotein analysis to freshly collected tissue (1, 13–17) emphasized that excised tissue is reactive. The guidelines below illustrate methods for the reducing preanalytical variables (adapted from (1)).

1. Tissue should be stabilized as soon as possible after excision. Taking into consideration the average time for procurement in a community hospital, the recommended maximum elapsed time is 20 min from excision to stabilization (e.g., flash freezing, thermal denaturation, or chemical stabilization).
2. Tissue stabilization and preservation methods should be compatible with the intended downstream analysis. Preservation of tissue histology and morphology is essential for verification of tissue type and cellular content.
3. For documentation, sample excision/collection time, elapsed time to preservation/stabilization, and length of fixation time are essential data elements for sample quality control.
4. Kinase pathway stabilization methods should block both sides of the kinase/phosphatase kinetic reaction. Blocking only phosphatases can cause false elevation of an analyte's phosphorylation level.

In this chapter, we describe tissue collection and processing for analysis of phosphoproteins by reverse phase protein microarray (RPMA).

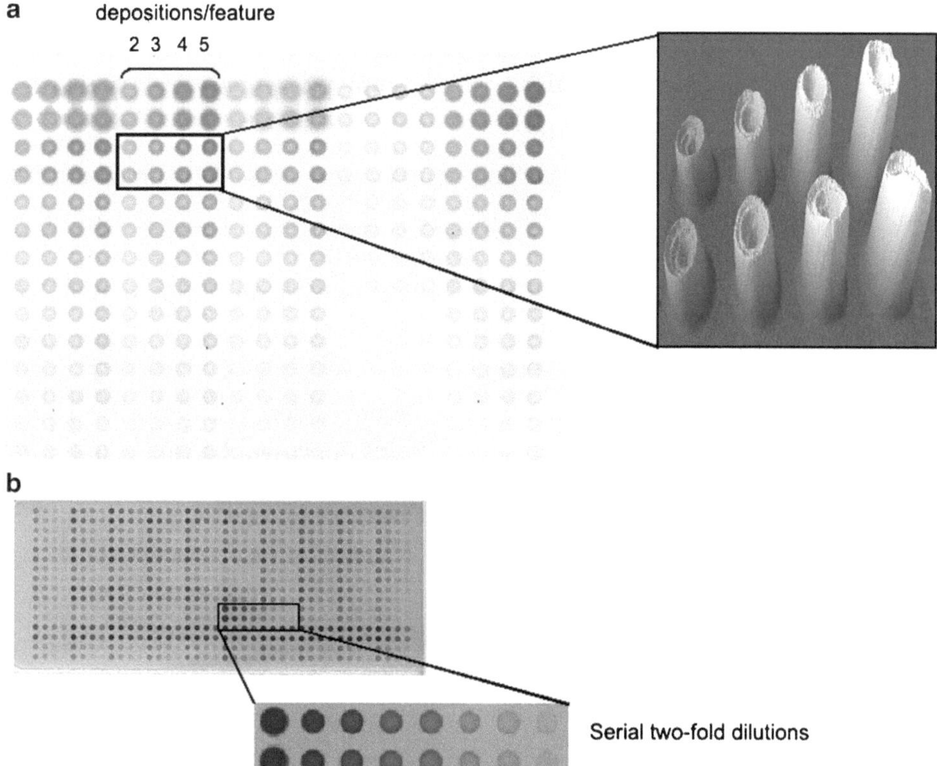

Fig. 2. Reverse phase protein microarray (RPMA) format. Binding capacity of the nitrocellulose membrane can be determined empirically by varying the amount of sample deposited on the nitrocellulose. (**a**) Whole cell lysates were printed in the vertical direction in twofold serial dilutions. In the horizontal direction, the lysate was printed at 2, 3, 4, or 5 depositions/feature (hits/spot). Spot morphology for each set of depositions was assessed using MicroVigene spot analysis software (Vigene Tech). (**b**) Typical RPMA constructed with whole cell lsyates. Each sample, control and standard were printed in duplicate, serial twofold dilutions using an Aushon Biosystemes 2470 arrayer equipped with 350 μm pins. The dilutions were printed in the horizontal direction.

2. Materials

2.1. Tissue Procurement and Cell Lysis

1. Tissue samples obtained by surgical resection, fine needle aspiration, or biopsy, not to exceed 10 mm × 5 mm (see Note 1).
2. Tissue lysis buffer: 450 μL T-PER™ Tissue Protein Extraction Reagent (Pierce), 450 μL 2× SDS Tris–glycine loading buffer (Invitrogen), and 100 μL TCEP Bond Breaker™ (Tris(2-carboxyethyl)phosphine (Pierce)).
3. Mortar and pestle: for snap frozen tissue pulverization.
4. Tissue homogenizer: for fresh tissue disruption (see Note 2).

2.2. Frozen Section Preparation

Frozen sections may be used for laser capture microdissection prior to printing RPMAs. Alternately, frozen sections may be used to assess overall tissue morphology and to prepare whole slide lysates.

1. Cryomolds.
2. Optimal cutting temperature (O.C.T.) compound (Sakura Finetek).
3. Dry ice.
4. Cryostat with appropriate blade, temperature setting –20°C or colder, and chucks.
5. Plastic slide box.
6. Precleaned glass microscope slides.
7. Bitran storage bags, 2×4 cm (Fisher Scientific) and/or aluminum foil.

2.3. Reverse Phase Protein Microarray

1. Aushon 2470 arrayer (Aushon Biosystems, Billerica, MA, USA).
2. 384-Well microtiter plates with lids (see Note 3).
3. Nitrocellulose-coated slides (KeraFast FAST™ slides, Schott Nexterion® NC-C slides, or ONCYTE® Nitrocellulose Film Slides, Grace Bio-Labs) (see Note 4).
4. 70% Ethanol.
5. Commercial cell lysates, such as HeLa+Pervanadate or A431+EGF (see Note 5). A minimum of 3–20 μL of each individual lysate is needed to construct a twofold dilution sequence on the array, in a 384-well microtiter plate (see Note 6).
6. Desiccant (Drierite, anhydrous calcium sulfate).

2.4. Reverse Phase Protein Microarray Immunostaining

1. I-Block blocking solution: 2.0 g I-Block (Applied Biosystems), 1,000 mL phosphate-buffered saline 1× without calcium or magnesium, 1.0 mL Tween-20. Dissolve I-Block Protein Blocking powder in PBS on a hot plate with constant stirring (see Note 7). Cool the solution to room temperature and add Tween 20. I-Block solution can be stored at 4°C for 1 week.
2. Re-Blot™ Mild Antigen Stripping solution 10× (Millipore/Chemicon).
3. Primary antibody of choice.
4. Biotinylated secondary antibody, species matched to primary antibody.
5. Dako CSA kit (Dako).
6. Biotin blocking system (Dako).
7. Antibody diluent with background reducing components (Dako).

8. Tris-buffered saline with Tween (TBST, Dako).
9. DAB+, Liquid Substrate Chromogen System (Carcinogenic, contact hazard: wear gloves while handling) (Dako).

2.5. Image Acquisition

1. High-resolution flat-bed scanner.
2. Laser scanner or CCD scanner (excitation 280 nm, emission 618 nm).

2.6. Sypro Ruby Protein Blot Stain

1. Sypro Ruby Protein Fixative Solution (7% v/v acetic acid and 10% v/v methanol in deionized water). Close tightly and store at room temperature. Solution is stable for 2 months.
2. Sypro Ruby Protein Blot Stain (Invitrogen).

3. Methods

3.1. Tissue Procurement

Based on current best practices for protein preservation, the tissue sample should be frozen as soon as possible after procurement to minimize phosphoprotein fluctuations (1). While there appears to be great variation in the fluctuation times between tissue types due to intrinsic kinases, nucleases, proteases, and phosphatases, prompt freezing of the tissue limits these potential molecular changes. Freezing the tissue sample in an embedding media such as Sakura Finetek's O.C.T. compound prevents the formation of water crystals that can disrupt a tissue's cellular structure. In addition, this aqueous polyvinyl alcohol compound provides support to the tissue and aids in the cryo-sectioning process.

3.2. Tissue Lysis and Protein Extraction

1. The desired maximum final total protein concentration for a whole cell lysate is 0.5 μg/μL total protein. Snap frozen tissue (without cryopreservative): Weigh frozen tissue sample on an analytical balance. Pulverize frozen tissue. Place the frozen tissue in a microcentrifuge tube. Add 1,000 μL tissue lysis buffer for each 200 mg of tissue.

   Fresh tissue: Weigh tissue sample on an analytical balance. Place the fresh tissue in a microcentrifuge tube. Add 1,000 μL tissue lysis buffer for each 200 mg of tissue. Immediately homogenize tissue (see Note 2).
2. Briefly vortex the microcentrifuge tube containing the whole cell lysate.
3. Immediately heat the lysate at 100°C for 5–8 min. After tissue lysis, the lysates should be printed on the microarray as soon as possible. If a delay in printing is anticipated, the lysates may be stored at −80°C (see Note 8).

3.3. O.C.T. Embedded Frozen Tissue for Frozen Section Preparation

1. Prepare all supplies prior to the biopsy procedure to avoid delay once the specimen has been obtained.
2. Label the handle and the front surface of the cryomold with the sample identifying information.
3. Cover the bottom of the cryomold with O.C.T. to a depth of 2–4 mm.
4. Orient the specimen on top of the O.C.T. in the cryomold in the desired position, keeping in mind that the side facing down will be the first tissue surface cut.
5. Completely cover the tissue with O.C.T. and place immediately, in a horizontal position, in a container of dry ice. Covering the container of dry ice will speed freezing.
6. O.C.T. will appear white once it has been frozen. The embedded tissue may or may not be visible within the O.C.T. To add further protection, the cryomold can be wrapped in aluminum foil, placed inside a 2×4 cm Bitran plastic bag, or placed in a 50-mL conical Falcon tube.
7. The frozen tissue should be stored at –70°C to –80°C.

3.3.1. Frozen Section Preparation

1. Label glass microscope slides with a pencil. Place slides face-up on top of the cryostat.
2. Remove the cryomold containing tissue from the freezer and place it in a box with dry ice. Peel the cryomold from the O.C.T. tissue block. Place the tissue block either in the dry ice or in the cryostat to keep it frozen.
3. Place a small amount of O.C.T. on a room temperature chuck. Place the O.C.T. embedded tissue block directly on the room temperature O.C.T. on the chuck. Immediately place the chuck in the cryostat.
4. Allow the O.C.T. to freeze, forming a bond between the tissue block and the chuck.
5. Place a blade in the knife holder.
6. Place the chuck containing the tissue block in the chuck holder and tighten the holder.
7. Align the tissue face parallel with the blade.
8. Set the micrometer setting to the desired thickness (5–8 μm is optimal for laser capture microdissection).
9. Cut sections until a full tissue thickness is obtained. The micrometer may be adjusted to cut thicker sections until the tissue face is reached.
10. Place the tissue section on a room temperature glass slide. Hold the slide so the tissue will adhere to the clean, front surface of the slide (see Note 9).

11. Do not allow the frozen section slides to thaw on the slide or air dry at room temperature. Keep the slides frozen by placing them in the cryostat box or directly into a prechilled slide box kept on dry ice.
12. After all the frozen section slides for one sample have been cut, remove the blade from the knife holder, and discard the blade in a sharp container or blade holder.
13. Loosen the chuck from the chuck holder and remove the chuck.
14. Squirt a small amount of O.C.T. on top of the tissue block to cover the tissue.
15. Immediately place the tissue in the Pelletier to rapidly freeze the O.C.T. Allow to freeze for 2–4 min. Alternatively, a piece of prechilled smooth metal may be placed on top of the O.C.T. to freeze the O.C.T.
16. Remove the tissue block from the Pelletier.
17. Use a room temperature putty knife to pry the O.C.T. block away from the chuck.
18. Put the cryomold and tissue block in a Bitran bag, or wrap the cryomold and tissue block in aluminum foil. Label the bag/aluminum foil.
19. Store the tissue block and frozen section slides at −80°C.

3.4. Reverse Phase Protein Microarray Construction

RPMAs are a multiplexed proteomic platform used to evaluate cell signaling protein levels or phosphoprotein profiles in many samples printed on one array for one specific endpoint per array (18–23). Over 100 array slides can be printed with 40 μL of protein lysate and each array is probed with a single antibody. In addition to printing sample lysates, it is also essential to print control lysates such as commercial cell lysates, recombinant peptides, or peptide mixtures that are known to contain the antigens being investigated. All samples are printed in a dilution curve, which permits the selection of the optimal sample protein concentration for individual antibodies that have varying affinities.

The Aushon 2470 arrayer utilizes a solid pin format for the application of cell lysates or other protein containing fluids onto a matrix of nitrocellulose mounted on a glass microscope slide (see Note 10). Prior to printing cell lysates on a RPMA, the number of cells required should be optimized preceding the final array construction (see Note 11) (Fig. 3). The arrays are subsequently stained using a Dako CSA (Catalyzed Signal Amplification) System that includes blocking and signal amplification reagents that are compatible with chromogenic (DAB), chemiluminescent, or fluorescent (Li-Cor® IRDye680) detection reagents.

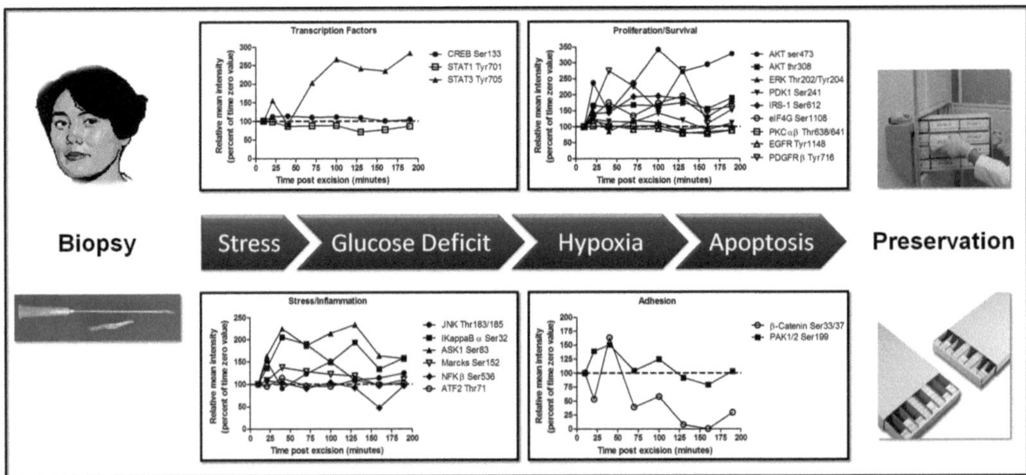

Fig. 3. Overview of tissue phosphoprotein preservation variablity.

1. Fill the Aushon 2470 arrayer water wash container with deionized water and empty the waste container. Care should be taken to verify that the tubing is sufficiently inserted into each container.
2. Fill the humidifier with deionized water (see Note 12).
3. Load nitrocellulose-coated slides onto slide platens with the frosted edge of the slide on the right. Insure the slides are securely held by the platen clips. The array slide printing order is top left to bottom left of the platen, followed by top right to bottom right. Place the slide platens into the Aushon 2470 arrayer.
4. If the lysates have been stored frozen, thaw the lysates and heat the lysates (do not heat commercial cell lysates samples) in a dry heat block or boiling water bath for 7 min at 100°C. Cool to room temperature.
5. Load samples into a 384-well plate, creating a four-point, two-fold dilution curve. Refer to Fig. 4 for an example plate map based on a 20-pin print head configuration (see Note 13).
6. Place the lid on top of the 384-well plates (see Note 14).
7. With the two metal clips on the plate holder open and A1 in the lower left hand corner, slide the 384-well plate to the back of the holder. Flip the metal clips to the closed position.
8. Place the plate holder in the elevator with A1 facing the outside of the instrument.
9. Turn on the power to the Aushon 2470 arrayer. Start the arrayer software by double-clicking on the "Aushon 2470" icon. Enter the username and password.

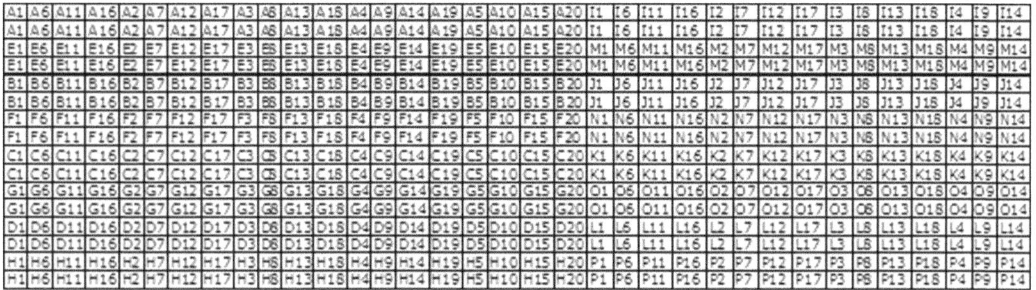

Fig. 4. Example 384-well platemap. The figure represents (1) which microtiter wells contain samples to construct a duplicate four-point, twofold dilution series and (2) the location of the spots on the RPMA from those wells. The microtiter plate well designations, such as A1, B1, C1, etc., represent which well contains a given sample. The position of the sample on the printed RPMA is depicted by the location of the well in the figure. For example, the sample in well A1 will be printed in the *top left corner* of the array, while the sample in A6 will be printed adjacent to A1. Samples in consecutive microtiter wells are not printed next to each other due to the pin head configuration, pin diameter, and microtiter plate well spacing.

10. The array program window will be displayed. Define the number of microtiter plates to be used for printing and the location of samples in each microtiter plate.
11. Double-click on a microtiter plate listed in the source well plate library. This automatically places a microtiter plate in the well plate hotel. Repeat this step for each microtiter plate that will be used in the print run.
12. Click on the first microtiter plate listed in the well plate hotel. The highlight color will change from green to blue.
13. Click on "overlay extractions" beneath the selected well plate image. This allows you to visualize which microtiter plate wells will be used for printing for a given plate. An extraction is equivalent to one dip of the print head into a set of wells. For a 20-pin format with 350 μm pins, one extraction corresponds to wells A1, A2, A3, A4, A5, B1, B2, B3, B4, B5, C1, C2, C3, C4, C5 and D1, D2, D3, D4, D5.

    To program the arrayer to print 80 samples, from four microtiter plates, in rows A–P, use the following format:

    Select four unique extractions, start at 1 for plate #1

    Select four unique extractions, start at 5 for plate #2

    Select four unique extractions, start at 9 for plate #3

    Select four unique extractions, start at 13 for plate #4
14. Designate the left offset and feature to feature spacing of the *x* and *y* axis (Table 1) (see Note 15).
15. Select three depositions per feature, which will print approximately 30 nL lysate per spot (see Note 16).

Table 1
Default spot parameters for the Aushon 2470 arrayer

Parameter	350 μm pins	185 μm pins
Top offset (arrayer specific)	4.5	4.5
Left offset	5.0	5.0
Deposited spot diameter	650 μm	250 μm
Feature-to-feature spacing *x*-axis	1,125	500
Feature-to-feature spacing *y*-axis	1,125	562.5
Number of depositions per feature	3	3
Max feature in Y-Dir	4	8
Max feature in X-Dir	4	9

16. Select the number of super arrays per substrate: (1) if using microtiter wells A–H; (2) if using microtiter plate wells I–P in addition to rows A–H.

    Replicate positioning: Linear (vertical)

    Number of replicates: (1) for duplicates, (2) for triplicates
17. Click "Next" to set the wash parameters.

    Submerged dwell time: 4 s (see Note 17).
18. Click "Next" to select the number of slides to be printed. One to ten slides may be printed on any platen.
19. If the humidity is less than the 50%, set the humidity control to 50%.
20. Verify that all instrument preparation steps and programming steps have been completed.
21. Click the green "Start Deposition" icon. The microtiter plate door and the slide platen door will automatically lock. The system will begin initialization by homing all components and taking inventory of the microtiter plates and slide platens.
22. Deposition is complete when the arrayer software displays the message "Quit or Continue." Select "quit" to terminate printing. Select "continue" to unload the printed array slides and load additional slides for printing (see Note 18).
23. To thoroughly clean the pins after each print run, load 20 μL of 70% ethanol into wells A1-D20 of a 384-well microtiter plate. Load one nitrocellulose-coated slide into the arrayer. Program the arrayer as outlined above to print three depositions/feature on one slide. Start the deposition. At the end of the deposition process, select "quit" to terminate printing. Remove and discard the nitrocellulose slide and the microtiter plate.

24. Turn the Aushon 2470 arrayer power off, and place the printed array slides in a slide box. Store the slide box in a plastic storage bag with desiccant at −20°C (see Note 19).

3.5. Reverse Phase Protein Microarray Immunostaining

3.5.1. Array Slide Pretreatment

RPMA technology allows the simultaneous detection and analysis of signal intensity among a group of samples. This method requires a single antibody–epitope interaction with the protein of interest. All RPMA slides, with the exception of the one probed with the fluorescent Sypro Ruby staining solution, should be blocked prior to the staining procedure.

1. Allow frozen RPMA slides to warm at room temperature for approximately 5–10 min. Leave the slides in the box with dessicant during this time.
2. Prepare a 1× solution of Mild Re-Blot (stock is 10×) in deionized water.
3. Incubate the microarray slides that are to be stained with antibodies in 1× Mild Re-Blot™ solution for 15 min on a rocker/shaker (see Notes 20 and 21). Do not use Re-Blot™ for arrays printed with serum or low molecular weight (LMW) serum fractions, or for arrays to be stained with Sypro Ruby Total Protein Blot stain. For serum/LMW serum fraction arrays, place the slides directly in I-Block solution.
4. Remove the Re-Blot™ solution and wash the microarray slides with 1× PBS (calcium and magnesium free) twice for 5 min each.
5. Decant the last PBS wash and immediately place the slide in blocking solution (I-Block solution). Incubate in I-Block at room temperature with constant rocking for a minimum of 60 min (see Note 22).

3.5.2. Microarray Immunostaining

The Dako Autostainer allows simultaneous staining of 48 slides (see Note 23). The number of slides to be stained is chosen in relation to the number of endpoints of interest and the number of species used to generate the primary antibodies. Antibodies from different animal species can be used during the same Autostainer run. However, to quantify the nonspecific background signal generated from the interaction between the secondary antibody and samples, it is essential to include in each staining run one slide that is probed with secondary antibody only for each species of secondary antibody used. The secondary antibody control slides must be matched to the primary antibody species. For example, if the primary antibodies selected consist of mouse and rabbit antibodies, then two secondary antibody control slides are required, one for the rabbit antibodies and one for the mouse antibodies. The signal intensity of the slide probed with secondary antibody only is subtracted from the signal intensity of the primary + secondary antibody stained slide.

Table 2
Dako Autostainer programming grid for reverse phase protein microarrays using a catalyzed signal amplification (CSA) method

Reagent	Time (min)	Reagent category
Buffer		Rinse
Hydrogen peroxide block	5	Endogenous enzyme block
Buffer		Rinse
Avidin block	10	Auxiliary
Buffer		Rinse
Biotin block	10	Auxiliary
Buffer		Rinse
Protein block	5	Protein block
Blow air		Rinse
Primary antibody	30	Primary antibody
Buffer		Rinse
Buffer	3	Auxiliary
Buffer		Rinse
Secondary antibody	30	Secondary antibody
Buffer		Rinse
Buffer	3	Auxiliary
Buffer		Rinse
Streptavidin biotin complex	15	Secondary reagent
Buffer		Rinse
Buffer	3	Auxiliary
Buffer		Rinse
Amplification (biotinyl tyramide)	15	Secondary reagent
Buffer		Rinse
Buffer	3	Auxiliary
Buffer		Rinse
Streptavidin–HRP	15	Tertiary reagent
Buffer		Rinse
Buffer	3	Auxiliary
Buffer		Rinse

(continued)

Table 2 (continued)

Reagent	Time (min)	Reagent category
Switch to toxic waste		
DAB	5	Chromagen
Water		Rinse
Overnight water[a]	840	Auxiliary

[a] Optional step if operating the Autostainer overnight

1. Select unconjugated primary antibodies of interest (see Note 24).
2. Select biotinylated secondary antibodies corresponding to the species of the primary antibodies.
3. Program the Dako Autostainer (Table 2).
4. Prepare CSA solutions according to the manufacturer's directions.
5. Fill the buffer reservoir with 1× TBST and the water carboy with deionized water. Empty the waste container if necessary.
6. Load the reagents and slides on the Autostainer. Prevent the nitrocellulose from drying during slide loading. If necessary, rinse the slides with 1× TBST buffer during the slide loading process (see Note 25).
7. Prime the water first and then the buffer before starting the run.
8. At the end of the Autostainer run, remove the slides, rinse them with deionized water, and allow them to air dry.
9. Label the microarray slides specifying the date, study, and antibody that have been used in the staining procedure.

3.6. Colorimetric System Image Acquisition and Data Analysis

Any high-resolution scanner, provided with grayscale option, can be employed for image acquisition of diaminobenzedine (DAB)-stained microarrays, providing it generates 14- or 16-bit scanned images.

1. Adjust the image appearance (inverted/not inverted) as required by the image analysis software. Save the adjusted image as a TIFF file (see Note 26). Tiff images can be imported to a variety of data analysis software programs.
2. The pixel intensity of each spot is proportional to the amount of measured analyte per spot. Final intensity values for the RPMAs are obtained after subtraction of the negative control intensity value/spot (secondary antibody alone) and normalization to the total protein value/spot or to another, stable protein.

3.7. Sypro Ruby Total Protein Stain

The concentrations of total, phosphorylated, and cleaved proteins present in different samples can be determined by RPMA. The signal intensity normalization process can be based on total protein values, allowing the comparison of samples with different protein concentrations (see Note 27). Sypro Ruby Protein Blot Stain is a reversible fluorescent dye that binds to primary amino groups on proteins in an acidic environment. The dye has two excitation maxima at ~280 nm and at ~450 nm and an emission maximum near 618 nm (24, 25). Sypro Ruby Blot stain has a sensitivity of 1.0 ng to 1.0 μg protein per microliter of sample. Images of Sypro Ruby stained slides can be acquired with a laser scanner or a CCD camera (see Note 28).

1. If the array slides were stored frozen (−20°C), allow the selected slide(s) to room temperature.
2. Wash array slide(s) in deionized water for 5 min with constant rocking/shaking.
3. Incubate array slides in Sypro Ruby Protein Blot fixative solution at room temperature for 15 min with constant shaking.
4. Discard Sypro Ruby Protein Blot fixative solution and wash slides with deionized water four times for 5 min each.
5. Incubate slides with Sypro Ruby blot stain for a minimum of 30 min. Sypro Ruby is a photosensitive dye; therefore, protect the array slides from light by covering the container with aluminum foil.
6. Discard Sypro Ruby Blot stain. Rinse slides with deionized water 4× for 1 min each. Protect from light.
7. Allow slides to air dry. Protect the stained microarray slides from light.
8. Acquire slide images with a laser scanner, such as Revolution® 4550 Scanner (VIDAR) or a CCD camera such as the NovaRay (Alpha Innotech).

4. Notes

1. The tissue should be cut to a size no greater than one half the area of the cryomold so that it will fit into the cryomold without touching the sides of the mold. For the standard cryomold, specimen samples should not exceed 1 cm in height or width, or a thickness of more than 0.5 cm.
2. Any type of manual or automated tissue disruptor may be used that is compatible with tissue lysis buffer. Automated tissue disruptors may also be used such as pressure cyclers (Barocycler®, Pressure BioSciences Inc), instruments containing a lysing

matrix such as glass beads (Fast Prep®, MP Bio), or combination of pressure/ultrasound disruption instruments (Adaptive Focused Acoustics® series, Corvaris, Inc.).

3. Each Aushon 2470 arrayer is supplied with microtiter plate holders (referred to as source plates) that are designed to hold a specific vendor's microtiter plate. An example microtiter plate compatible with the Aushon 2470 arrays is the Genetix polystyrene 384-well plate, catalog number X6003. Microtiter plate lids are required for compatibility with the Aushon 2470 arrayers. The arrayer is equipped with a suction cup to remove the lid of the microtiter plate.
4. Each lot number of nitrocellulose-coated slides should be thoroughly examined prior to use including visual macroscopic examination of the membrane surface. Examine the nitrocellulose for defects such as scratches, holes, and alignment of the pad on the glass surface. Nitrocellulose-coated slides are available in a variety of formats including single pad and multipad configurations.
5. Every printed array slide should include lysates of known total protein concentration and performance with the detection system, such as commercial cell lysates, homebrew cell lysates, and/or peptides or phosphopeptides. These samples are for process control, indicating adequate deposition of protein and recognition by the primary antibody.
6. The lysate volume determines the number of arrays that can be printed. 20 μL of lysate is sufficient to construct 30–40 arrays in serial twofold dilutions. To prepare whole slide lysates for RPMA, add 10–20 μL of tissue lysis buffer per tissue section based on the area of the section. For microdissected tissue, add 1–3 μL of tissue lysis buffer per 1,000 cells.
7. Avoid boiling the I-Block solution. Heating the solution at low/mid heat levels for 10–15 min is usually adequate to completely solubilize the I-Block powder. I-Block is a casein-based protein solution. Boiling will cause protein degradation and potential alterations in blocking efficiency.
8. Whole cell protein lysates that have been stored frozen should be heated at 100°C for 5–8 min prior to preparing the lysate for microarray printing.
9. The tissue should be in the center of the slide. This is of particular importance for cells that will be procured by laser capture microdissection, as tissue that is too close to the end of the slide, or the sides, cannot be microdissected (26).
10. The Aushon 2470 arrayer employs a proprietary pin technology for printing samples. The pin design/manufacturing process limits fluid from adhering to the shaft of the pin. The pins are

positioned in a print head which moves in the *z*-axis only, while the microtiter plate and slides move in the *x* and *y* axis.

11. For arrays prepared from LCM procured tissue cells, a minimum of 15,000 cells are needed for printing multiple arrays (up to 50 arrays). LCM procured cells are lysed using tissue lysis buffer. Add 1–3 μL of tissue lysis buffer per 1,000 cells. The resulting lysates are diluted into twofold dilution curves in a 384-well microtiter plate (26).
12. Use ddH_2O in the humidifier to prevent mold/bacterial growth. Every 2 weeks, empty the humidifier's water chamber and clean the water chamber by rinsing with 70% ethanol, followed by several water rinses, and air-dry the water chamber.
13. Samples that fill an entire array (640 spots maximum from 350 μm pins) can be loaded into four individual 384-well plates to prevent significant evaporation during the pipetting and printing process. Samples in rows A–D can be loaded in plate 1, rows E–H in plate 2, rows I–L in plate 3, and rows M–P in plate 4.
14. The Aushon 2470 arrayer is equipped with suction cups to remove the lid from the microtiter plates. A lid MUST be placed on every microtiter plate that is loaded in the arrayer. The lid should be clean and dry, free of dust, adhesive, and liquid.
15. Top offset settings may vary slightly with each arrayer and/or slide manufacturer. Food dyes, used for baking or egg coloring, can be diluted in PBS or water to substitute as "samples" for evaluating spot placement by the robotic arraying device. Clean the arrayer pins thoroughly following printing of food dyes. 70% (v/v) ethanol dispensed into a microtiter plate can be used to effectively clean the pins. Dispense 20 μL of 70% ethanol into wells 1–20 in rows A, B, C, and D of a 384-well microtiter plate. Load one nitrocellulose-coated slide into the arrayer. Program and execute a print run for one slide, from one microtiter plate with four unique extractions at three depositions/feature.
16. Samples with protein concentrations less than 0.5 μg/μL can be effectively concentrated on an array by printing more depositions per feature. If more than five depositions/spot are necessary, first print at five depositions/spot, allow the spots to dry for 10 min, and then print additional depositions/spot. Nitrocellulose has a finite protein binding capacity based on its porosity and depth (Fig. 3) (27, 28). Therefore, printing more than five depositions/spot is generally not recommended as the nitrocellulose becomes saturated.
17. Carryover experiments should be conducted with each instrument to determine the optimal pin washing time for various sample matrices.

18. A gal file is generated with each print run and is saved in C:\Documents and Settings\user\My Documents\user\Array Data Files. It can be uploaded into software analysis packages for array layout spot identification.
19. Proteins immobilized on nitrocellulose slides are stable at –20°C for up to 3 years (personal experience) if stored in dry (with dessicant) conditions.
20. Do not use Re-Blot™ for arrays composed of serum or LMW fractionated serum samples. Re-Blot™ causes diffusion of the serum sample and/or buffer components beyond the printed spot resulting in a blurry, poorly defined spot.
21. The Re-Blot™ solution further denatures the protein immobilized onto the nitrocellulose slides thus improving the antibody–epitope recognition. Do not exceed the suggested incubation time (15 min). Re-Blot™ is a very basic solution (pH 14). Over-exposure to Re-Blot™ solutions may cause nitrocellulose alterations or nitrocellulose detachment (delamination) from the glass slide.
22. I-Block™ solution is a protein-based blocking reagent that is useful for blocking the nitrocellulose prior to immunostaining the array. A minimum blocking time of 1 h at room temperature, with gently rocking, is recommended while longer blocking times are not detrimental. If blocking must be performed overnight, block the slides at 4°C.
23. Although the Dako Autostainer has a maximum capacity of 48 slides, we have found it best to stain a maximum of 36 slides per staining run. The nitrocellulose slides have a tendency to dry out during extended staining runs. Paper towels soaked in water may be placed inside the sink area of the Autostainer to maintain humidity during the staining run. Alternatively, a shallow dish of water may be placed inside the left side of the Autostainer chamber.
24. Each primary antibody must be validated by Western blotting to confirm specific interaction between the protein of interest and the antibody, using complex samples similar to those which will be used on the array.
25. TBST contains a high concentration of salt. If the TBST is not rinsed from the stained nitrocellulose arrays, salt crystals may form on the nitrocellulose surface. Consequently, the Autostainer program includes a water rinse after DAB deposition. Moreover, it is possible to add a further water rinse (auxillary step) after the final water rinse in order to pause the instrument for 840 min. By doing this, the Autostainer is programmed to be in an idle status for 14 h, at which time the slides will be rinsed again with deionized water (Table 2).

26. Image adjustments should only include those adjustments that change all pixel intensities in an image in a linear, consistent manner. All image adjustments must be performed prior to spot analysis and must be consistent for all array slides. It is important to note that some image manipulation programs are capable of changing the actual pixels in the image.
27. The total protein concentration in a large set of printed array slides may vary from the initial slides printed to the last slides printed due in part to sample evaporation and the volume of fluid held by the pin. It is recommended to stain 1 of every 25 slides with Sypro Ruby Protein Blot stain. For example, if 120 microarrays have been printed it is suggested to use slides 25, 50, 75, and 100 for total protein quantification.
28. A UV transilluminator (~300 nm), a blue-light transilluminator, or a laser scanner that emits at 450, 473, 488, or 532 nm is appropriate for imaging a Sypro Ruby stained array. Example imaging systems are Kodak 4000 MM imager; Alpha Innotech NovaRay; Tecan Reloaded LS; Perkin Elmer ProScanArray HT; and Molecular Devices GenePix 4000B.

References

1. Espina, V., Edmiston, K. H., Heiby, M., Pierobon, M., Sciro, M., et al. (2008) A portrait of tissue phosphoprotein stability in the clinical tissue procurement process. *Mol Cell Proteomics* **7**, 1998–2018.
2. Li, X., Friedman, A. B., Roh, M. S., and Jope, R. S. (2005) Anesthesia and post-mortem interval profoundly influence the regulatory serine phosphorylation of glycogen synthase kinase-3 in mouse brain. *J Neurochem* **92**, 701–4.
3. Li, J., Gould, T. D., Yuan, P., Manji, H. K., and Chen, G. (2003) Post-mortem interval effects on the phosphorylation of signaling proteins. *Neuropsychopharmacology* **28**, 1017–25.
4. Becker, K. F., Schott, C., Hipp, S., Metzger, V., Porschewski, P., et al. (2007) Quantitative protein analysis from formalin-fixed tissues: implications for translational clinical research and nanoscale molecular diagnosis. *J Pathol* **211**, 370–8.
5. Fox, C. H., Johnson, F. B., Whiting, J., and Roller, P. P. (1985) Formaldehyde fixation. *J Histochem Cytochem* **33**, 845–53.
6. Helander, K. G. (1994) Kinetic studies of formaldehyde binding in tissue. *Biotech Histochem* **69**, 177–9.
7. Srinivasan, M., Sedmak, D., and Jewell, S. (2002) Effect of fixatives and tissue processing on the content and integrity of nucleic acids. *Am J Pathol* **161**, 1961–71.
8. Nassiri, M., Ramos, S., Zohourian, H., Vincek, V., Morales, A. R., et al. (2008) Preservation of biomolecules in breast cancer tissue by a formalin-free histology system. *BMC Clin Pathol* **8**, 1.
9. Devireddy, R. V. (2005) Predicted permeability parameters of human ovarian tissue cells to various cryoprotectants and water. *Mol Reprod Dev* **70**, 333–43.
10. He, Y., and Devireddy, R. V. (2005) An inverse approach to determine solute and solvent permeability parameters in artificial tissues. *Ann Biomed Eng* **33**, 709–18.
11. Goldstein, B. J. (2002) Protein-tyrosine phosphatases: emerging targets for therapeutic intervention in type 2 diabetes and related states of insulin resistance. *J Clin Endocrinol Metab* **87**, 2474–80.
12. Neel, B. G., and Tonks, N. K. (1997) Protein tyrosine phosphatases in signal transduction. *Curr Opin Cell Biol* **9**, 193–204.
13. Grellner, W., Vieler, S., and Madea, B. (2005) Transforming growth factors (TGF-alpha and TGF-beta1) in the determination of vitality and wound age: immunohistochemical study on human skin wounds. *Forensic Sci Int* **153**, 174–80.

14. Grellner, W. (2002) Time-dependent immunohistochemical detection of proinflammatory cytokines (IL-1beta, IL-6, TNF-alpha) in human skin wounds. *Forensic Sci Int* **130**, 90–6.
15. Grellner, W., and Madea, B. (2007) Demands on scientific studies: vitality of wounds and wound age estimation. *Forensic Sci Int* **165**, 150–4.
16. Ohshima, T. (2000) Forensic wound examination. *Forensic Sci Int* **113**, 153–64.
17. Oehmichen, M. (2004) Vitality and time course of wounds. *Forensic Sci Int* **144**, 221–31.
18. Paweletz, C. P., Charboneau, L., Bichsel, V. E., Simone, N. L., Chen, T., et al. (2001) Reverse phase protein microarrays which capture disease progression show activation of pro-survival pathways at the cancer invasion front. *Oncogene* **20**, 1981–9.
19. Petricoin, E. F., 3rd, Espina, V., Araujo, R. P., Midura, B., Yeung, C., et al. (2007) Phosphoprotein pathway mapping: Akt/mammalian target of rapamycin activation is negatively associated with childhood rhabdomyosarcoma survival. *Cancer Res* **67**, 3431–40.
20. VanMeter, A., Signore, M., Pierobon, M., Espina, V., Liotta, L. A., et al. (2007) Reverse-phase protein microarrays: application to biomarker discovery and translational medicine. *Expert Rev Mol Diagn* 7, 625–33.
21. Wulfkuhle, J. D., Speer, R., Pierobon, M., Laird, J., Espina, V., et al. (2008) Multiplexed cell signaling analysis of human breast cancer applications for personalized therapy. *J Proteome Res* 7, 1508–17.
22. Espina, V., Mehta, A. I., Winters, M. E., Calvert, V., Wulfkuhle, J., et al. (2003) Protein microarrays: molecular profiling technologies for clinical specimens. *Proteomics* **3**, 2091–100.
23. Belluco, C., Mammano, E., Petricoin, E., Prevedello, L., Calvert, V., et al. (2005) Kinase substrate protein microarray analysis of human colon cancer and hepatic metastasis. *Clin Chim Acta* **357**, 180–3.
24. Berggren, K., Steinberg, T. H., Lauber, W. M., Carroll, J. A., Lopez, M. F., et al. (1999) A luminescent ruthenium complex for ultrasensitive detection of proteins immobilized on membrane supports. *Anal Biochem* **276**, 129–43.
25. Berggren, K. N., Schulenberg, B., Lopez, M. F., Steinberg, T. H., Bogdanova, A., et al. (2002) An improved formulation of SYPRO Ruby protein gel stain: comparison with the original formulation and with a ruthenium II tris (bathophenanthroline disulfonate) formulation. *Proteomics* **2**, 486–98.
26. Espina, V., Wulfkuhle, J. D., Calvert, V. S., VanMeter, A., Zhou, W., et al. (2006) Laser-capture microdissection. *Nat Protoc* **1**, 586–603.
27. Stillman, B. A., and Tonkinson, J. L. (2000) FAST slides: a novel surface for microarrays. *Biotechniques* **29**, 630–5.
28. Tonkinson, J. L., and Stillman, B. A. (2002) Nitrocellulose: a tried and true polymer finds utility as a post-genomic substrate. *Front Biosci* 7, c1–12.

Chapter 4

Utilization of RNAi to Validate Antibodies for Reverse Phase Protein Arrays

Heiko Mannsperger, Stefan Uhlmann, Ulrike Korf, and Özgür Sahin

Abstract

Reverse phase protein arrays (RPPAs) emerged as a very useful tool for high-throughput screening of protein expression in large numbers of small specimen. Similar to other protein chemistry methods, antibody specificity is also a major concern for RPPA. Currently, testing antibodies on Western blot for specificity and applying serial dilution curves to determine signal/concentration linearity of RPPA signals are most commonly employed to validate antibodies for RPPA applications. However, even the detection antibodies fulfilling both requirements do not always give the expected result. Chemically synthesized small interfering RNAs (siRNAs) are one of the most promising and time-efficient tools for loss-of-function studies by specifically targeting the gene of interest resulting in a reduction at the protein expression level, and are therefore used to dissect biological processes. Here, we report the utilization of siRNA-treated sample lysates for the quantification of a protein of interest as a useful and reliable tool to validate antibody specificity for RPPAs. As our results indicate, we recommend the use of antibodies which give the highest dynamic range between the control siRNA-treated samples and the target protein (here: EGFR) siRNA-treated ones on RPPAs, to be able to quantify even small differences of protein abundance with high confidence.

Key words: Reverse phase protein arrays, Antibody validation, Antibody specificity, Epidermal growth factor receptor, RNAi, siRNAs

1. Introduction

Reverse phase protein arrays (RPPAs) have been developed as a promising tool for high-throughput screening of protein expression in large sample sets (1, 2). The basic principle of RPPAs follows the idea of a dot immunoblot where large numbers of samples are arrayed on numerous solid phase carriers in parallel and can thus be probed with a different monospecific antibody (Fig. 1). RPPAs provide a semiquantitative readout, meaning that the expression of a certain target protein can be compared among all samples printed per subarray.

Ulrike Korf (ed.), *Protein Microarrays: Methods and Protocols*, Methods in Molecular Biology, vol. 785,
DOI 10.1007/978-1-61779-286-1_4, © Springer Science+Business Media, LLC 2011

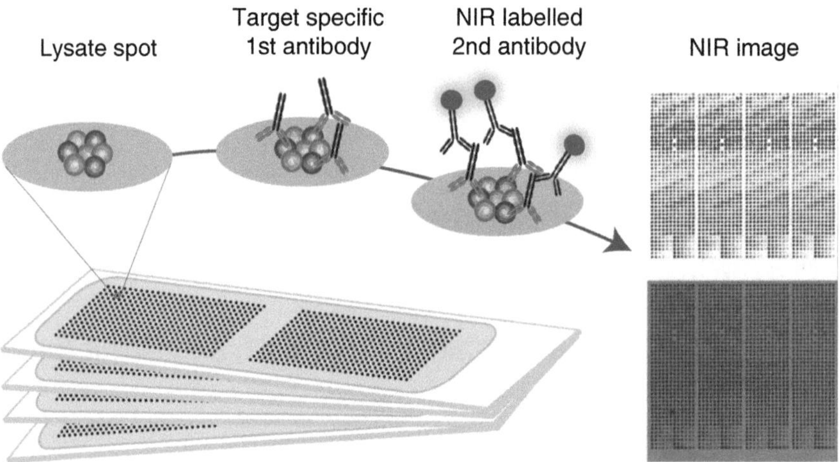

Fig. 1. The principle and workflow of reverse phase protein arrays (RPPAs). Cellular or tissue lysates are immobilized on a surface by spotting the lysates onto a nitrocellulose coated glass slide. Next, the target proteins contained in the lysate are detected with the help of monospecific primary antibodies. A NIR-labeled secondary antibody then detects the primary antibody and provides visualization and quantification of the signal (graphics by Frauke Henjes).

1.1. The Antibody Specificity for RPPA

For antibodies used in RPPA experiments, the specificity is more important than for antibodies used for Western blotting (WB) or immunohistochemistry (IHC). In WB experiments, unspecific binding of antibodies can be identified by comparing the molecular weight (MW) of the target protein with the MW of the detected proteins. Similarly, in IHC experiments, unspecific binding can be identified with respect to the cellular localization of the target protein for which certain localization is expected under certain conditions. In contrast, RPPA experiments provide no additional control of antibody specificity. Even minor unspecific binding leads to an increase of the signal and masks the target-specific signal. Testing antibodies on WB for monospecificity (Fig. 2a) is the most common method used to validate antibodies for RPPA experiments (3). However, even the antibodies showing monospecificity on WB as well as a linear correlation between protein concentration and signal intensity in serially diluted samples (Fig. 2b) are not always suitable to quantify their target protein on RPPAs.

1.2. RNA Interference

RNA interference (RNAi) is a biological process where small RNA molecules silence gene expression, either by inducing sequence-specific degradation of target mRNA or by inhibiting translation (4). After its first discovery by Fire and Mello in *Caenorhabditis elegans* (5) and proof that these mechanisms also work in mammalian cells (6), RNAi opened up a new era in reverse genetics and related fields enabling large-scale loss-of-function studies.

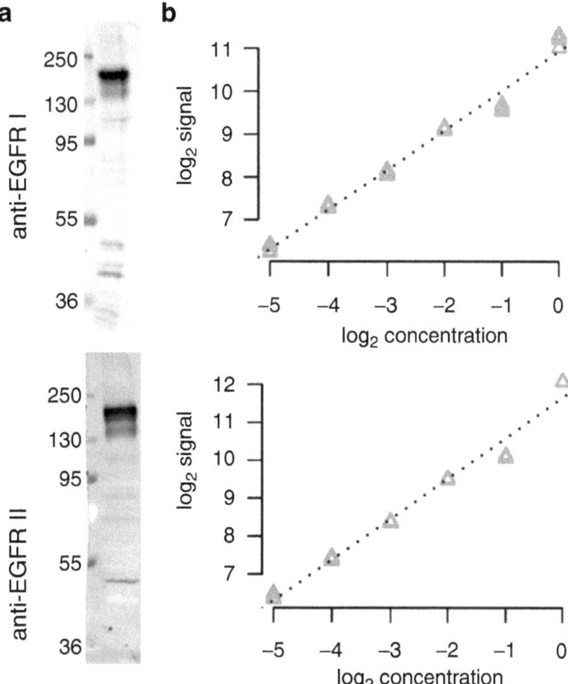

Fig. 2. Current approach to validate antibodies for use with RPPAs. (**a**) Antibody monospecificity is validated by Western blot and (**b**) the linear correlation of protein to signal intensities is shown as dilution series of cell lysate on RPPA.

Chemically synthesized small interfering RNA (siRNA) molecules have been shown to be potent effectors of post-transcriptional gene silencing, resulting in specific inhibition of protein expression (Fig. 3), and they are considered to be one of the most promising and time-efficient tools in dissecting several biological processes. We have previously applied RPPAs to quantify the residual protein expression levels after applying multiple siRNAs simultaneously (7), as well as to reconstruct protein networks by quantifying all of the proteins in the network after knocking down each protein (8, 9).

1.3. Use of RNAi to Validate Antibodies for RPPAs

As previously reported, positive controls to validate the phospho-specific antibodies against several proteins can be generated using various treatments like UV light or growth factors (10). This way, one can easily discriminate the highly phosphorylated samples from nonphosphorylated ones in any given sample with high confidence.

As we have stated in Subheading 1.1, a single band on WB as well as a linear signal may not be sufficient to validate antibodies for RPPAs. Alternatively, for the specific detection and sensitive quantification of the expression of target proteins, one would need a recombinant protein of the target being analyzed at different

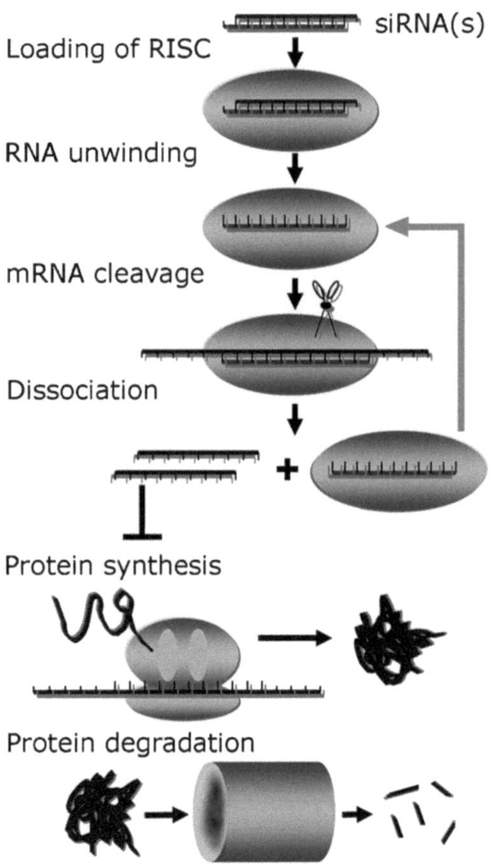

Fig. 3. Mechanism of siRNA-mediated protein knockdown. Chemically synthesized double-stranded RNA molecules are first loaded into a protein complex that is called RNA-induced silencing complex (RISC) whose helicase activity unwinds the double-stranded siRNAs in an ATP-dependent manner. The antisense strand then targets the specific mRNA and results in cleavage. In consequence, cleaved mRNAs of the target gene cannot be translated into the protein. Due to the turnover of proteins in the cell and the lack of supply with new protein product, a reduction in the residual protein level is achieved.

dilutions, which is a tedious and expensive approach. Therefore, we recommend including one more easy-to-perform and time-efficient validation step. Here, we report the use of siRNA-treated sample lysates as an indispensable step to validate antibodies against a specific target protein for RPPA using the knockdown of EGFR as an example. After validation of the knockdown efficiency of siRNAs by qRT-PCR at mRNA level (quantitatively) (Fig. 4a) and/or Western blotting at protein level (qualitatively or semiquantitatively) (Fig. 4b), the siRNA-treated samples are printed onto nitrocellulose-coated slides. The antibody that shows the largest difference between siRNA-treated sample and siRNA control corresponds to the antibody with the highest dynamic range, and is thus more suitable for use in RPPA experiments (Fig. 4c).

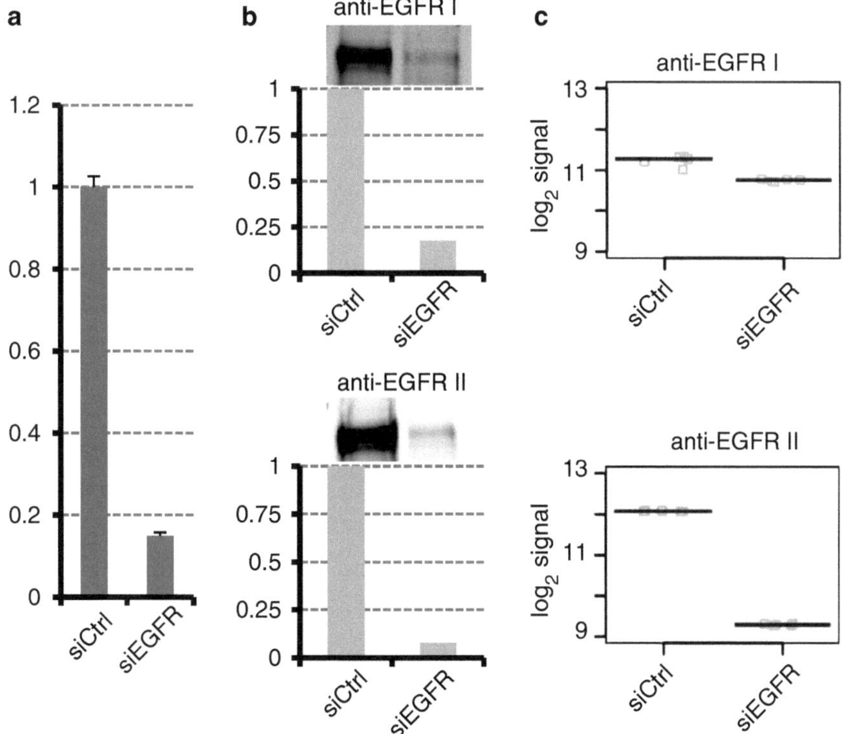

Fig. 4. Use of siRNAs for the validation of antibodies for use with RPPAs. (**a**) qRT-PCR validation of siRNA-mediated knockdown of EGFR. (**b**) Validation of two different EGFR antibodies (anti-EGFR I and anti-EGFR-II) in Western blot by assessing their capacity to detect a knockdown of EGFR protein after treatment with EGFR siRNA qualitatively and semiquantitatively. (**c**) Utilization of control siRNA- and EGFR siRNA-treated samples for the validation of EGFR antibodies which are suitable for RPPAs.

We report that if the antibody is able to detect the reduced target protein expression of the knockdown sample with a higher dynamic range, it can also detect minor and, of course, also major changes in protein expression in any biological and clinical application of RPPA.

2. Materials

2.1. Cell Culture

1. Human breast cancer cell line MDA-MB-231 (ATCC, Manassas, VA).
2. Leibovitz's L-15 medium (Sigma, St Louis, MO) supplemented with 1% nonessential amino acids, 10% fetal bovine serum (Gibco-BRL, Bethesda, MD), and 3 g/L sodium bicarbonate (AppliChem, Darmstadt, Germany).
3. Trypsin–EDTA solution with 0.25% trypsin (Sigma, St Louis, MO).

2.2. siRNAs and Materials for Transfections

1. Reduced-serum transfection medium Opti-MEM (Gibco-BRL, Bethesda, MD).
2. Lipofectamine 2000 (Invitrogen, CA).
3. siRNAs targeting EGFR (NM_005228.3) are obtained from Dharmacon (Lafayette, CO) and control siRNA Allstar Negative Control siRNA is from Qiagen (Hilden, Germany).

2.3. Preparation of Cell Lysates

1. Cell lysis buffer: M-PER Mammalian Protein Extraction Reagent (Thermo Scientific).
2. Protease inhibitor: mini Complete (Roche).
3. Phosphatase inhibitor: PhosStop (Roche).

2.4. Western Blotting

1. Protein gel system: BioRad (Munich, Germany).
2. Semidry protein blotting device: Trans-Blot SD Transfer Cell (BioRad, Munich, Germany).
3. Acrylamide Gel: BioRad (Munich, Germany).
4. Loading buffer: Roti-Load (Roth, Karlsruhe, Germany).
5. Running buffer: 25 mM Tris–HCl, 192 mM glycine, 0.1% SDS (w/v).
6. Blotting paper: Whatman.
7. PVDF membrane.
8. Cathode solution: 200 mL methanol, 5.2 g aminohexanoic acid, add ddH_2O 1 L.
9. Anode solution I: 200 mL methanol, 36.4 g Tris base, add ddH_2O 1 L.
10. Anode solution II: 200 mL methanol, 3 g Tris base, add ddH_2O 1 L.
11. Wash buffer: 0.2% Tween in PBS.
12. Blocking buffer: Odyssey blocking buffer 1:2 diluted in Wash buffer.
13. Buffer for second antibody: 0.01% (w/v) SDS in Wash buffer.

2.5. RPPA Materials

1. Contact spotter: Aushon 2470 (Aushon, Billerica, MA).
2. Nitrocellulose-coated glass slides: Oncyte Avid (Grace Biolabs, Bend, OR).
3. 384-Well microtiter plate.
4. Spotting buffer: 1% (w/v) Tween in PBS.
5. Needle wash solution: ddH_2O.

2.6. Analysis of RPPA Results

1. NIR Scanner: Odyssey (LI-COR, Lincoln, NB).
2. Image analysis software: GenePix Pro (Molecular Device, Sunnyvale, CA).
3. Data analysis: R statistical software.

3. Methods

3.1. Cell Culture

Seed 2×10^5 of MDA-MD-231 cells per well in a six-well format in their growth medium 24 h prior to the transfection and incubate at 37°C and 5% CO_2.

3.2. siRNA Transfections

The following procedure is designed for the transfection of one well in a six-well plate.

1. Prepare two mixes (mix A and B) with 250 μL of Opti-MEM, respectively.
2. Add 4 μL of Lipofectamine 2000 to mix A and 1.5 μL siEGFR (20 μM) to mix B to get a final concentration of 20 nM.
3. Vortex briefly and incubate both mixes (A and B) for 5 min at room temperature.
4. Afterward, transfer mix A to mix B and vortex thoroughly. Incubate again at room temperature, this time for 20 min.
5. During this 20-min incubation period, aspirate the medium and rinse with PBS.
6. Add 1 mL of medium without antibiotics to each well.
7. After this 20-min incubation step, transfer 500 μL of the transfection mix (combined mix of A and B) into one well of the six-well plate and incubate the cells for 48 h.

3.3. Preparation of Samples for Western Blotting

1. Trypsinize samples from cell culture dish.
2. Add one tablet PhosStop and one tablet mini Complete to 10 mL M-PER.
3. Lyse cell pellet from six-well dish in 30 μL lysis buffer.
4. Incubate lysate for 20 min at 4°C (see Note 1).
5. Centrifuge lysate for 8 min at 16,000 × *g*, 4°C.
6. Take supernatant and analyze protein concentration using BCA or Bradford assay.
7. Dilute samples with M-PER to adjust the same concentration in all samples, fill spotting source plate, and add Tween to a final concentration of 0.05%. Store samples at –20°C until spotting.

3.4. Western Blotting

1. Thaw the samples on ice and take subsequently the required sample volume in the labeled Eppi. Mix the sample with fourfold denaturating buffer (three parts sample, one part denaturating buffer) on the vortex mixer and centrifuge shortly. Denature the sample for 5 min at 95°C. Put the sample for 3 min on ice and centrifuge shortly afterward.
2. To adjust protein concentration in the lanes on the gel the sample concentration has to be normalized by adding diluted (onefold) denaturating buffer.

3. Mount the running chamber, fill in running buffer, and check buffer level after a few minutes.
4. Remove the comb from the top of the gel; rinse slots with running buffer; and load the required sample volume (see Note 2).
5. Connect power source and run the electrophoresis depending on the density of the gel and the molecular weight of the target protein (see Note 3).
6. For the semidry blot, assemble the layers as follows: four blotting papers soaked in Anode buffer I, two blotting papers soaked in Anode buffer II, PVDF-Membrane (steeped for 5 s in methanol), gel (stacking gel removed), six blotting papers in Cathode buffer. To remove air bubbles between the layers, roll the assembly with the thick end of a Pasteur pipette. Transfer the proteins onto the membrane for 1 h at 25 V.
7. Place the membrane in an incubation box after blotting. Add 10 mL of Blocking buffer and block the membrane for 1 h at RT on a rocking platform.
8. Dilute the target specific first AB in Blocking buffer, depending on the instruction manuals of the company (e.g., dilute ABs from Cell Signaling technology 1:1,000), and incubate over night on a rocking platform in the cold storage room. After incubation, wash the blot four times in 10 mL wash buffer.
9. Dilute the corresponding second AB 1:10,000 in second AB incubation buffer and incubate for 1 h maximum at RT on a rocking platform. Wash the blot four times in 10 mL wash buffer.
10. Store the membrane in wash buffer until the scanning. Place the membrane directly on the glass surface of the scanner and scan the membrane wet (avoid drying).

3.5. RPPA Spotting

1. Thaw spotting source plate on ice and centrifuge for 1 min at 360 × *g*. Place samples in the source plate carrier of the Aushon 2470.
2. Label slides and put them on the slide plate carrier.
3. Check wash buffer and waste container of the spotting device.
4. Design array layout and start print run.

3.6. Microarray Immunostaining

Incubate RPPAs according to the standard NIR detection method or apply antibody-mediated signal amplification (AMSA) as described in the previous chapter (see Note 4). A further method based on tyramide signal amplification is described by Spurrier et al. (3) (see Note 5).

3.7. Analysis of RPPA Results

1. Analyze the NIR image using the GenePix Pro software. Apply the GenePix array list (gal file) that is produced by the spotting device (see Note 6).

2. Prepare tables describing the samples (see Note 7) and the arrays (see Note 8) and put them together with the GenePix result files (gpr files) in one folder.
3. Read, merge, and normalize the data using the package RPPanalyzer implemented in the statistical software R (see Note 9).

4. Notes

1. Ensure that the lysate is gently mixed during the incubation. By using a thermomixer (Thermo scientific) or a rotating wheel. Smaller lysate volumes require vigorous mixing.
2. For the analysis of EGFR in MDA-MB-231 cells with NIR detection system, total protein amount of at least 10 μg per lane is required.
3. Use an SDS-gel with 7.5% acrylamide and run it for 60 min at 120 kV for the detection of EGFR.
4. For quality control include one blank array (incubated with second antibody only) in each incubation run and for normalization purposes one array incubated with total protein stain in every spotting run.
5. We recommend using a NIR-based detection method because of the high dynamic range of NIR signals.
6. Adjust gal file produced by the Aushon 2470 for the analysis of LiCor Oyssey NIR images by multiply all spacing parameter in the gal file with the factor 0.47166. To reduce the standard deviation between the replicate spots do not use the "Resize features during alignment" option, which is the default setting in GenePix.
7. The sample description is stored as a tab-delimited text file (ASCII code only) containing the columns describing the location in the source well plate: plate, row, column, and columns describing the details of each sample. The columns for sample type, sample, and concentration are obligatory but any other attribute of interest can be added.
8. The slide description file is stored as a tab delimited text file (ASCII code only) describing the array features. The columns gpr (with the names of the gpr files), pad (integer for the number of the pads), slide (slide number), incubation run, spotting run, target, and antibody ID. For further information any other attribute of interest can be added.
9. Read in the data using the read.Data command and correct for background using correctBG. In case of serially diluted samples,

calculate the concentration value before the normalization step (otherwise you will flatten the slope of the dilution series). If you are not familiar with data analysis using R it is possible to export the data as text file after the normalization step (with the command write.Data) and continue analysis using spreadsheet software.

References

1. Paweletz, C.P., et al., Reverse phase protein microarrays which capture disease progression show activation of pro-survival pathways at the cancer invasion front. Oncogene, 2001. 20(16): 1981–9.
2. Charboneau, L., et al., Utility of reverse phase protein arrays: applications to signalling pathways and human body arrays. Brief Funct Genomic Proteomic, 2002. 1(3): 305–15.
3. Spurrier, B., S. Ramalingam, and S. Nishizuka, Reverse-phase protein lysate microarrays for cell signaling analysis. Nat Protoc, 2008. 3(11): 1796–808.
4. Mittal, V., Improving the efficiency of RNA interference in mammals. Nat Rev Genet, 2004. 5(5): 355–65.
5. Fire, A., et al., Potent and specific genetic interference by double-stranded RNA in Caenorhabditis elegans. Nature, 1998. 391(6669): 806–11.
6. Elbashir, S.M., et al., Duplexes of 21-nucleotide RNAs mediate RNA interference in cultured mammalian cells. Nature, 2001. 411(6836): 494–8.
7. Sahin, O., et al., Combinatorial RNAi for quantitative protein network analysis. Proc Natl Acad Sci USA, 2007. 104(16): 6579–84.
8. Sahin, O., et al., Modeling ERBB receptor-regulated G1/S transition to find novel targets for de novo trastuzumab resistance. BMC Syst Biol, 2009. 3: 1.
9. Frohlich, H., et al., Deterministic Effects Propagation Networks for reconstructing protein signaling networks from multiple interventions. BMC Bioinformatics, 2009. 10: 322.
10. Spurrier, B., et al., Antibody screening database for protein kinetic modeling. Proteomics, 2007. 7(18): 3259–63.

Chapter 5

Antibody-Mediated Signal Amplification for Reverse Phase Protein Array-Based Protein Quantification

Jan C. Brase, Heiko Mannsperger, Holger Sültmann, and Ulrike Korf

Abstract

Reverse phase protein array (RPPA) techniques allow the quantitative analysis of signal transduction events in a high-throughput format. Sensitivity is important for RPPA-based detection approaches, since numerous signaling proteins or posttranslational modifications are present at low levels. Especially, the proteomic analysis of clinical samples exposes its own challenges with respect to sensitivity. Antibody-mediated signal amplification (AMSA) is a novel strategy relying on sequential incubation steps with fluorescently labeled secondary antibodies reactive against each other. AMSA is a simple extension of the standard quantification in the near-infrared range and is highly specific and robust. In this chapter, we present the amplification protocol and application examples for the time-resolved analysis of signaling pathways as well as protein profiling of clinical samples.

Key words: Reverse phase protein microarray, Signal amplification, Antibody specificity, Near-infrared detection, Sensitivity

1. Introduction

Reverse phase protein arrays (RPPAs) have emerged as a high-throughput technique for the analysis of signaling pathways and validation of biomarker candidates in biological and clinical samples (1–5). In RPPA-based experiments, protein lysates of cells or clinical tissue samples are commonly robotically spotted on nitrocellulose-coated glass slides. Sensitivity and specificity of signal detection are critical for RPPA-based measurements, since protein lysates are routinely spotted without any separation or purification steps. Proteins of interest can be present at low level requiring means for specific signal amplification.

Ulrike Korf (ed.), *Protein Microarrays: Methods and Protocols*, Methods in Molecular Biology, vol. 785,
DOI 10.1007/978-1-61779-286-1_5, © Springer Science+Business Media, LLC 2011

Near-infrared (NIR) fluorescence-based detection was reported as useful for reverse phase microarray applications (6, 7). Antibody-mediated signal amplification (AMSA) is a convenient and cost-effective NIR fluorescence-based approach for the robust and specific quantification of low abundant proteins on RPPAs. After target protein detection, amplification is carried out by two different NIR-labeled species-specific antibodies. Both secondary antibodies are designed to recognize the respective other species. Several rounds of incubation increase the density of fluorescent signals and result in a strong quantitative read-out signal. All working steps can be adapted to a fully automated procedure.

Typical applications of AMSA are the quantitative analysis of time-resolved measurements in systems-biology applications as well as protein profiling of clinical samples. Especially, the extraction of proteins from clinical samples from embedded tissue samples results frequently in lysates with low total protein concentrations. Besides that, assay costs for AMSA are low since all reagents are used highly diluted in the course of signal amplification.

2. Materials

2.1. Cell Culture and Time-Course Experiment

1. Cultured cells and medium (see Note 1).
2. Solution of trypsin (Sigma, Munich, Germany), store at –20°C.
3. Modified lysis buffer for cell lines: MPER (Pierce, Rockford, USA), store at RT.
4. 1 M *ortho*-sodium-vanadate stock solution (Sigma), store at RT.
5. 0.5 M NaF stock solution (Honeywell Riedel-de-Haën, Seelze, Germany), store at RT.
6. Complete mini protease inhibitor (Roche, Basel, Switzerland), store at 4°C.
7. 10 ml MPER containing 2 mM *ortho*-sodium-vanadate, 10 mM NaF, one complete mini protease inhibitor.
8. EGF (Sigma) dissolved in PBS (50 ng/ml) and stored in single use aliquots at –80°C.
9. Gefitinib (Astra Zeneca, London, UK), diluted in DMSO (Sigma) and stored in aliquots at 4°C.
10. Cetuximab (Merck Serono, Darmstadt, Germany) stored at –4°C.
11. Cell scrapers (TPP, Zurich, Switzerland).

2.2. Clinical Tissue Preparation

1. TPER buffer (Pierce).
2. 2,000 μM Staurosporine (Roche) store in aliquots at −80°C.
3. 0.5 M EDTA stock solution.
4. Modified lysis buffer for tissue extraction. T-PER buffer (Pierce), one PhosStop per 10 ml (Roche), 2 μM staurosporine (Roche), 1 mM EDTA, one complete mini per 10 ml (Roche).
5. Stainless steel beads (Qiagen, Hilden, Germany).
6. Tissuelyser (Qiagen).
7. Qiashredder (Qiagen).

2.3. Antibody Specificity: Western Blotting

Antibodies used for RPPA must be thoroughly characterized (6). A suitable method is also described in Chapter 4 by Mannsperger and coauthors.

2.4. Printing of Protein Microarrays

1. Spotting buffer: 50% glycerol/deionized water, 0.05% Triton X-100, store at 4°C in ddH_2O.
2. Nitrocellulose-coated glass slides (Grace Biolabs, Bent, OR, USA).
3. Protein printer (see Note 2).

2.5. Antibody-Mediated Signal Amplification on Protein Microarrays

1. PBS stock solution: 1.37 M NaCl, 27 mM KCl, 18 mM KH_2PO_4, 100 mM NA_2PO_4, pH 7.4.
2. Incubation chamber SN1000308003 (Metecon, Mannheim, Germany) (see Note 3).
3. Modified blocking buffer: 33% Odyssey blocking buffer (LI-COR, Lincoln, USA), 1% BSA, 0.02% NP40 (Igepal Ca-630; Sigma) in PBS, store at 4°C.
4. Buffer with background-reducing components (Dako, Glostrup, Denmark), store at 4°C.
5. Washing buffer: 1× PBS, 0.02% NP40, 0.02% SDS.
6. Robot Biomek FXP (Beckmann Coulter, Harbor Boulevard, USA).
7. Target-specific antibodies (rabbit or mouse).
8. Anti-rabbit Alexa680-labeled (raised in goat) antibody (Invitrogen).
9. Anti-goat Alexa680-labeled (raised in rabbit) antibody (Invitrogen).

2.6. Statistical Analysis

1. Odyssey NIR scanner (LI-COR Bioscience).
2. GenePix-Pro 5.1 (Axon Instruments, Sunnyvale, USA).
3. Statistical computing environment R.

3. Methods

3.1. Cell Culture and Time-Course Experiment

1. Cell lines are grown until 90% confluence and split three times per week when reaching confluence.
2. For the time-course experiment, 5×10^5 cells are seeded per well in six-well plates and cultivated for 24 h under serum-free conditions. After a 24-h period of starvation, EGF is applied at a concentration of 50 ng/ml.
3. For experiments with epidermal growth factor inhibitors, cells are cultivated with 2 μM Gefitinib or 100 nM Cetuximab under serum-free conditions for 30 min before EGF stimulation.
4. At each time point, the medium is removed and cells are washed with ice-cold PBS.
5. Cells are transferred to a collection tube after manual scraping on ice.
6. Lysis is performed for 15 min at 4°C with shaking.
7. Measure protein quantity with BCA assay, store tubes at -80°C.

3.2. Clinical Tissue Preparation

1. Place tubes with frozen tissue on dry ice and determine the weight of the tissue samples using a balance.
2. Open tubes and add 10 μl volume of tissue lysis buffer per mg of tissue sample.
3. Add one stainless steel bead per tube.
4. Incubate for 5 min on wet ice to thaw the tissue specimen.
5. Samples are homogenized 4 min with a tissuelyser (30 Hz).
6. Freeze samples on dry ice. Store all tubes at –80°C.
7. Thaw and mix samples in a thermomixer (300 rpm) for 15 min at 4°C.
8. Cell debris is subsequently pelleted (12 min, $>10{,}000 \times g$, 4°C).
9. Transfer supernatant to a shredder tube (Qiagen), centrifuge 1 min at $>10{,}000 \times g$, 4°C.
10. Store samples on wet ice, measure quantity with BCA, store tubes at –80°C.

3.3. Antibody Specificity: Western Blotting

Antibody specificity tests are described in detail in Chapter 4 by Mannsperger et al. Specificity test can be performed with standard NIR detection as well as AMSA (Fig. 1, also see Note 4).

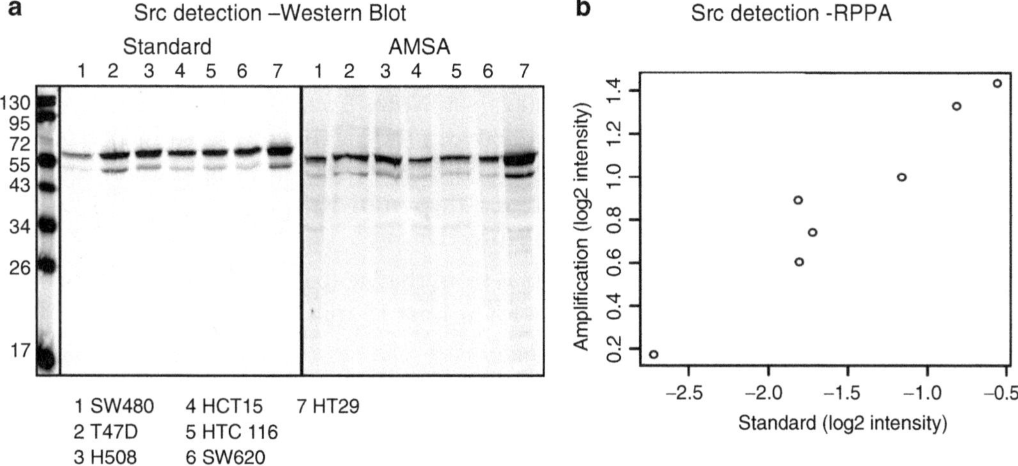

Fig. 1. Data comparison between standard near-infrared detection and antibody-mediated signal amplification of the protein SRC in seven cancer cell lines by Western blotting (**a**) and associated correlation analysis of SRC-specific RPPA detection (**b**).

3.4. Printing of Protein Microarrays

1. All protein lysates are normalized to an identical protein concentration with lysis buffer.
2. All samples are printed onto nitrocellulose slides using a non-contact or pin tool spotter.
3. Slides are stored at 4°C and used within a week (−20°C for long-time storage).

3.5. Antibody-Mediated Signal Amplification on Microarrays

All working steps can be performed as manual or automated procedure (see Note 5).

All washing and incubation steps are carried out at RT with gentle shaking.

1. Mount slides in incubation chambers.
2. Dispense 600 μl of modified blocking buffer and incubate for 1 h.
3. Primary antibodies are diluted in a buffer with background-reducing components (see Note 6).
4. Dispense 400 μl primary antibody dilution into each incubation chamber and incubate for 2 h.
5. Remove antibody solution by aspiration and wash slides in wash buffer (1.5 ml) four times for 5 min.
6. Slides are incubated with 400 μl anti-rabbit Alexa680-conjugated secondary antibody (dilution 1:8,000) for 30 min (see Note 7).
7. Slides are rinsed with wash buffer four times for 5 min as described before.

8. 400 μl anti-goat Alexa680-labeled (raised in rabbit) and anti-rabbit Alexa680-labeled (raised in goat) antibodies (both Invitrogen) are applied consecutively in a total of four cycles (dilution 1:8,000) (see Notes 8 and 9). Slides are washed four times for 5 min between the incubation steps. Each secondary antibody is incubated for 30 min. The amplification steps are sufficient to increase the sensitivity and S/N of the RPPA measurements (see Note 10).
9. Remove the slides from the incubation chamber and allow the slides to air-dry for at least 15 min at room temperature.

3.6. Statistical Analysis and Data Interpretation

1. Slides are scanned with the Odyssey Infrared Imaging System (scan settings: resolution: 21 μm, 700 or 800 nm dependent on the secondary antibody) (see Note 11).
2. Image analysis can be done using the GenePix Pro 5.1 Software.
3. Spot intensity is corrected for background and noise due to unspecific antibody binding.
4. (a) Protein profiling of clinical samples can be done using statistical computing environment R. The protein expression analysis of normal and prostate cancer tissue is shown as an example in Fig. 2 (also see Note 12). (b) RPPA in combination with AMSA detection can be applied for the time-resolved measurements in systems-biology type of experiments. One example is shown in Fig. 3 (also see Note 13).

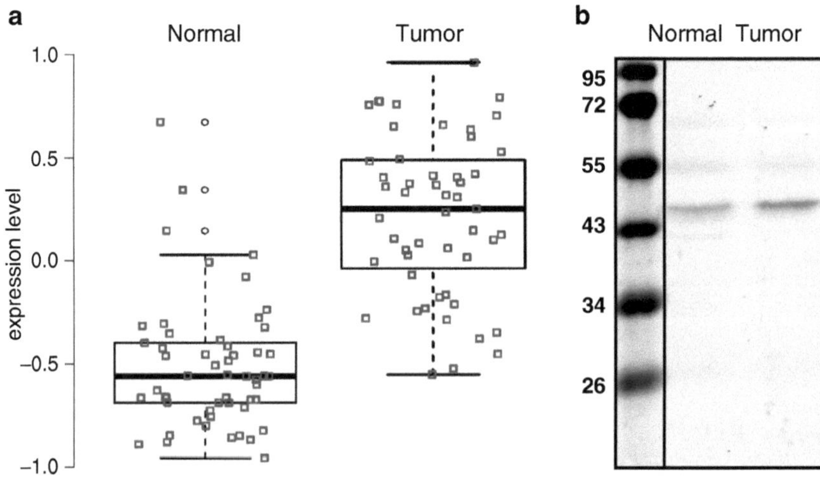

Fig. 2. (**a**) RPPA-based protein quantification of MEK in prostate cancer samples in comparison to benign samples. (**b**) Antibody specificity validation using Western blotting. Detection of MEK in pools of eight protein samples collected from benign prostate tissue or from eight cancer samples.

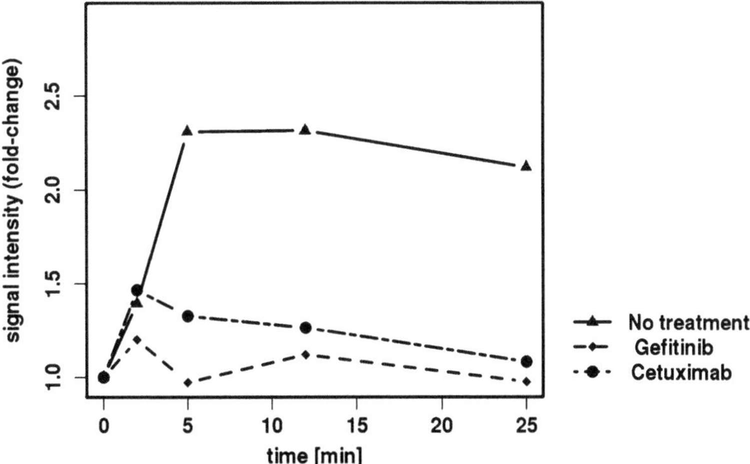

Fig. 3. Time-resolved analysis of SW480 cells stimulated with 50 ng/ml EGF. Signals were normalized to total protein concentration. The expression level of pERK was monitored and normalized to the nonstimulated sample (0 min). Blocking EGFR using gefitinib or cetuximab inhibits ERK phosphorylation. To ensure comparability, the signals are illustrated on the same scale using the fold-change to time point zero. Cetuximab treatment prior to EGF resulted in reduced ERK phosphorylation and inactivation was also more rapid. Gefitinib treatment revealed a more stringent inhibition of epidermal growth factor signaling.

4. Notes

1. Medium supplements depend on the particular cell line used, experimental set-up, and the biological question. The data shown in Fig. 3 were produced in the human colon cell line SW480 cells (ATCC), and RPMI 1640 medium supplemented with 10% FCS and 1% Pen/Strep.
2. Different printers can be used for the preparation of RPPAs. We used either a noncontact inkjetprinter Sprint (ArrayJet, Roslin, Scotland) or pin tool spotter, 2470 arrayer (Aushon, Billerica, MA, USA). The preparation protocol depends on the particular spotter. For instance, equal volumes of spotting buffer are added to each sample when the noncontact inketprinter is used.
3. Incubation chambers for protein microarray applications are available from various vendors. We prefer to use the incubation chamber SN1000308003 separating the nitrocellulose-coated glass slide into two independent subarrays. This allows incubating with two different antibodies on a single slide with two subarrays.
4. By directly comparing standard NIR and AMSA on Western blots, no artificial background signals were observed for the amplification routine. RPPA-based measurements also highly correlate with Western blot data as shown for a set of highly

abundant proteins (exemplified in Fig. 1). Since the AMSA protocol is more time consuming than the standard NIR detection we recommend the use of standard NIR detection for the validation of target-specific antibodies.

5. Although the AMSA protocol can in principle be performed manually. We recommend the use of a 96-well head robot to reduce hands-on time and to minimize experimental variation. We established a robotics protocol to increase and simplify the throughput: First, the slides are placed into incubation chambers and incubated with blocking buffer. Dilutions of the different first as well as the secondary antibodies are prepared in 96-well plates. A 96-channel head robot was set-up to use the prepared plates as required. The most efficient time plan: prepare the 96-well plate in the afternoon and start the robot program. The assay incubation steps require 8 h so that the slides can be analyzed the next morning.
6. The dilution of primary antibodies depends on a particular detection antibody and has to be tested ahead of the array-based detection.
7. Anti-mouse Alexa680 labeled (raised in goat) antibody is used in the first cycle for the detection of primary antibodies raised in mouse.
8. Secondary antibodies employed for signal amplification are from commercial suppliers and of highest purity. However, we confirmed the species specificity of all antibodies by printing dilution series of commercially available human immunoglobulin protein on nitrocellulose slides. RPPA-based signal detection according to the established protocol omitting incubation with a target-specific detection antibody yielded no signals. Thus, both species-specific antibodies revealed no cross-reactivity with human antibodies potentially present in clinical samples prepared from human material.
9. To determine the appropriate number of amplification cycles, we determined the best tradeoff between specific and background signal intensity gain using spike-in control. A continuous increase of signal intensities was observed from 1 to 4 amplification steps. The signal-to-noise ratio improved considerably, and the best results were obtained by performing four amplification cycles. Five cycles of amplification did not have further advantage. Therefore, we recommend four cycles of amplification after standard NIR detection.
10. We compared the sensitivity and signal-to-noise ratios between AMSA and standard NIR detection. Dilution curves of spike-in proteins showed that this amplification method yields lower detection limits and significantly increased signal-to-noise ratios. The limit of detection was reduced almost tenfold when compared to standard NIR detection procedure. Additional

experiments demonstrate that due to improved signal-to-noise ratios AMSA is especially useful for the RPPA-based quantification of proteins expressed at a low level and reveals a large linear signal range.

11. Depending on the applied first antibody the intensity may need to be adjusted. If saturation signals are visible, the intensity has to be reduced.

12. Useful antibodies for RPPA-based protein quantification are commonly characterized by a single band and low cross-reactivity in Western blotting. Each antibody should be validated using biological samples, which are analyzed with RPPAs. Therefore, we recommend the use of a representative sample pool from each biological or clinical group for Western blot validation (exemplified in Fig. 2b).

13. Time-resolved measurements of signaling events can efficiently be monitored by RPPAs in combination with AMSA. In our application example two drugs targeting the EGFR signaling pathway were applied to monitor the effect on downstream signaling events. SW480 human colon cancer cells were used to measure the effects of Gefitinib and Cetuximab on downstream protein phosphorylation in a time-resolved manner. All samples were collected after stimulation with EGF. A specific antibody directed against phosphorylated ERK was used to monitor intracellular signaling (Fig. 3).

Acknowledgments

We thank Maike Wosch and Annika Bittmann for their excellent technical assistance. This work was supported by the German Federal Ministry for Education and Science in the framework of the Program for Medical Genome Research (grants 01GS0890 and 01GS0864), the Program for Medical Systems Biology (grant 0315396B), as well as the Helmholtz Systems Biology Initiative (SBCancer).

References

1. Paweletz, C. P., Charboneau, L., Bichsel, V. E., Simone, N. L., Chen, T., Gillespie, J. W., Emmert-Buck, M. R., Roth, M. J., Petricoin, I. E., and Liotta, L. A. (2001) Reverse phase protein microarrays which capture disease progression show activation of pro-survival pathways at the cancer invasion front, *Oncogene 20*, 1981–1989.
2. Nishizuka, S., Charboneau, L., Young, L., Major, S., Reinhold, W. C., Waltham, M., Kouros-Mehr, H., Bussey, K. J., Lee, J. K., Espina, V., Munson, P. J., Petricoin, E., 3rd, Liotta, L. A., and Weinstein, J. N. (2003) Proteomic profiling of the NCI-60 cancer cell lines using new high-density reverse-phase lysate microarrays, *Proc Natl Acad Sci USA 100*, 14229–14234.
3. Sheehan, K. M., Gulmann, C., Eichler, G. S., Weinstein, J. N., Barrett, H. L., Kay, E. W., Conroy, R. M., Liotta, L. A., and Petricoin, E. F., 3rd. (2008) Signal pathway profiling of

epithelial and stromal compartments of colonic carcinoma reveals epithelial-mesenchymal transition, *Oncogene 27*, 323–331.

4. Haller, F., Lobke, C., Ruschhaupt, M., Cameron, S., Schulten, H. J., Schwager, S., von Heydebreck, A., Gunawan, B., Langer, C., Ramadori, G., Sultmann, H., Poustka, A., Korf, U., and Fuzesi, L. (2008) Loss of 9p leads to p16INK4A down-regulation and enables RB/E2F1-dependent cell cycle promotion in gastrointestinal stromal tumours (GISTs), *J Pathol 215*, 253–262.
5. Haller, F., Lobke, C., Ruschhaupt, M., Schulten, H. J., Schwager, S., Gunawan, B., Armbrust, T., Langer, C., Ramadori, G., Sultmann, H., Poustka, A., Korf, U., and Fuzesi, L. (2008) Increased KIT signaling with up-regulation of cyclin D correlates to accelerated proliferation and shorter disease-free survival in gastrointestinal stromal tumours (GISTs) with KIT exon 11 deletions, *J Pathol 216*, 225–235.
6. Loebke, C., Sueltmann, H., Schmidt, C., Henjes, F., Wiemann, S., Poustka, A., and Korf, U. (2007) Infrared-based protein detection arrays for quantitative proteomics, *Proteomics 7*, 558–564.
7. Calvert, V. S., Tang, Y., Boveia, V., Wulfkuhle, J., Schutz-Geschwender, A., Olive, D. M., Liotta, L. A., and III, E. F. P. (2004) Development of Multiplexed Protein Profiling and Detection Using Near Infrared Detection of Reverse-Phase Protein Microarrays, *Clinical Proteomics 1*, 81–89.

Chapter 6

Reverse-Phase Protein Lysate Microarray (RPA) for the Experimental Validation of Quantitative Protein Network Models

Satoshi S. Nishizuka

Abstract

Theoretical models describing complex biological phenomena have been accumulating. However, most of these models have been created with hypothetical parameter determination without seeing actual cell reactions. The parameter determination requires high-dimensional data monitoring, particularly at the protein level. It has been a difficult task to develop the standard model system because of the lack of an appropriate validation technique. Reverse-phase protein lysate microarray (RPA) is one of the most potent technologies for high-dimensional proteomic monitoring. Therefore, proteomic monitoring by RPA may contribute substantially to develop theoretical protein network models based on experimental validation.

Key words: Theoretical model, High dimension, Experimental validation, Perturbation model, Chemotherapy, Protein response, Network analysis

1. Introduction

The recent massive acquisition of data in molecular biology can contribute to the simulation of a biological event. However, it has been criticized because many of those quantitative models are not available in reality or the characterization is insufficient. One of the significances of the simulation is that a theoretical model enables one to predict biological consequences. However, the majority of simulation work has been done by using only computational approach, for which we must confirm the likelihood of the simulation results by independent molecular and cellular approaches (1).

The issue in experimental validation is that conventional molecular biology methods do not always provide sufficient resolution to

Ulrike Korf (ed.), *Protein Microarrays: Methods and Protocols*, Methods in Molecular Biology, vol. 785,
DOI 10.1007/978-1-61779-286-1_6,

learn about molecular interactions and cell signals. Herein, I present human in vitro models using cancer cell lines for quantitative biology at the protein level (2). Cultured cancer cells do not necessarily succeed all human cell entities but have played a major role in small molecular lead compounds screening in the context of anticancer drug discovery (3). However, the advantage to use cultured cell is that perturbation or simulation can be set by administration of anticancer drugs, gamma-irradiation, and gene knockout/knockdown in a simplest manner yet retaining cellular signal transduction property.

Signal transduction in the cell is initiated by many different types of stimulus or stress, and is subsequently transmitted from the cell membrane, cytoplasm, and finally the nucleus. To trace this transduction, events over time, and in order, have to be captured in a given time frame (4). Lysate-based assays such as Western blot and reverse-phase protein lysate microarray (RPA) can include all the molecular changes in a sample, but monitoring has to be conducted in a discrete manner. If the events only keep increasing or decreasing, a few time points will be enough to represent the entire dynamics; however, signal transduction can change in a second, minute, hour, or day order. Although the procedure is still labor intensive, the RPA should provide unique and essential information to interpret theoretical simulation models of cellular signal transduction.

2. Materials

2.1. Type of Cells

One of the major appropriate applications of signal transduction studies is a pair of stimulated and perturbed models (5–8). Both adherent and floating cultured cells are applicable; primary culture may be used in both types but is often challenging. If there is no necessity to use a particular cell type or tissues of origin, possessing faster growing clonality-confirmed, higher drug sensitivity is desirable.

2.2. Cell Pellets and Lysate Buffer

1. Cell scraper (BD Falcon) for adherent cells. Cells are counted by automated or manual cytometer (C-Chip, Digital Bio).
2. Ice-cold PBS.
3. Refrigerated microcentrifuge (ideally multi “vertical” cartridge; MX-305, TOMY).
4. Microtubes whose bottom is sharp (to be able to compare cell pellets by their volume).
5. Pink buffer (9 M urea, 4% CHAPS, 2% pH 8.0–10.5 Pharmalyte, and 65 mM DTT). Store at −80°C (see ref. 2).

2.3. Lysate Arrays

1. Nitrocellulose membrane-coated glass slide (51 mm×20 mm, AVID305170, Grace BioLabs).
2. V-bottom 384-well microplates (X6004, Genetix).
3. Multiwell chamber (i.e., Mini-Incubation Trays, Bio Rad).
4. Multichannel pipette for 4.5-mm interval (Thermo).
5. 0.67× Dilution buffer (6 M urea, 2.7% CHAPS, 1.3% pH 8.0–10.5 Pharmalyte, 43.3 mM DTT; see ref. 2).
6. Aushon 2470 microarrayer (Aushon BioSystems) (9).

2.4. Reagents for Signal Development

1. Colloidal gold (Total Protein Stain Blotting Grade, Bio Rad).
2. Blocking buffer. Dissolve 1 g i-Block (Tropix) in 1,000 ml TBST.
3. Avidin (Avidin Solution, #00-4303, Invitrogen).
4. Biotin (d-Biotin Solution, #00-4303, Invitrogen).
5. Primary antibodies.
6. Washing buffer (0.1% Tween-20 in TBS).
7. Catalyzed signal amplification (CSA) kit (K1500, DAKO Cytomation).

2.5. Apparatus for Quantitative Analysis of RPA

1. Flatbed scanner (GT-X970, EPSON).
2. Wedge Density Strip (Reflective Gray Step Wedge, OW20NIH, No. 0504, Danes Picta; see ref. 10).

2.6. Strip Western Blot

1. MIX lysate is made by adding a small amount of each cell lysate prepared individually prior to the Strip Western (11–13).
2. A 4–12% gradient gel with a prep well (NuPAGE 4–12% Bis–Tris Gel 1.5 mm×2D well, Invitrogen, see Note 1).
3. MIX lysate is added to an excess volume of Laemmli buffer (see Note 2).
4. Running buffer (NuPAGE MES SDS running buffer 20×, Invitrogen).
5. Transfer buffer (NuPAGE transfer buffer 20×).
6. Nitrocellulose membrane (iBlot NC Anode Stack, Invitrogen).
7. Blotting apparatus (iBlot Dry Blotting System, Invitrogen).
8. Washing buffer (see Subheading 2.5, step 4).
9. Multiwell chamber (i.e., Mini-Incubation Trays, Bio Rad).
10. Specificity-qualified primary antibodies (see Subheading 3.6).
11. SuperSignal (SuperSignal West Pico or Dura, Chemiluminescent Substrate; see Note 3).
12. Film (Amersham Hyperfilm, ECL, GE Healthcare).

2.7. Computer Programs

1. Entry to middle class computer models is sufficient. Intel CoreDuo CPU (2.66 GHz) mounted on Dell Inspiron/530S computer is used for all calculations.
2. Excel (Microsoft), JMP (JMP), and Matlab (Mathworks) software packages. P-SCAN and ProteinScan (executed on Matlab script language) can be downloaded from http://abs.cit.nih.gov/pscan/.
3. The JMP statistical package can be used for a quality control of each generated data file. Further advanced analysis for image intensity calculations is currently almost exclusively performed by P-SCAN and ProteinScan programs (12).

3. Methods

Unlike conventional Western blot, more than one data point is required to draw a dose–response curve of a protein to extract one most representative number from the linear range of a curve (12). Most of the current techniques, however, have not been able to accommodate multiple data points for quantitative measurement of a sample. As identified in the initial report by Paweletz et al. (14) RPA employs a twofold dilution series, which facilitates sophisticated quantitative assay. It is also important to assess the specificity of primary antibody prior to the quantitative assay.

3.1. Stimulation of the Cells for Time-Course Sample Collection

1. Although the more time points there are the better, it may not be realistic to collect ten samples within 10 s; or every 2 h over a week. It is essential to determine what the best time frame (we call it "experimental window") is. Moreover, prior to the sample collection, it is important to run the experiment effectively to confirm if any significant molecular reactions can be seen in a given time frame. Lower resolution (in terms of time) methods are highly recommended to obtain a rough but correct idea of whether or not the window is appropriate (see ref. 13; Fig. 1; Note 4).
2. Here, I describe sample collection in the case of drug stimulation to a pair of parental and acquired resistant cells. First culture both types of cells in separate T-150 flasks, and then split each type of cell into a handy size flask (i.e., T-25; see Note 5).
3. Add the stimulus (e.g., drug, otherwise, radiations, UVs, growth factors, etc.) time point by time point if the stimulus is in liquid form so that sample collection can be done altogether when the last stimulation time is up. Stimulate cells at once only when it is convenient (i.e., a high-dose irradiation).

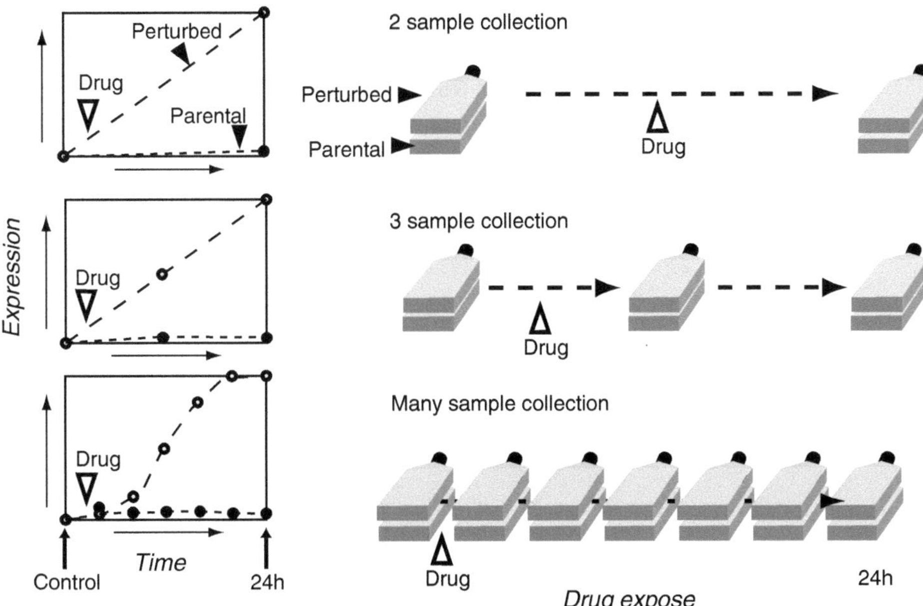

Fig. 1. Sample collection of a pair of parental and "perturbed" cells from a time-course experiment. RPA is suitable for monitoring protein expression from many samples. If a few samples are collected from a time course, protein expression as a function of time can be differently interpreted from the actual protein change (*upper panels* show protein increases in a linear fashion while more time points give a sigmoidal change). Fewer sample collections still provide a good idea of whether or not the time frame is appropriate for monitoring protein expression.

3.2. Preparation of Cell Lysate

1. At the end of the stimulation time period, cells are rinsed three times with ice-cold PBS (for adherent cells; wash with centrifuge for floating cells). Cells are then scraped (or counted for floating cells) to be transferred into the ice-cold PBS.
2. Single-suspension cells are centrifuged in a 1.8-ml microcentrifuge tube for 30 s with 3,000 rpm (900×*g*) to wash cells. Repeat this step once.
3. Carefully remove the PBS supernatant with aspiration and then fully remove the residual PBS with a manual micropipetter.
4. A cell pellet should be seen in the bottom of the tube. Estimate the volume of the pellet by comparing to "volume scale" (see Note 6).
5. Add the same amount of Pink buffer to the cell pellet so that the protein concentration will finally be diluted approximately 1.5 times. Mix well by finger tapping to avoid making bubbles. Continue until cell debris is hardly visible.
6. Finally, a 15,000 rpm (21,900×*g*) centrifuge is performed to transfer the lysate (supernatant) into a clean tube.
7. Store the lysate in a –80°C freezer. The lysate can be stored for up to 2 years.

3.3. Reverse-Phase Protein Lysate Arrays

1. The number of samples from multiple conditions may be more than a few hundred. RPA is a technology whose format is a microscale dot format Western blot followed by immunochemical signal detection by specific primary antibodies (Fig. 2). Recent technological developments allow the printing of more than 15,000 features of cell lysate on a single glass slide (see ref. 9; Fig. 3; Note 7).
2. To print a set of lysate samples, organize all lysates into a set of 384-well v-bottom microplates that require 20 μl per well. We divided the microplate into twelve 32-well sectors because our arrayer's pinhead alignment is a 4 × 8 configuration (see ref. 2; Fig. 2). Make a twofold dilution series with the set of 32 samples over sectors. In our standard procedure, we repeated the

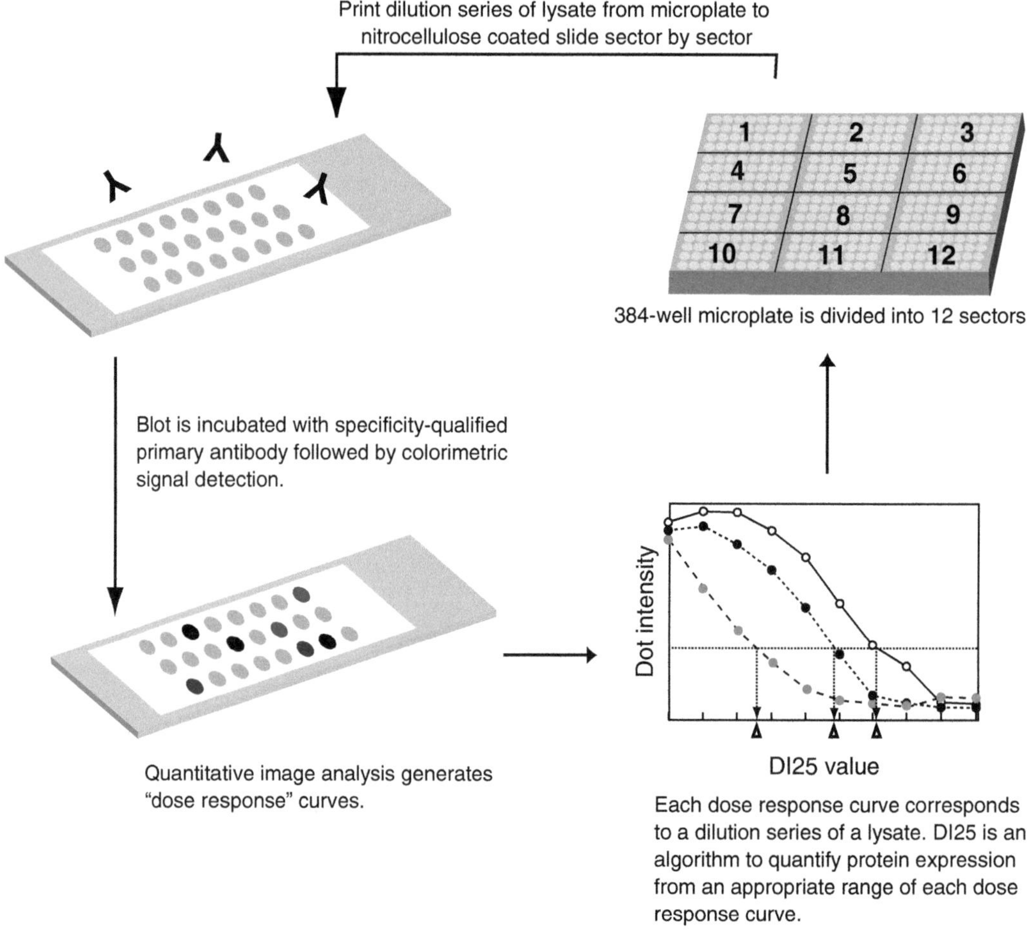

Fig. 2. Flow of RPA experiment and image analysis. Protein lysates are prepared in a 384-microplate in a twofold serial dilution. The resulting RPA gives a dose–response curve per sample per antibody. The quantitative protein expression value is given by DI25 (dose interpolation 25% algorithm 9), which is the value of lysate concentration necessary to reach the level of 25 percentile of entire dot density of a given RPA (*horizontal dashed line* is 25% level).

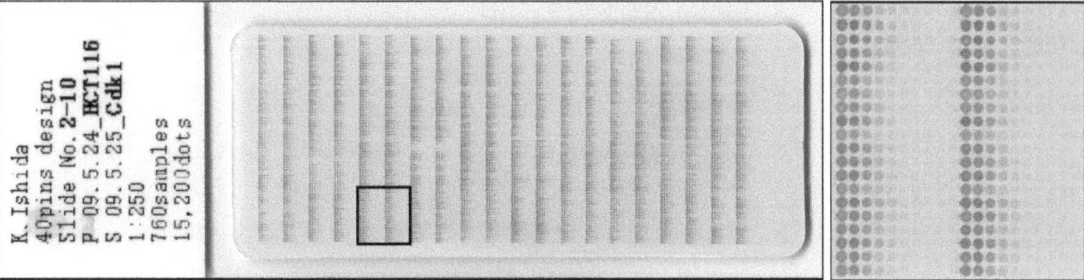

Fig. 3. High-density RPA. A total of 15,200 features in 51 × 20 mm area (*left*). The RPA was produced using Aushon 2470 with 32-pin configuration but a total of 40 pins were used by two times print runs. Lysate samples from multiple conditions were probed with specific primary antibody followed by CSA immunostaining. *Inset* on the RPA indicates the area printed per pin (*right*). Enlarged view of the *inset* printed by a single pin. Lysate samples are replicated in horizontal direction. Each row is of dilution series of a sample lysate.

dilution nine times (ten concentrations) but it can be altered depending on the final array layout up to 12 times.

3. Place the set of 384-well microplates into the Aushon 2470 microarrayer along with nitrocellulose-coated glass slides. The RPA density is constrained by the pin diameter and the number of rows, which is the number of microplates. Since our standard pin diameter is 110 μm, twenty 384-well microplates is the maximum in our current protocol.
4. Run the array printing according to the software GUI. Twenty microplates, 45 slides printing takes approximately 16 h.
5. With at least one of the 45 slides produced, Colloidal Gold staining should be performed to check if the printing process has been successfully completed, there should be uniformity, alignment, density variation of features. Subsequently, the rest of the arrays are stained with a specific primary antibody followed by tyramide-linked amplified immunochemistry (see ref. 15; Fig. 3).

3.4. Immunochemical Signal Development on RPA

Since the CSA system is an effective but complex method (see Note 8), previous papers have introduced the use of an autostainer. However, it has been found that manual staining achieves the equivalent quality of signal development, although the number of slides stained in a single run is 5–10 times less. Here as an alternative, the manual method is described (slide rinsing with TBST is required after every step).

1. Wash with ddH_2O and lightly agitate for 15 min × 2.
2. Incubate slides in I-block with a slight agitation (either at room temperature for 2 h or in a cold room for overnight).
3. H_2O_2 (5 min).
4. Avidin block (25 min).

5. Biotin block (25 min).
6. I-Block (10 min).
7. Primary antibody (30 min).
8. Secondary antibody (Mouse or Rabbit "link"; i.e., Biotinylated secondary antibody).
9. HRP (forming streptavidin–biotin–HRP complex; 15 min).
10. Biotin conjugated tyramide (15 min).
11. HRP (5 min).
12. DAB (5 min).

3.5. Acquisition of Quantitative Output

The tyramide-linked signal amplification is a colorimetric enzymatic reaction. Hence, it is not necessary to use a designated fluorescent microarray scanner. Most ordinary optical flatbed microarray scanners have high enough specifications for RPA scanning in terms of dynamic range and resolution. However, we use a Wedge Density Strip to adjust the range of density because image-digitizing algorithms of these scanners are generally not accessible (10). Scanned images are produced in a TIF format and subsequently quantified by the P-SCAN program, which gives a density score feature by feature with the *XY* grid address. The density score of each feature is subsequently plotted as a dose–response curve to facilitate calculation of the most representing number from a linear range (Fig. 2, DI25 value).

3.6. Strip Western Blot

1. Collected samples are loaded onto "prep well" SDS-PAGE apparatus. The gradient-type gel is recommended to see wider range of molecular weight of proteins. After transferring the gel onto a nitrocellulose membrane, the membrane is cut into a set of 4-mm strips. The mini-gel format produces 14–16 strips (Fig. 4; see Note 9).
2. Each strip is incubated on a multiwell chamber with one primary antibody for 24 h at 4°C. Primary antibodies in the chamber are then washed with TBST individually. Signal development is done using an HRP-catalyzed chemiluminescence method (Super Signal, see Note 3).
3. The band results are categorized into the following four groups: (1) Single predominant band at the expected molecular weight; (2) multiple bands; (3) predominant band at unexpected molecular weight; (4) no band. Only (1) is considered qualified for a quantitative analysis on RPA (several attempts are made to increase the chance). Statistics on screening results have been published previously (see refs. 11–13; Fig. 4).

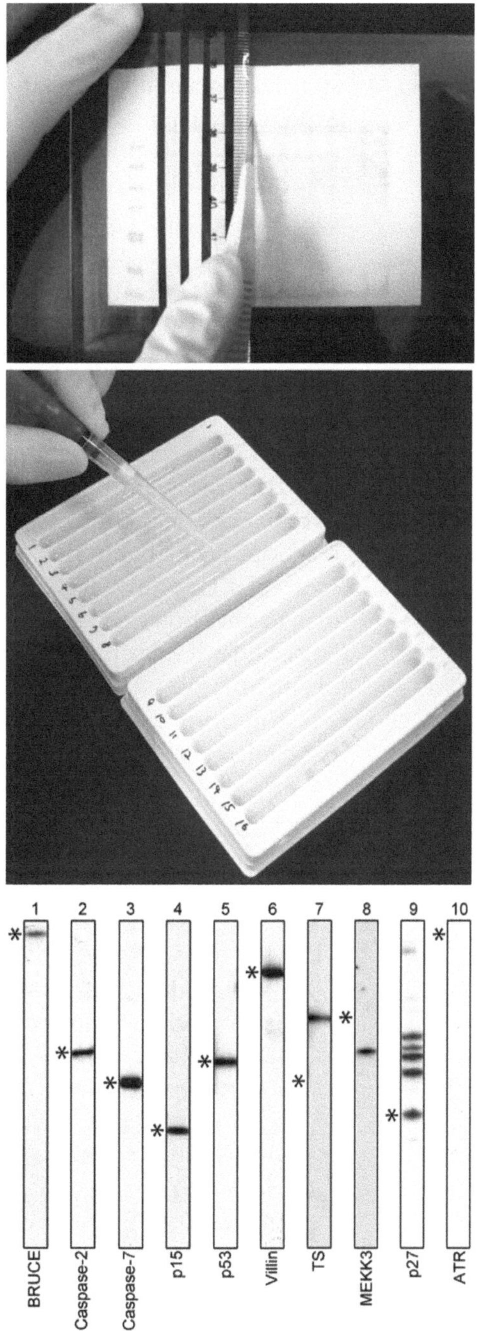

Fig. 4. Strip western blot. (*Top*) Cutting a nitrocellulose membrane into 14–16 strips. (*Middle*) Strip membranes are incubated with different primary antibodies in a multiwell incubation tray. (*Bottom*) Signal development of each antibody followed by chemiluminescent immunostaining. The name of target antigen of each antibody is listed at the bottom of the membrane. Strips 1–6 indicate "single band at expected range"; 7–8, "single band at wrong range"; 9, multiple bands; and 10, "no band".

3.7. Interpreting Cellular Signal Transduction by Theoretical Models

Signal transduction is a chain reaction of many molecules, which rarely uses an identical set of pathway components because the signal is determined by a large number of small parameters. Hence, experimental approaches are not necessarily always able to cover all parameters. Theoretical modeling has been another powerful approach to interpret the signal transduction (16). Most theoretical modeling has been done by a set of simple equations designed based on the literature. Although complete modeling requires not only the network structure but also reaction rates, concentrations, and spatial distribution of each molecule at any time point, quantitative analysis of signal transduction at the protein level provides essential information to predict the response or organize an extremely complex system (1) while the RPA technology offers a validation method for the models (5). Previous studies revealed that comparing protein reactions derived by ODEs and quantitative protein data could be an appropriate method to predict the reactions occurring in the "perturbed" model (5, 6). In fact, recent molecular targeting drugs are supposed to have the known molecular target; however, the cellular/molecular response at molecular/cellular level is not fully understood. Preclinical in vitro testing by quantitative proteomic analysis provides a concrete insight without drug administration to the patients when the drugs are used in the clinical setting.

4. Notes

1. RPA is a dot format Western blot followed by an immunochemical detection. It is important to evaluate the specificity of specific primary antibodies using higher dimensional (than a dot) methods such as Western blot. However, the configuration of an ordinary SDS-PAGE Western blot is for testing a given protein expression against a few to a dozen samples per run whereas antibodies for RPA have to be a set of qualified antibodies against samples printed on the RPA. Hence, we used the "Strip Western" method in which each of the many antibodies can be tested on a 4-mm width nitrocellulose membrane strip.
2. Each sample collected by the procedure mentioned in the chapter as "high dimensional sample collections" should be kept in separate tubes. Using the collected samples, a MIX sample for strip Western is made. Protein expression changes over time depending on the degree of stimulus. For instance, phosphorylation of EGFR takes place immediately after its EGF stimuli and diminishes after the spike (7). A set of high-resolution time-course samples should be able to represent the phosphorylation but it cannot be seen if one time point sample

or a lysate from static states is used for antibody testing. Hence, we collect samples from the time course and mix each sample.

3. Depending on the strength of the signal on initial development, the substrate reagents' strength can be increased (Pico, Dura, Fempto).
4. After adding the stimulation, it can be considered that a network level reaction may be taking place if a key protein change is seen by an ordinary Western blot of six hourly samples within a time frame of 24 h. Samples of the actual experiment are collected every 2 h.
5. For instance, we cultured one T-150 and split it into 15 T-25s with the ratio of 1:3 and then cultured the set of T-25 until each becomes about 80% confluent. It is critical to adjust cell pellet volume using the naked eyes.
6. Since handling many (>100) samples is generally required, it is critical to keep the cell pellet volume constant. If sample collection is properly performed by an experienced individual, the deviation can be kept within twofold. To make sure of the cell pellet volume estimate, prepare 5 volume ranges of food color so that each volume of the cell pellet is more precisely estimated.
7. This high-density lysate spot provides an opportunity to measure protein expression in a more rigorously quantitative manner, because a multiple dilution series of a cell lysate can be printed in a row, which gives a dose–response curve.
8. In contrast to the chemiluminescent signal detection method for Strip Western blot, a colorimetric detection namely CSA (DAKO) has been almost exclusively used for RPA signal detection. It has been reported that with the CSA method, the detection limit is 1,000 times more sensitive than the conventional system (15). Hence, it is suitable for detecting extremely small amounts of targets such as those in a small amount of complex lysate printed on RPA. In addition, the final product of the detection, a pigment, can be scanned by an ordinary CCD-coupled flatbed scanner, which is a convenient method when many slides (RPAs) need to be scanned because the scanners are much faster than other microarray fluorescent scanners while maintaining a satisfactory quality in terms of resolution and dynamic range for quantitative analysis. We found that a manual staining method performs at least equally or better than machine staining although the number of slides stained is smaller than the autostainer (4–8 vs. 20–40 slides per run).
9. This is a process for screening antibodies if they produce a single band at the expected molecular weight. The SDS-PAGE step is used for separating the full range of molecular weight proteins – there is no particular target molecular weight.

Acknowledgments

The author particularly thanks Kazushige Ishida, Sundhar Ramalingam, and Brett Spurrier for their contribution for the development of the experimental procedures; and Lynn Young for completing the computational part of the analysis. The support of Teppei Matsuo, Hironobu Noda, Takeshi Iwaya, Miyuki Ikeda, and Go Wakabayashi is greatly appreciated. This work was supported by KAKENHI (Grant-in-Aid for Scientific Research (C), 50316387 and 50453311).

References

1. Kollmann, M., and Sourjik, V. (2007) In Silico Biology: From Simulation to Understanding. *Curr Biol*, **17**, R132–134.
2. Spurrier, B., Ramalingam, S., and Nishizuka, S. (2008) Reverse-Phase Protein Lysate Microarrays for Cell Signaling Analysis. *Nat Protoc*, **3**, 1796–1808.
3. Weinstein, J. N., Myers, T. G., O'Connor, P. M., Friend, S. H., Fornace, A. J., Jr., Kohn, K. W., Fojo, T., Bates, S. E., Rubinstein, L. V., Anderson, N. L., Buolamwini, J. K., van Osdol, W. W., Monks, A. P., Scudiero, D. A., Sausville, E. A., Zaharevitz, D. W., Bunow, B., Viswanadhan, V. N., Johnson, G. S., Wittes, R. E., and Paull, K. D. (1997) An Information-Intensive Approach to the Molecular Pharmacology of Cancer. *Science*, **275**, 343–349.
4. Nishizuka, S., and Spurrier, B. (2008) Experimental Validation for Quantitative Protein Network Models. *Curr Opin Biotechnol*, **19**, 41–49.
5. Ramalingam, S., Honkanen, P., Young, L., Shimura, T., Austin, J., Steeg, P. S., and Nishizuka, S. (2007) Quantitative Assessment of the P53-Mdm2 Feedback Loop Using Protein Lysate Microarrays. *Cancer Res*, **67**, 6247–6252.
6. Sahin, O., Lobke, C., Korf, U., Appelhans, H., Sultmann, H., Poustka, A., Wiemann, S., and Arlt, D. (2007) Combinatorial Rnai for Quantitative Protein Network Analysis. *Proc Natl Acad Sci USA*, **104**, 6579–6584.
7. Winters, M. E., Mehta, A. I., Petricoin, E. F., 3rd, Kohn, E. C., and Liotta, L. A. (2005) Supra-Additive Growth Inhibition by a Celecoxib Analogue and Carboxyamido-Triazole Is Primarily Mediated through Apoptosis. *Cancer Res*, **65**, 3853–3860.
8. Rudelius, M., Pittaluga, S., Nishizuka, S., Pham, T. H., Fend, F., Jaffe, E. S., Quintanilla-Martinez, L., and Raffeld, M. (2006) Constitutive Activation of Akt Contributes to the Pathogenesis and Survival of Mantle Cell Lymphoma. *Blood*, **108**, 1668–1676.
9. Spurrier, B., Honkanen, P., Holway, A., Kumamoto, K., Terashima, M., Takenoshita, S., Wakabayashi, G., Austin, J., and Nishizuka, S. (2008) Protein and Lysate Array Technologies in Cancer Research. *Biotechnol Adv*, **26**, 361–369.
10. Nishizuka, S., Washburn, N. R., and Munson, P. J. (2006) Evaluation Method of Ordinary Flatbed Scanners for Quantitative Density Analysis. *Biotechniques*, **40**, 442, 444, 446 passim.
11. Major, S. M., Nishizuka, S., Morita, D., Rowland, R., Sunshine, M., Shankavaram, U., Washburn, F., Asin, D., Kouros-Mehr, H., Kane, D., and Weinstein, J. N. (2006) Abminer: A Bioinformatic Resource on Available Monoclonal Antibodies and Corresponding Gene Identifiers for Genomic, Proteomic, and Immunologic Studies. *BMC Bioinformatics*, 7, 192.
12. Nishizuka, S., Charboneau, L., Young, L., Major, S., Reinhold, W. C., Waltham, M., Kouros-Mehr, H., Bussey, K. J., Lee, J. K., Espina, V., Munson, P. J., Petricoin, E., 3rd, Liotta, L. A., and Weinstein, J. N. (2003) Proteomic Profiling of the Nci-60 Cancer Cell Lines Using New High-Density Reverse-Phase Lysate Microarrays. *Proc Natl Acad Sci USA*, **100**, 14229–14234.

13. Spurrier, B., Washburn, F. L., Asin, S., Ramalingam, S., and Nishizuka, S. (2007) Antibody Screening Database for Protein Kinetic Modeling. *Proteomics*, 7, 3259–3263.
14. Paweletz, C. P., Charboneau, L., Bichsel, V. E., Simone, N. L., Chen, T., Gillespie, J. W., Emmert-Buck, M. R., Roth, M. J., Petricoin, I. E., and Liotta, L. A. (2001) Reverse Phase Protein Microarrays Which Capture Disease Progression Show Activation of Pro-Survival Pathways at the Cancer Invasion Front. *Oncogene*, **20**, 1981–1989.
15. Nishizuka, S. (2006) Profiling Cancer Stem Cells Using Protein Array Technology. *Eur J Cancer*, **42**, 1273–1282.
16. Di Ventura, B., Lemerle, C., Michalodimitrakis, K., and Serrano, L. (2006) From in Vivo to in Silico Biology and Back. *Nature*, **443**, 527–533.

Chapter 7

Characterization of Kinase Inhibitors Using Reverse Phase Protein Arrays

Georg Martiny-Baron, Dorothea Haasen, Daniel D'Dorazio, Johannes Voshol, and Doriano Fabbro

Abstract

Using the reverse protein array platform in combination with planar waveguide technology, which allows detection of proteins in spotted cell lysates with high sensitivity in a 96-well microtiter-plate format for growing, treating, and lysing cells was shown to be suitable for this approach and indicates the usefulness of the technology as a screening tool for characterization of large numbers of kinase inhibitors. In this study, we have used reverse protein arrays to profile kinase inhibitors in various cellular pathways in order to unravel their MoA. Multiplexing and simultaneous analysis of several phospho-proteins within the same lysate allows (1) the estimation of inhibitor concentrations needed to shut down an entire pathway, (2) the estimation of inhibitor selectivity, and (3) the comparison of inhibitors of different kinases within one assay. For example, parallel analysis of p-InsR, p-PKB, p-GSK-3, p-MEK, p-ERK, and p-S6rp in insulin treated A14 cells allows profiling for inhibitors of the InsR, PI3K, PKB, mTor, RAF, and MEK. Selective kinase inhibitors revealed different specific inhibitory pattern of the analyzed phospho-read outs. Altogether, multiplexed analysis of reverse (phase) protein arrays is a powerful tool to characterize kinase inhibitors in a semi-automated low to medium throughput assay format.

Key words: Antibodies, Kinase inhibitors, Phospho-proteomics, Protein-arrays, Signaling-pathways

1. Introduction

Protein kinases play key roles in cellular signaling being involved in a variety of diseases accompanied by deregulation of cell growth, viability, and differentiation. Aberrant activation leads to deregulation of cellular signaling cascades and has been shown to be associated with cancer (1, 2). The investigation and development of selective kinase inhibitors are currently a major focus of drug discovery efforts in the pharmaceutical industry (3–7). Understanding

Ulrike Korf (ed.), *Protein Microarrays: Methods and Protocols*, Methods in Molecular Biology, vol. 785,
DOI 10.1007/978-1-61779-286-1_7,

the selectivity of kinase inhibitors has become a major issue, in particular for indications that require chronic administration.

Different assay systems are used for selection and characterization of kinase inhibitors. The most economic approach with highest throughput is the biochemical, cell-free assays using purified recombinant kinases and artificial peptide substrates. As this assay relies on recombinant enzymes and artificial substrate, the SAR obtained using the biochemical assay requires confirmation and validation at the cellular level. The cell-based assays for kinases offer the advantage of testing selected compounds, if they are cell permeable, against the native kinases in their natural environment by measuring the phosphorylation status of signaling pathway components using antiphosphoprotein antibodies that specifically recognize the most proximal substrates of the target kinase (7, 8). In addition, cell-based assay also allows to measure phosphorylation events that are downstream of the target allowing an approximate assessment of the potency and selectivity of the kinase inhibitors. Thus, the activity status of multiple signaling pathways can be probed through parallel phospho-specific analysis. Besides the laborious Western blot, which allows only a limited throughput, the current gold standard for this purpose is the sandwich-ELISA, which is available in many custom or commercial formats. The latter usually comprises the detection of one or a few phosphorylated proteins directly related to the action of the targeted kinase. However, because of the inherent potential promiscuity of kinase inhibitors, a much more extensive characterization of compound activities across a wide range of signaling pathways and their components is desirable to select the inhibitors with the appropriate profile.

Recently, reverse (phase) protein arrays (RPAs) have emerged as an alternative to the sandwich (or forward) assay formats (9). This type of array, in which a protein extract is immobilized and queried with antibodies or other reagents that bind to a specific protein in the sample, is often referred to as reverse (phase) protein array (RPAs). RPAs have been used for several years in their most basic form, the dot blot, in which drops of cell or tissue extract are applied to a membrane or a coated glass slide (9). Among the different proteomics technologies that are suitable for that purpose, we describe here a reverse array platform based on using the planar waveguide technology with that significantly improved sensitivity (10). Planar waveguide reverse protein arrays make it feasible to obtain reproducible and quantitative protein expression information about the dynamic aspects of cell signaling. As the quality of the antibodies is key to the successful application of reverse arrays, a significant effort is required to their validation before the antibodies are applied to the RPA platform.

In this chapter, we describe the use of RPA

(a) As a standardized process.
(b) To measure phopho-pathways that are in good agreement with other technologies (WB and ELISA).
(c) To be multiplexed with RGA to deliver specific information on the selectivity of kinase inhibitors allowing the dissection of delineation and compound screening in a cellular model as well as comprehensive analysis of the transduction of signals.

2. Materials

2.1. Kinase Inhibitors

A panel of kinase inhibitors was prepared as 10 mM stock solutions in DMSO and stored at −20°C (Table 1). For the cell-based assays, serial dilutions of the compounds were prepared in starvation medium. The final DMSO concentration in the cellular assay was kept constant at 0.1%.

2.2. Consumables

1. 96-Well deep-well plates (Greiner, Cat. No. 780215).
2. 96-Well deep-well plates (Thermo Scientific Matrix, Cat. No. 4221).
3. 384-Well V-bottom polypropylene plates (Greiner, Cat. No. 781201).
4. 96-Well V-bottom polypropylene plates (Thermo Scientific Matrix, Cat. No. 4919).
5. Amicon Ultrafree-MC 1.5 ml tube 0.22 μm filter unit (Millipore, Cat. No UFC30GV00).

Table 1
List of kinase inhibitors

Kinase inhibitor	Target
NVP-TAE226	InsR/IGF1R/FAK (11)
NVP-BEZ235	PI3K/mTor (12)
NVP-RAD001	mTor (13)
PD0325901	MEK1/2 (14)
AZD6244	MEK1/2 (15)
CP-690,550	Jak1/2/3 (16)
NVP-AEE788	EGFR/VEGFR (17)

6. Cell culture-treated 96-well plates, F-bottom (Costar, Cat. No. 3596), filter, 0.22 μm pore width (Millipore, Cat. No. SCGP00525).
7. Maxisorp 96-well F-bottom black ELISA plates (Nunc, Cat. No. 437111).
8. Black-wall, clear-bottom, 384-well cell culture assay plates (with low fluorescence background), Corning Life Sciences, Catalogue #3712.
9. 75 cm^2 cell culture flasks, Corning, Catalogue #430641.
10. LiveBLAzer™-FRET B/G Substrate (CCF4-AM), Invitrogen K1030.
11. Cell culture freezing medium, Invitrogen 11101-011.
12. DMEM (high-glucose), Invitrogen 11965-092.
13. DMSO Fluka 41647.
14. Opti-MEM® reduced serum medium, Invitrogen 11058-021.
15. Fetal bovine serum (FBS), dialyzed, tissue-culture grade, Invitrogen 26400-044.
16. Nonessential amino acids (NEAA), Invitrogen 11140-050.
17. Penicillin/streptomycin, Invitrogen 15140-122.
18. Phosphate-buffered saline without calcium and magnesium [PBS (−)], Invitrogen 14190-136.
19. HEPES (1 M, pH 7.3), Invitrogen 15630-080.
20. Sodium pyruvate, Invitrogen 11360-070.
21. Epidermal growth factor (EGF), Invitrogen 13247-051.
22. 0.05% Trypsin/EDTA, Invitrogen 25300-054.
23. Blasticidin, Invitrogen R210-01. Solution D, Invitrogen K1157.
24. BSA fraction V, Sigma A-9418.

2.3. Cell Lines

1. The human epidermoid carcinoma cell line A431 (ATCC No.: CRL-1555) was cultivated in DMEM high glucose supplemented with 10% heat inactivated fetal FCS and 1% sodium pyruvate.
2. A14, mouse fibroblast NIH3T3 cells overexpressing the human insulin receptor have been described (18).
3. The cervix adenocarcinoma cell line HeLa (ATCC No.: CCL-2) was cultured in DMEM high glucose containing 10% heat-inactivated FBS and 1% sodium pyruvate.
4. The human epidermoid carcinoma CellSensor™ AP-1-bla ME-180 cell line, which contains a beta-lactamase reporter gene under the control of the AP-1 response element that has been stably integrated into ME-180 cells, was purchased from

Invitrogen (Cat. No. K1185). The cell line was cultivated in DMEM high glucose supplemented with 10% dialyzed FCS, 1% sodium pyruvate, 1% NEAA, 5 mg/ml blasticidin, and 25 mM HEPES (complete medium for ME-180).

5. The AP-1-bla ME180 cell line was purchased from Invitrogen (Cat. No. K1185).
6. ME180 cells (ATCC # HTB-33™) are derived from a cervix epidermoid carcinoma.
7. The AP-1-bla ME180 cell line is stably transfected with a plasmid containing the beta-lactamase gene under the control of a 5-mer of a consensus AP-1 response element.

2.4. Instruments and Software

1. 5810 R Centrifuge (Eppendorf, Germany).
2. CyBi-Well vario (CyBio, Germany).
3. ELx405CW washer (Biotek, Germany).
4. Heraeus BBD6220 CO_2 incubator (Thermo Electron Corporation, USA).
5. Multidrop 384 dispenser (Thermo Scientific, USA).
6. Nano-Plotter NP2.1 (GeSiM, Germany).
7. PlateMate Plus (Thermo Scientific Matrix, USA).
8. STAR (Hamilton, Switzerland).
9. XLfit Excel Add-In 4.2 (IDBS Limited, UK).
10. ZeptoCARRIER (Zeptosens, Switzerland).
11. ZeptoFOG (Zeptosens, Switzerland).
12. ZeptoREADER (Zeptosens, Switzerland).
13. ZeptoVIEW software package (Zeptosens, Switzerland).

2.5. Antibodies

See Table 2.

3. Methods

3.1. RGA Assay

1. Cells were routinely passaged three times per week maintaining a confluence between 20 and 80% in growth medium DMEM containing 10% dialyzed FBS, 0.1 mM NEAA, 1 mM sodium pyruvate, 25 mM HEPES (pH 7.3), 100 U/ml penicillin, 100 μg/ml streptomycin, and 5 μg/ml blasticidin.
2. For the assays, 7,500 cells were seeded per well in 32 μl Opti-MEM containing 0.5% dialyzed FBS, 0.1 mM NEAA, 1 mM sodium pyruvate, 100 U/ml penicillin, and 100 μg/ml streptomycin.

Table 2
Antibodies

Antigen	Epitope	Source	Provider	Cat. no.	Lot no.	Dilution
pEGFR	pTyr1173	Rabbit polyclonal	Biosource	44-794G	502	1:50,000
pERK	pThr202/pTyr204	Rabbit monoclonal	Dako	ECA 297	00021G	1:2,000
pPLCg (pPLCγ1)	pTyr783	Rabbit polyclonal	CST	2821	3	1:1,000
pS6rp	pSer235/236	Rabbit polyclonal	CST	2211	11	1:1,000
pSTAT3	pTyr705	Rabbit polyclonal	CST	9131	9	1:1,000
pMEK	pSer217/221	Rabbit monoclonal [41G9]	CST	9154	3	1:500
Prohibitin	Full-length protein	Mouse monoclonal [II-14-10]	Abcam	ab1836	325291 355651	1:100 1:250
Anti-rabbit AF647	Rabbit antibodies[a]	Goat IgG (H + L)	Invitrogen	A21245	49626A	1:2,000
Anti-mouse AF555	Mouse antibodies[b]	Goat IgG (H + L)	Invitrogen	A21424	52319A	1:2,000
EGFR	Extracellular domain	Mouse monoclonal [H11]	NeoMarker	MS-316-P1ABX	316X61OF	1:500
PY20(AP)	pTyr	Mouse monoclonal [PY20]	ZYMED	03.7722	50699113	1:10,000

CST cell signaling technologies, *AF* Alexa fluor, *AP* alkaline phosphatase

[a]Reacts with IgG heavy chains and all classes of immunoglobulin light chains from rabbit, highly cross-adsorbed against bovine IgG, goat IgG, mouse IgG, rat IgG, and human IgG

[b]Reacts with IgG heavy chains and all classes of immunoglobulin light chains from mouse, highly cross-adsorbed against bovine IgG, goat IgG, rabbit IgG, rat IgG, human IgG, and human serum; IgG (H + L): whole antibody

3. Cells were grown for 16 h, 4 μl of the stimulus (in assay medium containing 5% DMSO) at the indicated concentration was added to the appropriate wells, they were incubated for 5 h followed by the addition of 8 μl 6×-substrate mix per well (12 μl of 1 mM LiveBLAzer™-FRET B/G Substrate (CCF4-AM) plus 60 μl Solution B plus 898 μl Solution C, and 30 μl Solution D) and an incubation time of 2.5 h at room temperature.
4. Plates for RGA were red on a Biotek Synergy 2 reader (Witec AG, Switzerland) with an excitation filter 400/30 and an emission filter for the green channel of 520/25 and 460/40 for the blue channel. Cells were plated using a Multidrop dispenser (Bioconcept, Switzerland), substrate and stimuli were added to cells manually or by using a Multidrop combi (Promega, Switzerland), compounds were added to cells manually or by using a Matrix PlateMate 2×2 with a 384-tips head (Thermo Fischer, Switzerland).
5. Background wells (no cells) were subtracted from the values of the control and stimulated cells as measured in the green (520/25) and blue channel (460/40) and the ratio blue/green was formed. EC50 and IC50 determination was carried out using the XLfit Dose–Response One Site 205 model (4 Parameter Logistic Model or Sigmoidal Dose–Response Model fit $= (A+((B-A)/(1+((C/x)^\wedge D))))$.

3.2. Cell Culture and Treatment for RPA

The compounds selected for testing were stored at −20°C as 10 mM stock solutions in pure DMSO. Compound dilutions were prepared in 96-well V-bottom polypropylene plates from compound stock solutions by serial dilution in 90% (v/v) DMSO using a Hamilton STAR liquid handler. Controls (stimulated) in 90% (v/v) DMSO low controls (nonstimulated) were supplemented with 0.09 mM.

1. For compound testing, cells were seeded in 96-well plates at an appropriate density of about 45,000 cells/well. After 30 h, the culture medium was replaced by starvation medium for appropriate times.
2. Starved cells were treated with serial dilutions of compounds for 30 min (37°C, 5% CO_2) for 1 h followed by stimulation with growth factors (insulin: Invitrogen, 150 ng/ml final concentration) or EGF (Invitrogen, 50 ng/ml final concentration) for the indicated times.
3. Incubations were terminated by aspirating the medium, followed by two brief washes with cold PBS and lysis in 50 μl CLB96 buffer (Zeptosens, Witterswil, Switzerland) using two freeze/thaw cycles and the lysates were stored frozen at −80°C until further use following the 96-well plate RPA procedure (see below).

4. The cell lysate was cleared from debris by centrifugation for 10 min at 3,200×*g* in an Eppendorf 5810 R centrifuge and was diluted into a 384-well plate in a two-step dilution protocol using a 96-well pipetting head of a Matrix PlateMate Plus or a CyBi-Well vario instrument.
5. First, 70 μl per well of the cell lysate was diluted to 50% (v/v) in lysate dilution buffer I into a 96-well V-bottom polypropylene plate, giving a final concentration of 0.5% octylglucoside.
6. Second, the diluted cell lysate from each well was transferred directly (lysate concentration c1) or was again diluted to 75% (v/v) (c2), 50% (v/v) (c3), and 25% (v/v) (c4) in lysate dilution buffer II into the four corresponding quadrants of a 384-well V-bottom polypropylene plate.
7. A solution of fluorescently labeled BSA was prepared as a reference solution for microarray generation. 5 mg/ml stock solutions of Alexa fluor 647-conjugated BSA (red reference) and of Alexa fluor 555-conjugated BSA (green reference) were prediluted 1:5,000 and 1:2,500, respectively, in Zeptosens reference dilution buffer RDB1.
8. In a second step, these solutions were again diluted 1:40 in filtered 12% (v/v) Zeptosens reference dilution buffer RDB1 in Zeptosens reference spotting buffer CSBR1.

3.3. RPA Spotting of Reverse Phase Microarrays

1. For the production of reverse phase microarrays, the cell lysates were spotted onto Zeptosens planar waveguide chips (ZeptoMARK® hydrophobic chips ZeptoCHIPs, Zeptosens, Witterswil, Switzerland) using the piezo-based inkjet technology from GeSiM (Grosserkmannsdorf, Germany).
2. Spots were produced by single droplet depositions of 400 pl using the Nano-Plotter NP2.1.
3. The predefined layout of the ZeptoCHIP consists of six microarrays with an 9 mm pitch, allows a maximal number of four jets (piezo pipettes) with a 9 mm pitch which can be mounted to the pipetting head of the Nano-Plotter.
4. The architecture of an SBS standard 384-well microtiter plate with 16 rows and 24 columns with a 4.5-mm pitch between wells allows simultaneous spotting of samples from a full 384-well plate using three (jets) only if the plate positions on the Nano-Plotter is changed from landscape to portrait.
5. A separate position on the chip was implemented for the fluorescent reference solution and detection controls. For unattended spotting, runs with four 384-well plates two additional plate positions were introduced on the Nano-Plotter chip tray.
6. The modification of the hardware also required adaptation at the software level. The TransferSi mMultiPlates arraying

program for the Nano-Plotter was used for the array generation by three jets with or without duplicate spots per sample resulting in arrays with 32 samples or 64 samples in four lysate concentrations each. The spotting scheme was defined by the respective newly defined transfer files. Replica ZeptoCHIPs were generated as required for the number of multiplex readouts.

7. A solution of fluorescently labeled BSA was prepared as a reference solution for microarray generation. 5 mg/ml stock solutions of Alexa fluor 647-conjugated BSA (red reference) and of Alexa fluor 555-conjugated BSA (green reference) were prediluted 1:5,000 and 1:2,500, respectively, in Zeptosens reference dilution buffer RDB1. In a second step, these solutions were again diluted 1:40 in filtered 12% (v/v) Zeptosens reference dilution buffer RDB1 in Zeptosens reference spotting buffer CSBR1.

3.4. RPA Detection of Reverse Phase Microarrays

1. For the detection of the RPA, the microarrays were dried for 1 h at 37°C after spotting.
2. The surface of the ZeptoCHIPs was then blocked for 30 min in the gently flowing aerosol of blocking buffer of using the ZeptoFOG (Zeptosens, Witterswil, Switzerland, Fig. 2) blocking station that ensures homogenous deposition of blocking agent without deterioration of spot morphology.
3. The ZeptoCHIPs were rinsed with H_2O, dried under nitrogen, and stored at 4°C if not processed immediately.
4. For binding of antibodies, ZeptoCHIPs were inserted into a ZeptoCARRIER (Zeptosens, Witterswil, Switzerland) that provides a separate microfluidic chamber for each microarray, accommodating up to six ZeptoCHIPs.
5. After equilibration with antibody dilution buffer, 100 μl of primary antibody solution was added onto each microarray chamber, and the ZeptoCARRIER was incubated for 4 h or overnight at room temperature in the dark. Upon three washes with 100 μl of antibody dilution buffer, 100 μl of secondary antibody solution was added to each microarray chamber, followed by incubation for 1 h at room temperature in the dark.
6. After three final washes with 100 μl of antibody dilution buffer per microarray chamber, the ZeptoCARRIER was subjected to planar waveguide imaging at 635 nm (red) and 532 nm (green) excitation by the ZeptoREADER (Zeptosens, Witterswil, Switzerland). The fluorescence signal was integrated over a period of 1–10 s, depending on the signal intensity.
7. Array images were stored as 16-bit TIFF files and analyzed with the ZeptoView Pro (Zeptosens, Witterswil, Switzerland) software package (version 2.0, Zeptosens). Relative intensities were obtained by plotting net spot intensities against protein

concentrations, analogous to described procedures (9, 17) and renormalized for small variations in protein content using prohibitin, a mitochondrial marker, as internal standard.

8. A linear regression of the four lysate concentrations of each sample (c1–c4) was performed to give a single referenced fluorescence intensity (RFI) value for each lysate sample.

3.5. RPA EC50 Determination and Statistical Data Analysis

1. The percent of high (stimulated) control values (% CTL) were calculated as follows:

$$\% \text{ CTL} = \frac{100}{(\text{Mean}_{\text{high}} - \text{Mean}_{\text{low}}) \times (x - \text{Mean}_{\text{low}})}$$

2. Using the IDBS XL*fit* Excel Add-In 4.2, dose–response curves were fitted according to the Sigmoidal Dose–Response Model #205 by the equation:

$$f(x) = \text{Min} + \frac{\text{Max} - \text{Min}}{1 + \left(\frac{\text{EC50}}{x}\right)^{\text{Hill}}}$$

Min is the minimum $f(x)$, Max is the maximum $f(x)$, x is the concentration of the test compound, and Hill indicates the Archibald Hill coefficient.

3. Z' values were calculated as described previously (19).

$$Z' = 1 - \frac{(3 \times \text{SD}_{\text{high}}) + (3 \times \text{SD}_{\text{low}})}{\text{Mean}_{\text{high}} - \text{Mean}_{\text{low}}}$$

SD_{high} is the standard deviation of the "high values" (stimulated controls), $\text{Mean}_{\text{high}}$ is the mean value of the "high values" (stimulated controls), SD_{low} and Mean_{low} are the respective numbers for the "low values" (nonstimulated controls).

4. The signal-to-background (S/B) ratio was calculated as follows:

$$\text{S/B} = \frac{\text{Mean}_{\text{high}}}{\text{Mean}_{\text{low}}}$$

3.6. General Conclusions

1. The data presented here demonstrate that RPA can be used to set up cell-based profiling assays for kinase inhibitors in a multiplexed manner. For our experiments, we started from cells grown in 96-well plates which allow a medium throughput. In principle, multiplexing of different antibodies within the same cell lysate is nearly unlimited due to the low amount of protein lysate which is spotted on the chip.

2. For our assays, we focused the selection of our antibodies by a few key criteria. First, the selectivity of antibodies should be

proven by Western analysis and should at least show a threefold increase of signal compared to background in RPA experiments using treated and untreated cells. Second, if possible we always choose two different antibodies which monitor the activity of a target kinase. For example, we either monitored phosphorylation of the InsR by a specific antibody which recognizes the phosphorylated Tyr1334 of the receptor as well as with a generic antiphosphotyrosine antibody or monitored Janus kinase activity by measuring the phosphorlyation of two different distinct proximal substrates of the kinase, namely Stat1 and Stat3. Third, we tried to include measurements of key nodes of the pathways which were triggered by the cell stimulation. As exemplified by the InsR pathway we included phospho-targets of the PI3K-mTor pathway as well as of the RAF/MEK/ERK pathway. By multiplexing these key nodes, we could successfully discriminate inhibition of various kinases, including the InsR kinase, the PI3Ks, PKB, MEK as well as mTor within the same experiment.

3. A prerequisite for the successful application of RPA in the analysis of cellular signaling pathways is the availability of validated antibodies. Since detection and quantification are done in crude cell lysates, a careful validation of the antibodies, which are used, is needed to ensure high quality data. For the validation of phosphosite-specific antibodies, we used several validation criteria in our cell-based experiments: (1) increase in the phospho-signal upon cell treatment with appropriate growth factors or cytokines, (2) decrease in the phospho-signal upon treatment with appropriate kinase inhibitor, (3) Western blots which ensures specificity under the conditions which have been applied for RPA, and (4) decrease of phospho-signal due to siRNA-mediated knock down of the target protein.

4. Depending on the equipment the throughput of the RPA method is comparable to standard capture ELISAs but offers the advantage of nearly unlimited multiplexing. Thus, the method is clear superior to Western blot techniques. We have not extensively tried to adapt to a 384-well format due to limitations of the array layout to enhance the throughput of the method.

5. For RPA technology in a 96-well cell culture format, the reliability of the obtained results had to be confirmed using another assay system. For that reason, we compared standard capture ELISAs with RPA for compound profiling. Correlation factors of 0.9 and higher were usually observed in this side-by-side comparisons. Furthermore, the high signal-to-background ratio achieved with the evanescent field technology allows detection of target proteins present in the spotted cell lysates with extremely high sensitivity. The detection limits are in the range of femto- to zeptomoles of the target protein in the spotted lysates, depending on the experimental conditions

(cell types, lysis and spotting conditions, antibody concentrations) and on the quality of the target-specific antibodies (affinity, specificity).

6. One drawback of the method might be due to assay conditions which do not allow using SDS during lysate preparation and thus might result in limited solubilization of some of the target proteins. In some cases, we failed to measure a reasonable signal by RPA although Western blots and ELISA indicated the presence of the respective phospho-protein.
7. For our purposes, we wanted to illustrate that (1) that kinase inhibitors can be profiled with a reasonable throughput, (2) several inhibitors targeting different kinase can be profiled within one assay, and (3) how cross-talk between pathways can be delineated using a combination of pathway readouts and tool compounds which can block a specific "branch" of the signaling tree. The data presented here clearly show that this can be achieved by RPA. For example, we were able to characterize specific inhibitors of the InsR, PI3K, MEK1/2, and mTor within one and the same assay. In addition, the parallel measurements of different phospho-nodes within the assay allowed to discover a hyper-phosphorylation of MEK due to the binding of an allosteric inhibitor.
8. To profile kinase inhibitors in complex cellular systems, RPA data can be further combined with other methods such as RGA. One example is given for EGF-driven AP-1 reporter gene assay. A431 cells lack total inhibition of EGF-induced ERK phosphorylation after treatment with the highly potent MEK-inhibitor although complete inhibition of pERK was observed by RPA (data not shown). No dose–response, but a plateau around 30–50% inhibition was obtained with this compound between 0.01 and 10 μM.

4. Notes

1. Cell lysis and protein extraction in 96-well format for RPA
   (a) The 96-well plate format for cell growth and treatment was an essential prerequisite to obtain a reasonable throughput for testing of compounds with the reverse protein microarray technology.
   (b) Cell lysis was done with 100 μl volume of CLB96 lysis buffer mixed with spotting buffer (1 + 9) per well. Beneficial effect of the repeated freezing and thawing of the samples after addition of the CLB96 lysis buffer to the cells was observed, especially regarding the solubilization of

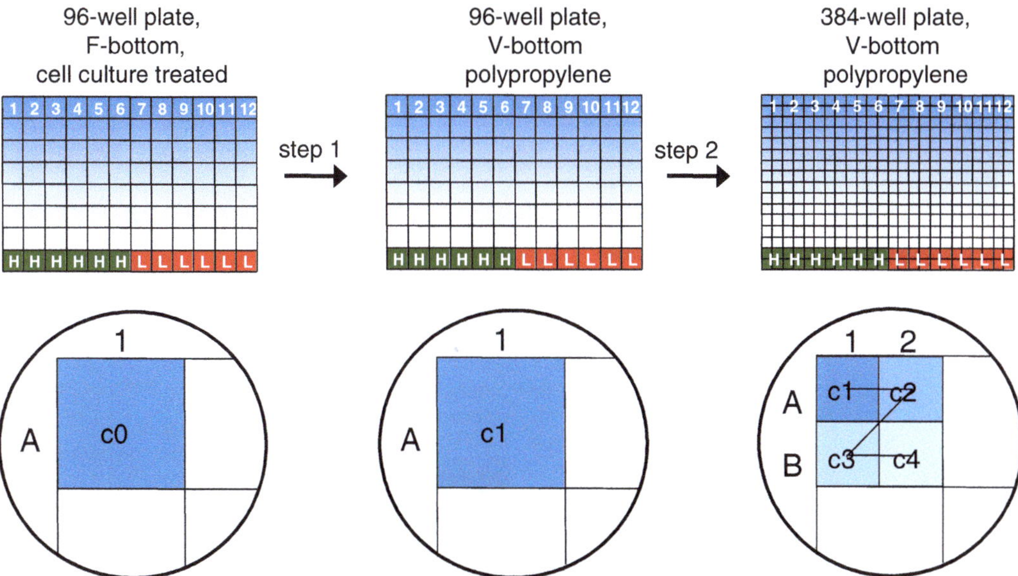

Fig. 1. Scheme of cell lysate dilution using a 96-well pipetting head. The cleared cell lysate (lysate concentration c0) was diluted to 50% (v/v) in lysate dilution buffer I into a 96-well V-bottom polypropylene plate. The diluted cell lysate from each well was transferred directly (lysate concentration c1) or was again diluted to 75% (v/v) (c2), 50% (v/v) (c3), and 25% (v/v) (c4) in lysate dilution buffer II into the four corresponding quadrants of a 384-well V-bottom polypropylene plate. 1–12: compounds; H: high (stimulated) controls; L: low (unstimulated) controls.

membrane-associated proteins like EGFR. Prior spotting, protein lysates were diluted in serial steps resulting in 100, 75, 50, and 25% of the original concentration (Fig. 1).

(c) After spotting, and development with the appropriate phospho-site specific antibody and fluorescence-labeled secondary antibody, the spot intensities were documented with the ZeptoReader at three different exposure times (1, 5, and 10 s). In many cases, the 1-s exposures yielded already sufficient signal-to-noise ratios for the analysis (Fig. 2).

2. Timing and time-consuming steps in RPA

(a) Although cell culture and lysis can be done in a 96-well format, the throughput of RPA is limited. One drawback of the equipment which we used for this study is the lack of SBS standard in the array layout. In addition, serial protein lysate dilutions from the 96-well to the 384-well require automated liquid handler.

(b) The most time-consuming steps in the overall procedure are the duration of spotting and the incubation time for the primary and secondary antibodies. For routine testing, we processed four 96-well plates in parallel. Since an entire experiment including cell seeding and data analysis takes roughly 4 days, one Nano-Spotter can pursue

Semi-automated RPA assays

Spotting – detection – image analysis – QC / IC50 calculation

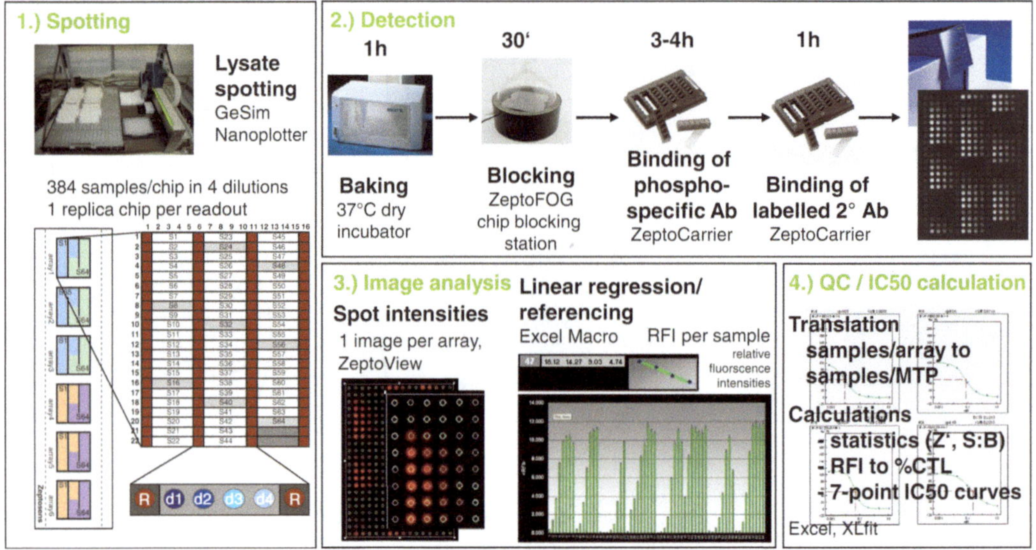

Fig. 2. Spotting scheme with 32 or 64 samples per array. The spotting scheme was either adapted to the original setup from Zeptosens with 32 samples (four lysate concentrations as duplicate spots) per array (**a** and **b**) or was designed for a new setup with 64 samples (four lysate concentrations as single spots) per array (**c** and **d**). Depending on the number of MTP the samples were spotted once onto each chip (**a** and **c**) or in duplicate, generating replicate arrays 1 & 4, 2 & 5, and 3 & 6 (**b** and **d**). *S* sample; *R* reference, c1–c4: lysate concentrations. The assay procedure for the detection of the reverse phase microarrays was essentially as described by Zeptosens (10). After spotting, the microarrays were dried for 1 h at 37°C. The unoccupied surface of the ZeptoCHIPs was blocked for 30 min in the gently flowing aerosol of blocking buffer of the ZeptoFOG blocking station that ensures homogenous deposition of blocking agent without deterioration of spot morphology. ZeptoCHIPs were rinsed with H_2O, dried under nitrogen, and stored at 4°C if not processed immediately. For binding of antibodies, ZeptoCHIPs were inserted into a ZeptoCARRIER that provides a separate microfluidic chamber for each microarray, accommodating up to six ZeptoCHIPs. After equilibration with antibody dilution buffer, 100 μl of primary antibody solution was added onto each microarray chamber, and the ZeptoCARRIER was incubated for 4 h or overnight at room temperature in the dark. Upon three washes with 100 μl of antibody dilution buffer, 100 μl of secondary antibody solution was added to each microarray chamber, followed by an incubation for 1 h at room temperature in the dark. After three final washes with 100 μl of antibody dilution buffer per microarray chamber, the ZeptoCARRIER was subjected to planar waveguide imaging at 635 nm (*red*) and 532 nm (*green*) excitation by the ZeptoREADER. The output was three 16-bit TIFF images per laser line of 1, 5, and 10 s exposure. The pixel intensities for the red and the green reference were around 5,000, 18,000, and 27,000 for the 1, 5, and 10 s images. *Image analysis.* TIFF images were manually selected for the optimal exposure time, avoiding saturating pixel intensities (>60,000). One image per array was subjected to image analysis by the ZeptoView software. Briefly, a grid structure with masks corresponding to single spots was superimposed on each array, and the average pixel intensity within each mask was determined. In a second step using an Excel Macro provided by Zeptosens, the intensity value of each lysate sample spot was normalized to the intensities of the corresponding reference spots that have been corrected for their position within the array. Last, a linear regression of the four lysate concentrations of each sample (c1–c4) was performed to give a single referenced fluorescence intensity (RFI) value for each lysate sample.

approximately 16–20 96-well plates per week. Comparison of data obtained from traditional pELISAs with data obtained by RPA revealed acceptable robustness of the RPA assay with a good correlation between the two assay formats.

(c) For specific analysis of a single read out in cell-based assays, other methods have been described which allow higher throughput than RPA, but due to the advantage of multiplexing of the assay, RPA offers a unique opportunity to profile different pathways phospho-nodes simultaneously within one assay.

3. Antibody validation for RPA
   (a) Antibody validation is an essential part of a reverse protein array platform since it is not straightforward to check the specificity of the antibody signal at the level of the arrays. In most cases, it will be sufficient to confirm specificity of the antibodies by Western blotting.
   (b) Ideally, a two-step validation should be used to build up a collection of prevalidated antibodies a standardized screening process. Antibodies are either tested on a number of different cell lines, if necessary after treatment of cells with generic stimuli such as Ser/Thr- (e.g., calyculin) and Tyr-phosphatase (e.g., orthovanadate) inhibitors in the case of antibodies to phospho-proteins.
   (c) This process can and should also be used to test batch-to-batch consistency, especially in the case of polyclonal antibodies.
   (d) Since the standardized screening process can only cover a limited number of different samples and conditions, it is crucial to confirm specificity by using Western blots on (a subset of) the actual cell line or tissue extracts that were analyzed on the RPAs. This can be done by performing side-by-side extractions for arrays and blots, or simply by using the exact same samples, since the CLB1 buffer is compatible with commonly used buffer systems for 1D gel electrophoresis (see Subheading 3).
   (e) All primary antibodies used in this study have been validated on western in parallel to the RPA studies. For antibody validation, samples in CLB1 buffer were diluted with the appropriate concentrated SDS sample buffer and left at room temperature for 30 min prior to application to the 1D gel.
   (f) For the reported RPA array experiments, antibodies were used at dilutions between 1:500 and 1:2,000 in CeLyA assay buffer CAB1 (Zeptosens). Alexa Fluor 647-labeled anti-rabbit IgG Fab fragments (Molecular Probes and/or Invitrogen, 1:500 dilution in CAB1) was used as a secondary antibody to generate the fluorescence signal.
4. Analysis of cellular signaling by RPA of the InsR
   (a) Activation of the InsR pathway

      The InsR controls blood glucose levels and is therefore responsible for glucose homoeostasis. Defective insulin production or desensitization of InsR signaling has been shown to result in diabetes (20, 21). Activated InsR undergoes autophosphorylation at multiple tyrosine residues which serve as docking sites for SH2 domain containing proteins such as the insulin receptor substrate (IRS) protein: IRS-1 and IRS-2. Recruitment and phosphorylation of IRS proteins result in the activation of the PI3K and of the RAF/ERK pathway. In order to assess the insulin pathway by RPA, we used antibodies raised against pInsR (Tyr1343), PKB (Akt) (Ser473), pGSK3α/β (Ser21/Ser9), pMEK1/2 (Ser217/Ser221), pERK1/2 (Thr202/Tyr204), pS6rp (Ser235/Ser236), and a generic antiphosphotyrosine antibody (4G10) (Fig. 3).

The recognition epitope of the pInsR Tyr1334 antibody is located at the carboxy terminus of the InsR-β-chain,

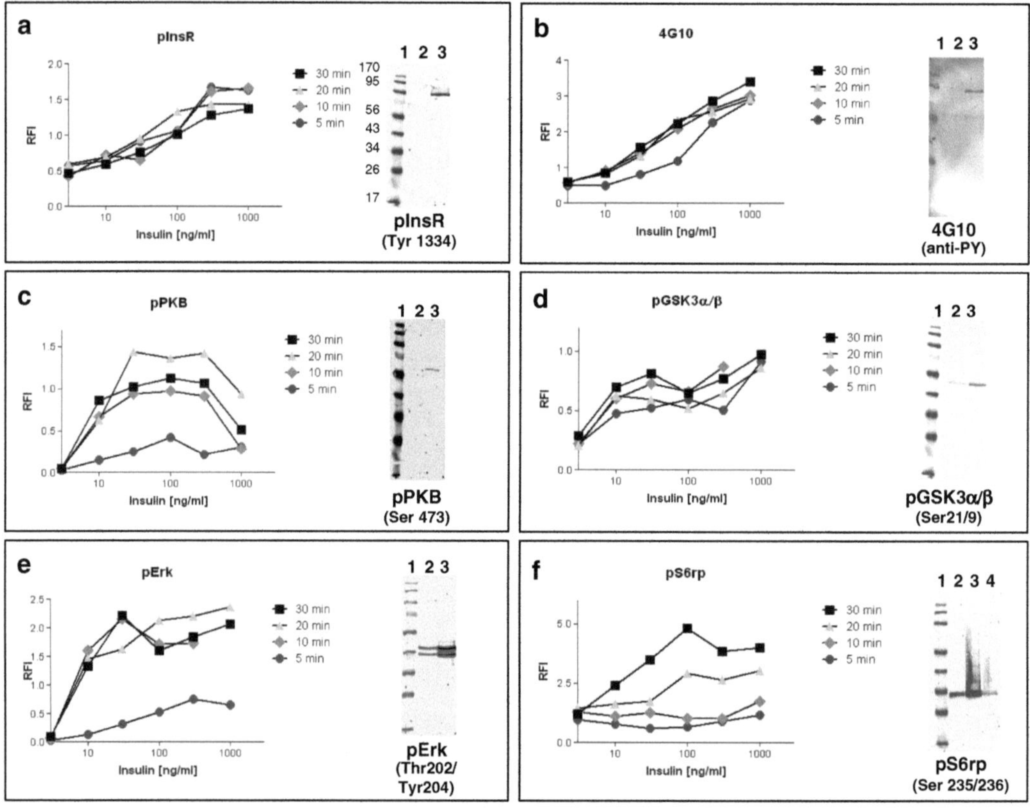

Fig. 3. Kinetic analysis of insulin-dependent protein phosphorylation in A14 cells. A14 cells have been cultured on 96-well plates, starved and stimulated with different amounts of insulin for 5 min (*circle*), 10 min (*rhombus*) 20 min (*triangle*), and 30 min (*square*). Cell lysates for RPA have been prepared as described, spotted on ZeptoCHIPs and analyzed for phosphorylated proteins ((**a**) pInsR; (**b**) 4G10; (**c**) pPKB; (**d**) pGSK3; (**e**) pERK; (**f**) pS6rp). All experiments have been repeated three to four times and a representative example is shown. For Western blot analysis, A14 cells have been starved (*lane 2*) and treated with 100 ng/ml insulin for 30 min (*lane 3*). Cells have been lysed in RIPA buffer prior SDS-PAGE. To reduce basal levels of pS6rp, the mTor inhibitor NVP-BEZ235 was added ((**f**) *lane 4*).

C-terminal to the kinase domain. The region around Tyr1334 of the InsR shows only limited homology to the corresponding IGF1R sequence and has been implicated in the regulation of MAPK phosphatase1 (22).

PKB (Akt) is activated by phospholipid binding and activation loop phosphorylation at Thr308 by PDK1 and by phosphorylation within the carboxy terminus at Ser473. Several kinases including mTor are targeting Ser473 of PKB (23).

GSK3 is a ubiquitously expressed serine/threonine protein kinase that phosphorylates and inactivates glycogen synthase. GSK3 is a critical downstream element of the PI3 kinase/PKB cell survival pathway, and its activity can be inhibited by PKB-mediated phosphorylation at Ser21 of GSK3α and Ser9 of GSK3β (24).

MEK1 is a member of the MAP kinase kinase family, involved in cell growth and differentiation. Ser217 and Ser221 are located within the activation loop of MEK and phosphorylation of these serine residues by RAF like kinases leads to activation of MEK kinase activity (25).

ERK 1/2 are members of the MAP kinase family and are direct targets of MEK. MEK activates ERK1 and ERK2 through phosphorylation of activation loop residues Thr202/Tyr204 and Thr185/Tyr187, respectively (26). The antibody used here does not discriminate between the two ERK isoforms.

For cell growth, growth factors have to initiate protein translation. Thus, many growth factors including insulin activate p70-S6 kinase, which phosphorylates S6rp at the carboxy terminus. S6rp is part of the ribosomal translational machinery and phosphorylation of Ser235 and Ser236 correlates with an increase in translation, particularly of mRNAs with an oligopyrimidine tract in their 5′ untranslated region. Within this group of mRNAs, proteins are encoded which are involved in cell cycle progression or are themselves part of the translational machinery (27).

Phosphotyramine has been used as an immunogen to generate the monoclonal Antibody 4G10. This antibody specifically recognizes phosphorylated tyrosine residues (28).

On Westerns, only low or undetectable levels of phospho-protein was observed for the InsR, PKB, MEK, ERK, and GSK3 in the control lysates derived from starved A14 cells. In contrast, relative high phosphorylation levels of S6rp were observed even in the absence of insulin. The reason for this might be due to the fact that the mTor pathway is not only triggered by activated PKB but can be also activated by other stimuli such as nutrients, thus resulting in partial activation of the mTor/p70-S6 kinase axis and thereby increases S6rp phosphorylation even in the absence of insulin.

This idea is supported by the notion, that addition of mTor inhibitors such as NVP-BEZ235 or NVP-RAD001 decreased S6rp phosphorylation even in the absence of insulin. The generic antiphosphotyrosine antibody 4G10 was also characterized on Western blots. Like for the pInsR antibody only a single major band of about 95 kDa, the size of the InsR-β chain, was detected by 4G10 upon insulin treatment of the cells. This indicates that due to the high expression levels of InsR in this cell line the InsR is the major tyrosine phospho-protein in the cell line used here.

(b) Pharmacological modulation of InsR activation

To analyze the InsR pathway we used A14 cells, a murine fibroblast which has been engineered to overexpress the human InsR (18). As shown in Fig. 3, phosphorylation levels of all analyzed proteins increased upon insulin stimulation in a time- and dose-dependent manner. Roughly 300–1,000 ng/ml insulin were needed for optimal InsR phosphorylation. While the signal increased approximately threefold for the pInsR antibody, a sevenfold increase was seen with the 4G10 antibody under optimal conditions. One reason for this observation might be due to the fact that InsR is phosphorylated on multiple tyrosine residues upon insulin stimulation. While the anti-pInsR Tyr1334 can only detect this specific phospho-residue, 4G10 can bind to the multiple sites and thereby increases the signal-to-background ratio. Increase of InsR phosphorylation was already seen after 5 min incubation with insulin and was stable up to 30 min (Fig. 3). To further demonstrate that both the pInsR antibody and the 4G10 mab measures InsR phosphorylation, we stimulated A14 cells with insulin as well as with IGF1 and EGF (data not shown). Only insulin increased the signal after immunostaining with pInsR and 4G10 antibodies while IGF1 activates the IGF1R, a close homologue of the InsR, and EGF, which activates EGFR. Neither IGF1 nor EGF was able to increase significantly the signals obtained with the pInsR ab and 4G10 ab (data not shown).

Reference compounds inhibiting the InsR, like NVP-TAE226, inhibited also the phosphorylation in a dose-dependent manner of all other tested proteins, demonstrating that inhibitors of the InsR pathway could be profiled using this RPA system (Fig. 4).

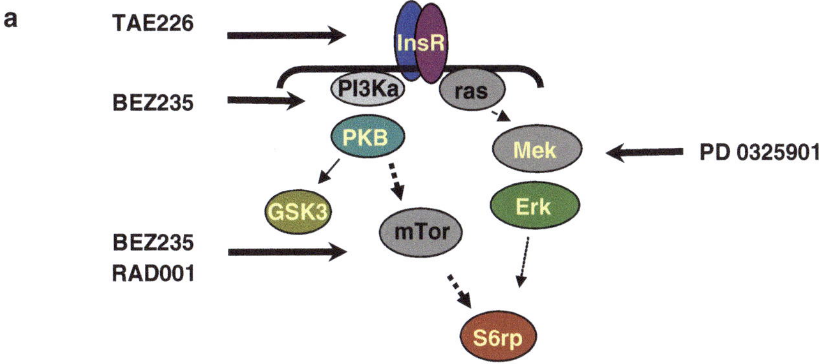

b

compound	target	4G10 EC50	pInsR EC50	pPKB EC50	pGSK3 EC50	pS6rp EC50	pErk EC50
TAE226	InsR	0.289	0.135	0.583	1.0256	0.562	0.38
BEZ235	PI3K/mTor	>10	>10	0.006	0.085	0.002	>10
PD 0325901	Mek	>10	>10	>10	>10	0.004	0.005
Rad 001	mTor	>10	>10	>10	>10	<0.0006	>10

Fig. 4. Profiling of reference compounds with inhibitory activity against InsR, PI3K isoforms, MEK, and mTor in insulin stimulated A14 cells. A14 cells have been cultured and starved on 96-well plates as described. Cells were than incubated with reference compounds (0.0006–10 μM) for 1 h and stimulated with 300 ng/ml insulin for another 30 min (plate scheme is exemplified in Fig. 1). Cell lysates for RPA were prepared, spotted on ZeptoCHIPs, and analyzed for protein phosphorylation as described in Subheading 3. (**a**) Schematic view of the InsR pathway. Phospho-proteins which have been analyzed by RPA are indicated in *yellow*, target kinases of reference compounds are indicated by *arrows*. (**b**) EC50s from RPA data have been calculated as described in Subheading 3 and are shown color coded (<1 μM, *orange*; >1 μM to <10 μM, *yellow*; >10 μM, *green*). All EC50 values are given in μM.

Inhibitors of PI3K, like NVP-BEZ235, which also inhibits catalytically the mTOR kinase and thus the mTORC2 complex that is responsible for the pS473 of PKB selectively inhibited phosphorylation levels of PKB, GSK3, and S6rp without affecting the InsR autophosphorylation and ERK phosphorylation. On the other hand, an allosteric and highly selective inhibitor of MEK like PD0325901, selectively inhibited pERK and S6rp, having no effect on all the other phospho-proteins in the InsR signaling pathway. The NVP-RAD001 rapalog targets specifically and allosterically the rapamycin-sensitive mTORC1 complex and selectively inhibited phosphorylation of S6rp without affecting any of the other read outs in the InsR pathway (Fig. 5).

These data demonstrate that by analyzing the multiple pathway components, in this case downstream of the InsR can lay the basis for the characterization of InsR pathway inhibitors including inhibitors of InsR, PI3K, and ERK pathway.

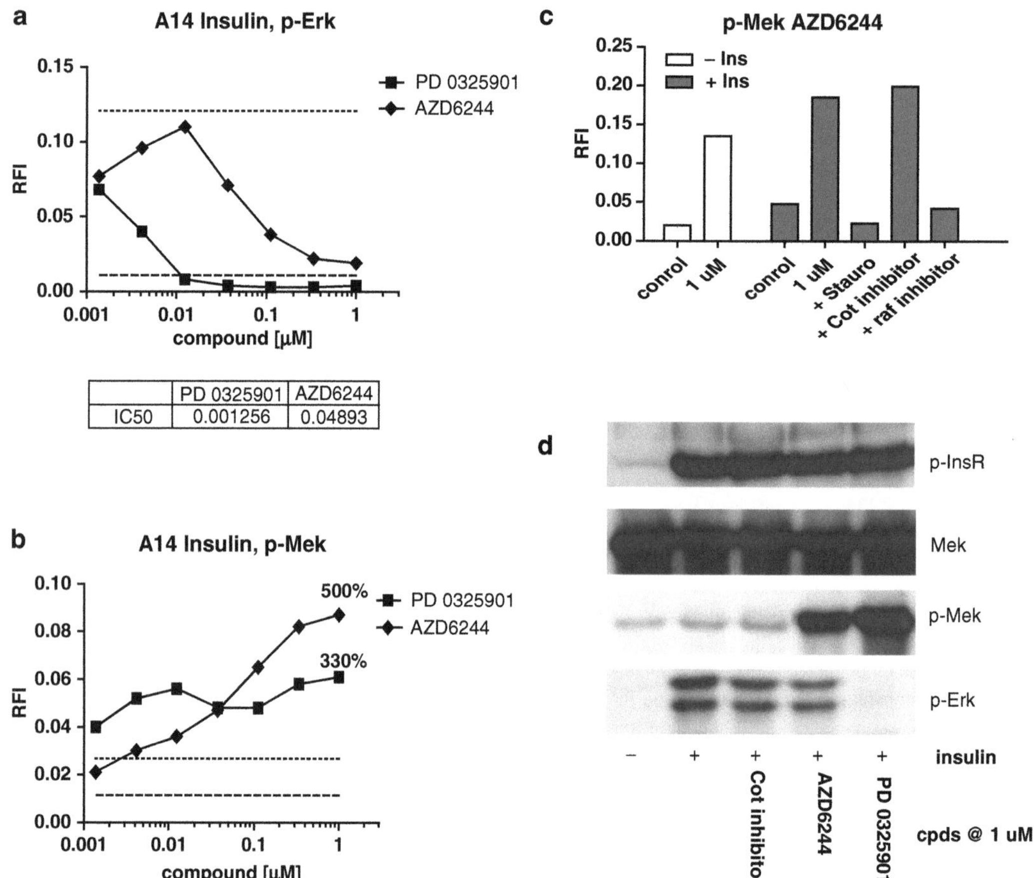

Fig. 5. Allosteric inhibitors of MEK induce hyper-phosphorylation of MEK. (**a**, **b**) A14 cells have been cultured and starved on 96-well plates as described. Cells were than incubated with allosteric inhibitors of MEK (0.0006–10 μM) for 1 h and stimulated with 300 ng/ml insulin for another 30 min. RFI values of starved negative controls and insulin-stimulated positive controls are indicated by *dotted lines*. RFI values for pERK (**a**) and pMEK (**b**) are shown. (**c**) A14 cells have been cultured and starved on 96-well plates as described. Cells have been treated with 1 μM AZD6244 for 1 h and either left for another 30 min in starvation medium (*open bars*) or treated with 100 ng/ml insulin (*gray bars*). (**d**) Western blot analysis of A14 cell lysates which have been starved or stimulated with 100 ng/ml insulin in the absence and presence of allosteric MEK inhibitors as well as a COT kinase inhibitor. Antibodies which have been used for Western blot analysis are indicated.

(c) Analysis of pharmacological modulation of InsR activation

Analysis of InsR phosphorylation and downstream targets of the InsR in a multiplexed manner is much more informative compared to a single assessment of InsR autophosphorylation assay. Partial inhibition of InsR does not necessarily lead to partial inhibition of the entire pathway. Signal amplification between the receptor and effector molecules might cause an activation of the pathway even with a reduced number of activated receptors. The RPA assay described allows assessing which concentrations are needed to inhibit individual downstream targets of the InsR. In addition, redundant signaling and cross-talk between different signaling pathways increase the complexity and make the interpretation of these data difficult. The simultaneous measurement of different signaling molecules within the same cell lysate allows unraveling of redundancy and cross-talk of pathways. For example, insulin-mediated phosphorylation of the S6rp involves both, the ERK and the PI3K pathway, indicating that these two pathways merge at the level of p70-S6 kinase.

This allows the characterization of the cellular profile of various kinase inhibitors on given signaling pathways. In addition, there is a good correlation between the RPA and other cell-based assay formats such as ELISAs, SureFire/Alphascreen technology, or Western blots.

5. Analysis of cellular signaling by RPA of the Jak/Stat

(a) Activation of the Jak/Stat pathway

Janus kinases (Jaks) are nonreceptor tyrosine kinases with four members, Jak1, Jak2, Jak3, and Tyrosine kinase 2 (Tyk2) (for review see refs. 29–31). Each protein has a kinase domain and a catalytically inactive pseudo-kinase domain, and they each bind cytokine receptors through amino-terminal FERM (Band-4.1, ezrin, radixin, moesin) domains. Upon binding of cytokines to their receptors, Jaks are activated and phosphorylate the receptors, creating docking sites for signaling molecules, especially members of the signal transducer and activator of transcription (Stat) family (for review see refs. 31, 32). The family of Stat proteins comprises seven members all critical involved in cytokine signaling. They consist of a coiled coil domain, a DNA binding domain, a transactivator domain as well as SH2 domain, which is critical involved in dimerization upon phosphorylation of Stat proteins through the Jaks (33). Phosphorylated and activated dimeric Stat proteins are translocated to the nucleus where they initiate transcription via binding to either GAS elements (γ-activated sequence: TTN5-6AA) or the IFN-α/β-stimulated response elements (ISRE).

Jak1 and Tyk2 are widely expressed. Jak1 mainly associates with cytokine receptors containing gp130 and IFN receptors. Jak1$^{-/-}$ mice die perinatally most likely due to neurological defects and show a SCID phenotype similar to the Jak3$^{-/-}$ mice. Tyk2$^{-/-}$ mice are viable but show subtle defects in IFNα/β signaling (33).

The phenotype of Jak1 knockout mice suggests that inhibition of Jak1 might cause severe side effects also in humans. Thus, kinase inhibitors which are developed as therapeutic agents should avoid targeting this kinase. Therefore, a cellular assay suited for inhibitor profiling which allows to specifically assess the activity of Jak1 and/or Tyk2 would be highly desired. The type I Interferon (IFN) receptor is expressed as a heterodimeric complex on many cell types and recognizes the IFN subtypes alpha and beta. In the receptor complex, two members of the cytokine receptor superfamily, IFNAR1 and IFNAR2, contribute to ligand binding and to activation of the associated Tyk2 and Jak1 (34–36).

(b) Pharmacological modulation of activation of the Jak/Stat pathway

In this study, we used antibodies raised against pStat1 (Tyr701) and pStat3 (Tyr705). Stat1 is activated by many cytokines but is critical for Interferon signaling since cells derived from Stat1 knockout mice are unresponsive to Interferon (37). Phosphorylation of Tyr701 of Stat1 induces protein dimerization which is a prerequisite for transcriptional activity of Stat1. Two Stat1 isoforms derived from differential splicing of the same gene have been reported which result in the transcription of a 84- and 91-kDa protein (38). Like Stat1, Stat3 is also activated by many cytokines and is especially important for fetal development since Stat3 knockout mice during early embryonic development (39). Tyr705 phosphorylation of Stat3 is associated with Stat3 dimerization that is as for Stat1 a prerequisite for Stat3 transcriptional activity. As described before, antibodies for RPA analysis were initially characterized by Western blot analysis. For this, Hela cells were seeded on six-well plates, starved for 5 h and treated with 100 ng/ml IFNα for 10 min. Analysis of cell lysates revealed that upon IFNα treatment of Hela cells the pStat1 antibody recognized specifically a double band of approximately 84/91 kDa, the sizes of the Stat1 full-length protein isoforms (Fig. 6). When the same samples were analyzed for pStat3 a single band of approximately 80 kDa, was observed (Fig. 6) which corresponds to the size of full-length Stat3 protein. Therefore, we decided to study the effects of IFNα on phosphorylation of both Stat proteins by RPA in order to develop a method for kinase inhibitor profiling.

In a first set of experiments, we studied dose and time dependency of Stat protein phosphorylation upon IFNα treatment. For this we stimulated Hela cells with various amounts of IFNα for 10, 20, 30, and 40 min and analyzed the levels of phospho-Stat1 and phospho-Stat3 in the lysates (Fig. 6). IFNα induced an approximately 10- to 20-fold increase in Stat phosphorylation for both, Stat1 and Stat3. To ensure that increase of Stat phosphorylation is dependent on Jak kinase activity we used CP-690,550.

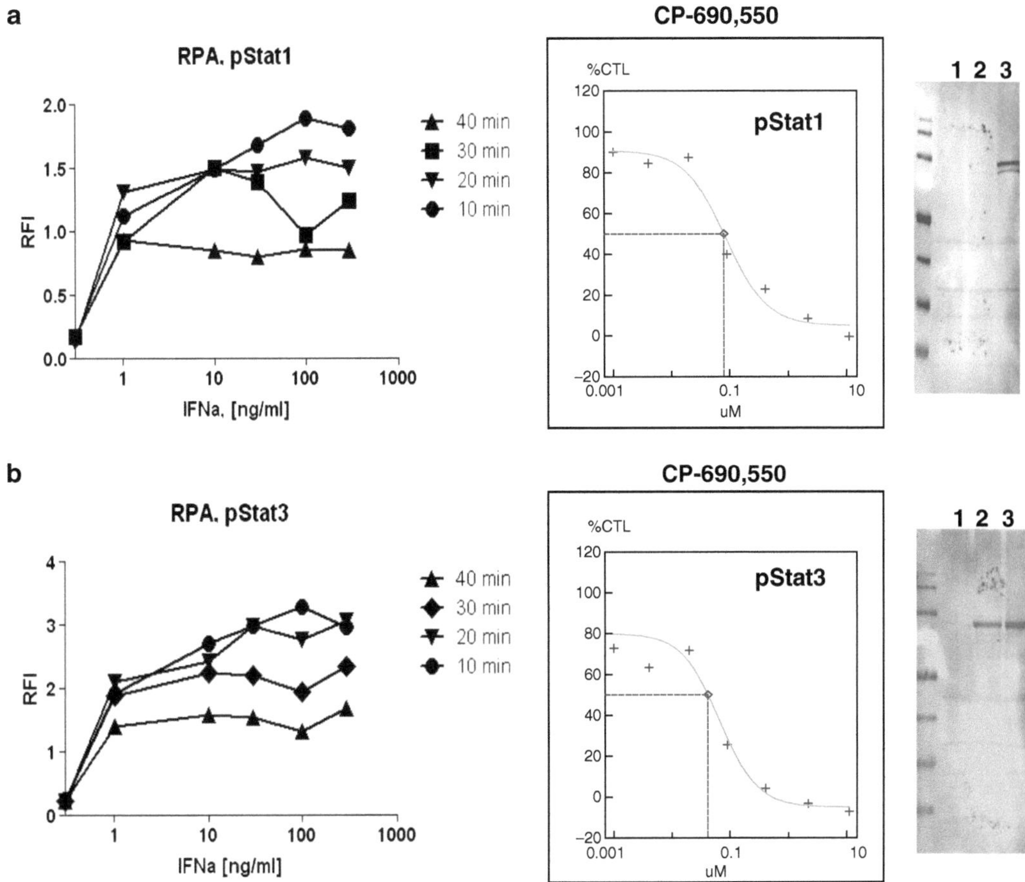

Fig. 6. Interferon alpha stimulated Stat1 and Stat3 phosphorylation. Hela cells have been seeded in 96-well plates, starved as described in Subheading 3 and stimulated with different amounts of IFNα and different incubation times as indicated. Cell lysates for RPA have been prepared as described, spotted on ZeptoCHIPs and analyzed for phosphorylated Stat1 (**a**) and Stat3 (**b**). Reference compound CP-690,550 has been tested in Hela cells which have been stimulated with 100 ng/ml IFNα for 10 min. For Western analysis Hela cells were grown in six-well plates until they reached approximately 90% confluency. Then cells were starved for 5 h in DMEM supplemented with 0.1% BSA and stimulated for 10 min either with 100 ng/ml IL6 or with IFNα. 50 μg of protein lysate were analyzed on Western blots using antibodies raised against pStat1 (Tyr701) or pStat3 (Tyr705). *Lane 1*, starved controls; *lane 2*, IL6 = 100 ng/ml IL6, 10 min; *lane 3*, IFNα = 100 ng/ml IFNα, 10 min. All experiments have been repeated three to four times and a representative example is shown.

This compound had been developed as a Jak3 inhibitor but subsequently shown to inhibit all Jak kinases in the low nanomolar range (16). CP-690,550 completely abolished Stat phosphorylation when added to cells which had been treated with 100 ng/ml IFNα for 10 min. Next we characterized a panel of kinase inhibitors of the various Jak kinases for Stat1 as well as Stat3 phosphorylation in Hela cells. As shown in Fig. 7, inhibition of Stat1 phosphorylation correlated very well with inhibition of Stat3 phosphorylation, indicating that both phospho-read outs are inhibited by the same up-stream Janus kinase.

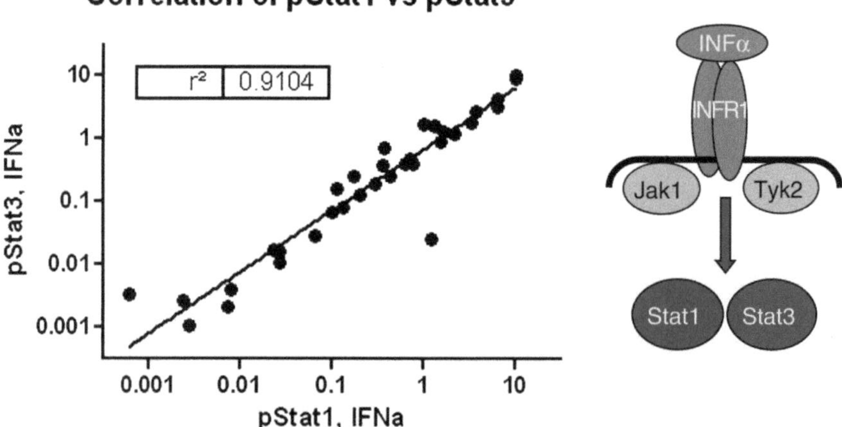

Fig. 7. Correlation of pStat1 and pStat3 in IFNα stimulated Hela cells. Hela cells have been seeded in 96-well plates, starved and stimulated with 100 ng/ml IFNα for 10 min. Thirty-four compounds have been analyzed for inhibition of Stat1 and Stat3 phosphorylation. IC50 for inhibition was calculated as described in Subheading 3 and results were plotted against each other. R2 obtained by linear regression is indicated. A schematic view of the IFNα pathway is shown on the *right*.

(c) Analysis of pharmacological modulation of Jak/Stat activation

A final validation experiment for the RPA assay was done by transfecting Hela cells with short interfering RNAs (siRNA) targeting either Jak1 or Tyk2 (Fig. 8). Seventy-two hours after siRNA transfection Hela cells were stimulated with different amounts of IFNα for 10 min. As expected, in cells which had been transfected with unrelated control siRNAs, a robust increase of both pStat1 and pStat3 was measured by RPA. In contrast, both siRNAs targeting either Jak1 or Tyk2 diminished the signal for phosphorylated Stat proteins. SiRNAs targeting Tyk2 were slightly more effective than siRNAs against Jak1. Combination of both siRNAs against Jak1 and Tyk2 completely abolished the response to IFNa, indicating that indeed Stat phosphorylation is mediated by both, Jak1 and Tyk2 and can be reliably measured by RPA (Fig. 8a, b). Again, knock down of Jak1 and Tyk2 as well as pStat protein levels were monitored by Western in order to confirm RPA data (Fig. 8c).

6. Unexpected findings for kinase inhibition of signaling pathways

It is often assumed that pathways are linear and that kinase inhibitors targeting central kinases of a given pathway should shut down-stream effectors of a given pathway. However, redundancy and feedback mechanisms have been reported which are often not taken into consideration.

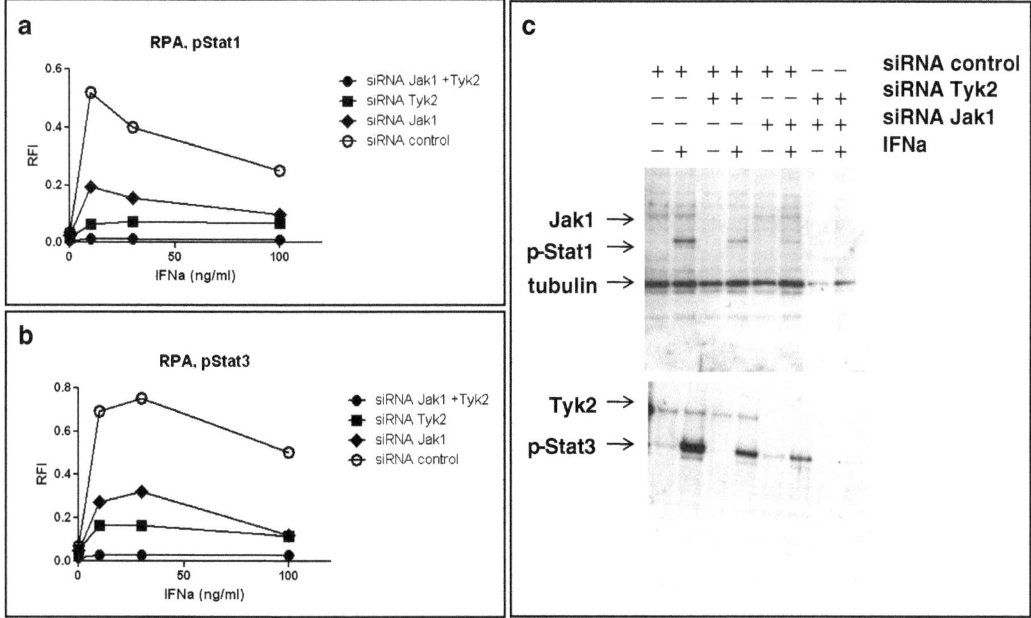

Fig. 8. Validation of IFNα RPA assay using siRNA. Validated siRNA targeting Jak1, Tyk2, or unrelated controls, respectively, have been purchased from Invitrogen. Hela cells have been transfected with siRNA (10 nM f.c.) as indicated in 96-well plates following the protocol provided by the supplier and grown for 72 h prior stimulation with IFNα. Phosphorylation of Stat1 (**a**) and Stat3 (**b**) was analyzed by RPA as described. Target knock down as well as inhibition of Stat phosphorylation was confirmed by Western (**c**).

In our studies, we used pERK as a read out to characterize inhibitors of the ERK pathway. For a more detailed analysis of allosteric inhibitors of MEK, we included phosphorylation of MEK (Ser217/Ser221) in the analysis. As MEK inhibitors we choose PD-0325901 and AZD6244, two well-described inhibitors of MEK1/2 (14, 15). Ser217/221 are located on the activation loop of MEK and needs to be phosphorylated by up-stream MAP3-kinases including RAF and Cot/Tpl-2. As expected, we observed potent dose-dependent inhibition of ERK phosphorylation in insulin-treated A14 cells by both MEK inhibitors (Fig. 5a). Surprisingly, this was not accompanied by the inhibition of MEK phosphorylation but by a dose-dependent increase of MEK phosphorylation (Fig. 5b). Interestingly, increase of pMEK was already visible when the compounds were added to starved cells, but was further enhanced in the presence of insulin (Fig. 5c). We speculated that binding of the allosteric inhibitor to MEK stabilizes the kinase in a conformation which makes it more accessible to up-stream kinases and thereby increases the amount of phosphorylated MEK protein. To prove this hypothesis we performed a combination experiment incubating AZD6244 together with either the unspecific kinase inhibitor staurosporine or more specific inhibitors of either Cot/Tpl-2 kinase or RAF kinase prior to insulin stimulation of A14 cells. In insulin-treated cells, MEK should be activated via the Ras and RAF which couple to the phosphorylated receptor as well as to IRS-1 and IRS-2. As shown in Fig. 5, both, staurosporine and a specific inhibitor of RAF were able to potently inhibit MEK hyper-phosphorylation while the specific inhibitor of Cot/Tpl-2 was inactive. Furthermore, findings of the RPA experiments could be recapitulated by Western blot analysis (Fig. 5d) indicating that allosteric inhibitors of MEK indeed induce MEK hyper-phosphorylation by their up-stream kinase RAF. MEK hyper-phosphorylation induced by allosteric inhibitors of MEK has also been observed by others (40). Like in our experiments, hyper-phosphorylation did not prevent inhibition of down-stream signaling such as ERK phosphorylation, although a mechanism for MEK hyper-phosphorylation was not elucidated. Increased phosphorylation of kinases by specific inhibitors has been described before and termed kinase priming. More recently, similar findings have been described for PKB (41). This report describes hyper-phosphorylation of two known regulatory phosphorylation sites of PKB, namely Ser473 and Thr308, as a direct consequence of binding of an ATP-competitive inhibitor to PKB. Similar has been described for protein kinase C (PKC) (42). PKC needs priming phosphorylation on multiple activation sites for full activity, and binding of inhibitors to the nucleotide binding site enhances priming phosphorylation of PKC independent of PKC autophosphorylation. Unraveling unexpected increase in protein phosphorylation by kinase inhibitors can be detected by the multiplexed pathway analysis using RPA. Even more, by multiplexed routine testing of several analytes it offers the opportunity to study these phenomena in a more systematic fashion.

7. Combining RGA with RPA

Reporter gene assays (RGAs) are widely applied to study signaling cascades in cellular systems. RGAs deliver an integrated readout, with a high validation state, downstream of the signaling cascades offering an excellent system to profile protein kinase inhibitors. Using RGAs, target-specific as well as nontarget-related effects of kinase inhibitors can be measured. There are a number of different reporter genes that can be used for this purpose. The luciferase reporter gene has been introduced in 1987 (43) and thereafter has revolutionized RGAs and other high-throughput screening (HTS) methods allowing highly sensitive and homogenous assay formats (44). However, the luciferase gene readout is based on lyses of cells defining an end point measurement. Only very recently, live-cell luciferase-based substrates became available. For the analysis of pathways and signaling in cells via RGAs, it is an advantage to measure in live cells allowing kinetic reads and multiplexing together with other readouts like toxicity. The beta lactamase (bla) reporter gene-based technology offers such an advantage. A cell permeable

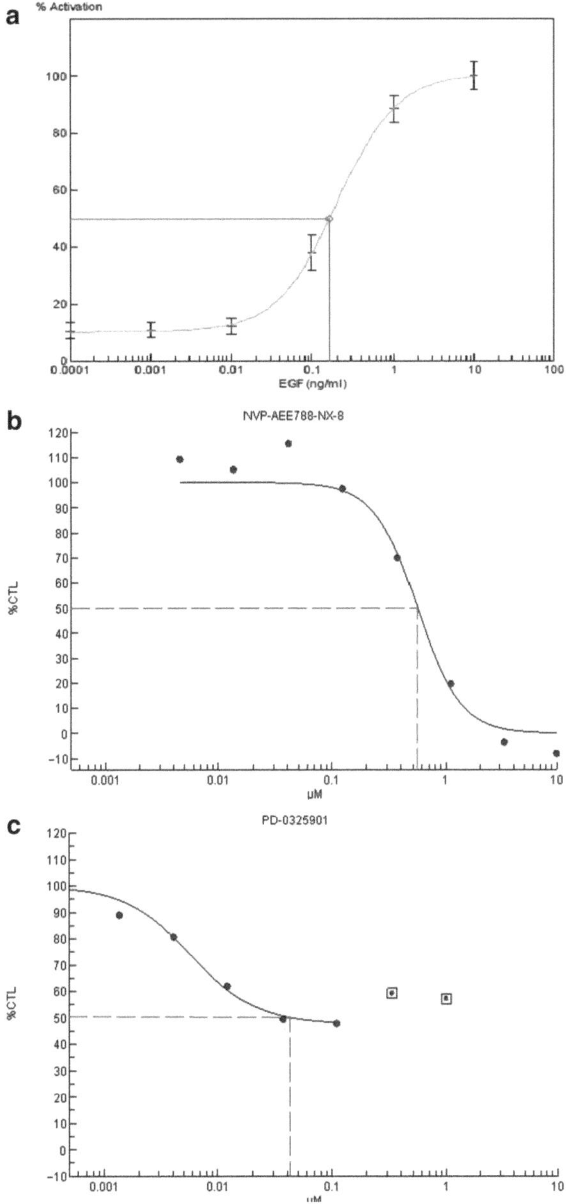

Fig. 9. Effect of EGFR- and MEK-1 inhibitors on EGF-stimulated AP-1-bla ME180 cells. The ME180, AP-1 reporter gene cell line was stimulated with a serial dilution of EGF ranging from 0.0001 to 10 ng/ml. The beta-lactamase reporter gene product was measured as described in Subheadings 2 and 3 (**a**). The specific EGFR inhibitor (NVP-AEE788) was added 1 h prestimulation at the indicated concentrations (**b**). The specific allosteric MEK-1 inhibitor (PD0325901) was added 1 h prestimulation at the indicated concentrations (**c**).

substrate is added to the medium that emits a green fluorescence based on an intramolecular FRET. Upon cleavage of the substrate by the expressed bla, the cleaved product emits a blue fluorescence allowing for a ratiometric measurement of substrate versus product (45). This technology is ideal for the profiling of compounds in a 384-well format but the assays can be further miniaturized allowing HTS (46).

Stimulation of the AP-1-bla ME180 cells using EGF leads to the potent activation of the reporter gene via the MEK–ERK- and the MEKK1–JNK1-signaling pathways with an EC50 of 1.6 ng/ml (Fig. 9a). Inhibition of the EGF receptor (EGFR) with the EGFR inhibitor NVP-AEE788 leads to a complete and potent inhibition of the reporter gene at an IC50 of 0.53 μM (Fig. 9b). In contrast, the potent MEK-1 inhibitor PD-0325901 leads to a 55% inhibition of the reporter gene at an IC50 of 0.042 μM (Fig. 9c). The incomplete inhibition presumably results from the activity transmitted via the MEKK1–JNK1 signaling path.

References

1. Fishman, M.C. & Porter, J.A. Pharmaceuticals: a new grammar for drug discovery. *Nature* **437**, 491–3 (2005).
2. Inoki, K., Corradetti, M.N. & Guan, K.L. Dysregulation of the TSC-mTOR pathway in human disease. *Nat Genet* **37**, 19–24 (2005).
3. Butcher, E.C. Can cell systems biology rescue drug discovery? *Nat Rev Drug Discov* **4**, 461–7 (2005).
4. Apic, G., Ignjatovic, T., Boyer, S. & Russell, R.B. Illuminating drug discovery with biological pathways. *FEBS Lett* **579**, 1872–7 (2005).
5. Butcher, E.C., Berg, E.L. & Kunkel, E.J. Systems biology in drug discovery. *Nat Biotechnol* **22**, 1253–9 (2004).
6. Sevecka, M. & MacBeath, G. State-based discovery: a multidimensional screen for small-molecule modulators of EGF signaling. *Nat Methods* **3**, 825–31 (2006).
7. Cho, C.R., Labow, M., Reinhardt, M., van Oostrum, J. & Peitsch, M.C. The application of systems biology to drug discovery. *Curr Opin Chem Biol* **10**, 294–302 (2006).
8. Mendes, K.N. et al. Analysis of signaling pathways in 90 cancer cell lines by protein lysate array. *J Proteome Res* **6**, 2753–67 (2007).
9. Sheehan, K.M. et al. Use of reverse phase protein microarrays and reference standard development for molecular network analysis of metastatic ovarian carcinoma. *Mol Cell Proteomics* **4**, 346–55 (2005).
10. Pawlak, M. et al. Zeptosens' protein microarrays: a novel high performance microarray platform for low abundance protein analysis. *Proteomics* **2**, 383–93 (2002).
11. Wang, Z.G. et al. TAE226, a dual inhibitor for FAK and IGF-IR, has inhibitory effects on mTOR signaling in esophageal cancer cells. *Oncol Rep* **20**, 1473–7 (2008).
12. Maira, S.M. et al. Identification and characterization of NVP-BEZ235, a new orally available dual phosphatidylinositol 3-kinase/mammalian target of rapamycin inhibitor with potent in vivo antitumor activity. *Mol Cancer Ther* **7**, 1851–63 (2008).
13. Schuler, W. et al. SDZ RAD, a new rapamycin derivative: pharmacological properties in vitro and in vivo. *Transplantation* **64**, 36–42 (1997).
14. Barrett, S.D. et al. The discovery of the benzhydroxamate MEK inhibitors CI-1040 and PD 0325901. *Bioorg Med Chem Lett* **18**, 6501–4 (2008).
15. Huynh, H., Soo, K.C., Chow, P.K. & Tran, E. Targeted inhibition of the extracellular signal-regulated kinase kinase pathway with AZD6244 (ARRY-142886) in the treatment of hepatocellular carcinoma. *Mol Cancer Ther* **6**, 138–46 (2007).
16. Changelian, P.S. et al. Prevention of organ allograft rejection by a specific Janus kinase 3 inhibitor. *Science* **302**, 875–8 (2003).

17. Traxler, P. et al. AEE788: a dual family epidermal growth factor receptor/ErbB2 and vascular endothelial growth factor receptor tyrosine kinase inhibitor with antitumor and antiangiogenic activity. *Cancer Res* **64**, 4931–41 (2004).
18. Burgering, B.M. et al. Insulin stimulation of gene expression mediated by p21ras activation. *Embo J* **10**, 1103–9 (1991).
19. Zhang, J.H., Chung, T.D. & Oldenburg, K.R. A Simple Statistical Parameter for Use in Evaluation and Validation of High Throughput Screening Assays. *J Biomol Screen* **4**, 67–73 (1999).
20. De Meyts, P. & Whittaker, J. Structural biology of insulin and IGF1 receptors: implications for drug design. *Nat Rev Drug Discov* **1**, 769–83 (2002).
21. Kanzaki, M. & Pessin, J.E. Signal integration and the specificity of insulin action. *Cell Biochem Biophys* **35**, 191–209 (2001).
22. Kusari, A.B., Byon, J.C. & Kusari, J. Substitution of two insulin receptor carboxy-terminal tyrosines with phenylalanine impairs the expression of MAP kinase phosphatase-1 (MKP-1) mRNA. *Mol Cell Biochem* **211**, 27–37 (2000).
23. Sarbassov, D.D., Guertin, D.A., Ali, S.M. & Sabatini, D.M. Phosphorylation and regulation of Akt/PKB by the rictor-mTOR complex. *Science* **307**, 1098–101 (2005).
24. Srivastava, A.K. & Pandey, S.K. Potential mechanism(s) involved in the regulation of glycogen synthesis by insulin. *Mol Cell Biochem* **182**, 135–41 (1998).
25. Alessi, D.R. et al. Identification of the sites in MAP kinase kinase-1 phosphorylated by p74raf-1. *Embo J* **13**, 1610–9 (1994).
26. Murphy, L.O. & Blenis, J. MAPK signal specificity: the right place at the right time. *Trends Biochem Sci* **31**, 268–75 (2006).
27. Peterson, R.T. & Schreiber, S.L. Translation control: connecting mitogens and the ribosome. *Curr Biol* **8**, R248-50 (1998).
28. White, M.F., Maron, R. & Kahn, C.R. Insulin rapidly stimulates tyrosine phosphorylation of a Mr-185,000 protein in intact cells. *Nature* **318**, 183–6 (1985).
29. Foxwell, B.M., Barrett, K. & Feldmann, M. Cytokine receptors: structure and signal transduction. *Clin Exp Immunol* **90**, 161–9 (1992).
30. Yamaoka, K. et al. The Janus kinases (Jaks). *Genome Biol* **5**, 253 (2004).
31. Kisseleva, T., Bhattacharya, S., Braunstein, J. & Schindler, C.W. Signaling through the JAK/STAT pathway, recent advances and future challenges. *Gene* **285**, 1–24 (2002).
32. Murray, P.J. The JAK-STAT signaling pathway: input and output integration. *J Immunol* **178**, 2623–9 (2007).
33. O'Shea, J.J., Gadina, M. & Schreiber, R.D. Cytokine signaling in 2002: new surprises in the Jak/Stat pathway. *Cell* **109 Suppl**, S121-31 (2002).
34. Stark, G.R., Kerr, I.M., Williams, B.R., Silverman, R.H. & Schreiber, R.D. How cells respond to interferons. *Annu Rev Biochem* **67**, 227–64 (1998).
35. Mogensen, K.E., Lewerenz, M., Reboul, J., Lutfalla, G. & Uze, G. The type I interferon receptor: structure, function, and evolution of a family business. *J Interferon Cytokine Res* **19**, 1069–98 (1999).
36. Yeh, T.C. & Pellegrini, S. The Janus kinase family of protein tyrosine kinases and their role in signaling. *Cell Mol Life Sci* **55**, 1523–34 (1999).
37. Durbin, J.E., Hackenmiller, R., Simon, M.C. & Levy, D.E. Targeted disruption of the mouse Stat1 gene results in compromised innate immunity to viral disease. *Cell* **84**, 443–50 (1996).
38. Schindler, C., Fu, X.Y., Improta, T., Aebersold, R. & Darnell, J.E., Jr. Proteins of transcription factor ISGF-3: one gene encodes the 91-and 84-kDa ISGF-3 proteins that are activated by interferon alpha. *Proc Natl Acad Sci USA* **89**, 7836–9 (1992).
39. Takeda, K. et al. Targeted disruption of the mouse Stat3 gene leads to early embryonic lethality. *Proc Natl Acad Sci USA* **94**, 3801–4 (1997).
40. Vogel, S. et al. MEK hyperphosphorylation coincides with cell cycle shut down of cultured smooth muscle cells. *J Cell Physiol* **206**, 25–34 (2006).
41. Okuzumi, T. et al. Inhibitor hijacking of Akt activation. *Nat Chem Biol* **5**, 484–93 (2009).
42. Cameron, A.J., Escribano, C., Saurin, A.T., Kostelecky, B. & Parker, P.J. PKC maturation is promoted by nucleotide pocket occupation independently of intrinsic kinase activity. *Nat Struct Mol Biol* **16**, 624–30 (2009).
43. de Wet, J.R., Wood, K.V., DeLuca, M., Helinski, D.R. & Subramani, S. Firefly luciferase gene: structure and expression in mammalian cells. *Mol Cell Biol* **7**, 725–37 (1987).
44. Fan, F. & Wood, K.V. Bioluminescent assays for high-throughput screening. *Assay Drug Dev Technol* **5**, 127–36 (2007).
45. Zlokarnik, G. et al. Quantitation of transcription and clonal selection of single living cells with beta-lactamase as reporter. *Science* **279**, 84–8 (1998).
46. Chin, J. et al. Miniaturization of cell-based beta-lactamase-dependent FRET assays to ultra-high throughput formats to identify agonists of human liver X receptors. *Assay Drug Dev Technol* **1**, 777–87 (2003).

Chapter 8

Use of Formalin-Fixed and Paraffin-Embedded Tissues for Diagnosis and Therapy in Routine Clinical Settings

Daniela Berg, Katharina Malinowsky, Bilge Reischauer, Claudia Wolff, and Karl-Friedrich Becker

Abstract

Formalin-fixed and paraffin-embedded (FFPE) tissues are used routinely everyday in hospitals world-wide for histopathological diagnosis of diseases like cancer. Due to formalin-induced cross-linking of proteins, FFPE tissues present a particular challenge for proteomic analysis. Nevertheless, there has been recent progress for extraction-based protein analysis in these tissues. Novel tools developed in the last few years are urgently needed because precise protein biomarker quantification in clinical FFPE tissues will be crucial for treatment decisions and to assess success or failure of current and future personalized molecular therapies. Furthermore, they will help to conceive why only a subset of patients responds to individualized treatments. Reverse phase protein array (RPPA) is a very promising new technology for quick and simultaneous analysis of many patient samples allowing relative and absolute protein quantifications. In this chapter, we show how protein extraction from FFPE tissues might facilitate the implementation of RPPA for therapy decisions and discuss challenges for application of RPPA in clinical trials and routine settings.

Key words: Reverse phase protein array, FFPE, Personalized therapy, Urokinase-typ plasminogen activator, Plasminogen activator inhibitor type 1, HER2, Signalling

Abbreviations

DTT	Dithiothreitol
EGFR	Epidermal growth factor receptor
ELISA	Enzyme-linked immunosorbent assay
ERK	Extracellular signal-regulated kinase
FFPE	Formalin-fixed and paraffin-embedded
FISH	Fluorescence in situ hybridisation
H&E	Hematoxylin and Eosin
HER2/neu/ErbB2	Human epidermal growth factor receptor 2

Ulrike Korf (ed.), *Protein Microarrays: Methods and Protocols*, Methods in Molecular Biology, vol. 785, DOI 10.1007/978-1-61779-286-1_8, © Springer Science+Business Media, LLC 2011

IHC	Immunohistochemistry
MAPK	Mitogen-activated protein kinase
PAI1	Plasminogen activator inhibitor 1
RPPA	Reverse phase protein array
SDS-PAGE	Sodium dodecyl sulfate polyacrylamide gel electrophoresis
uPA	Urokinase-type plasminogen activator

1. Formalin-Fixed and Paraffin-Embedded Tissues

Formalin-fixation and paraffin-embedding (FFPE) is the standard method for histological tissue preparation in most hospitals around the world. FFPE tissues are used for everyday routine diagnosis of many diseases, including cancer. The routine formalin-fixation process stabilizes proteins via cross-linking of macromolecules which keeps the tissues in an excellent condition for further histopathological analysis (1). Formalin is 37–40% formaldehyde in water, stabilized by 10% methanol. Tissues are routinely fixed in a phosphate buffered 10% solution of formalin. Formaldehyde reacts with amino groups of basic amino acids such as lysine, asparagine, arginine, histidine, and glutamine, leading to the formation of highly reactive methylol adducts. A subsequent condensation reaction of adducts occurs through Schiff base formation, resulting in the formation of methylene bridges with amine, guanidyl, phenol imidazol, and indole groups of several other amino acids such as arginine, asparagine, glutamine, histidine, tryptophan, and tyrosine. The results of these chemical reactions are inter- and intramolecular cross-linking of proteins (2).

Several reports have shown that not only proteins but also protein modifications (e.g., phosphorylations, glycosylations) are preserved during fixation and can be analysed even years later, for example by immunohistochemistry (IHC) (3, 4). Although IHC can provide valuable information on the abundances of proteins in tissues, this method is not suitable for the analysis of subtle quantitative changes in multiple classes of proteins taking place simultaneously within a cell or tissue (5). In addition, the quantification of protein expression based on IHC is difficult and depends at least in part on the observer.

In this chapter, we want to show how new methods of protein extraction from FFPE tissue and the application of RPPA using protein lysates from FFPE tissues might well allow the identification and quantification of protein biomarkers related to diagnosis, prognosis, monitoring, and treatment selection in the future.

2. Protein Extraction from FFPE Tissues

For the use in clinical diagnostics a protein extraction protocol should not be too complex and has to be compatible with downstream analysis such as Western-blot, RPPA, or mass spectrometry. FFPE tissues were long believed not to be suitable for study of proteins beyond immunohistochemistry. Formalin-induced cross-linking of proteins was thought to be irreversible and protein extraction of those tissues was impossible. Furthermore, proteins from FFPE tissues could be difficult to re-suspend and hard to separate and analyse. Nevertheless, only recently it became possible to successfully extract full-length, immunoreactive proteins from FFPE tissues (1, 6–13) (see below and Chap. 27 of this book).

Typically, biopsies or tissues taken during surgery are fixed in formalin as soon as possible, dehydrated in an alcohol series, and finally embedded in paraffin. From the resulting paraffin blocks 3 μm sections are prepared for H&E staining used for routine diagnosis. The same H&E stained slides may also be reference slides for the protein extraction procedure. For protein extraction 10 μm sections are prepared from the same blocks from which the sections for routine diagnosis were cut. To reduce cellular heterogeneity and to avoid selection of necrotic regions, tissue areas of interest are defined by histological inspection and, therefore, must be selected by a pathologist on the H&E-stained reference slide. It is not recommended to procure the tissues directly from the H&E stained slide as histological stains can decrease the yield of the extracted proteins (8). Tumour areas are dissected and proteins are extracted as described in detail in Chap. 27 using a commercial kit. A list of recently reported protocols for protein extraction is available in the same chapter. After protein extraction from FFPE tissues, the lysates from each patient can be analysed by Western blot or reverse phase protein microarrays (Fig. 1).

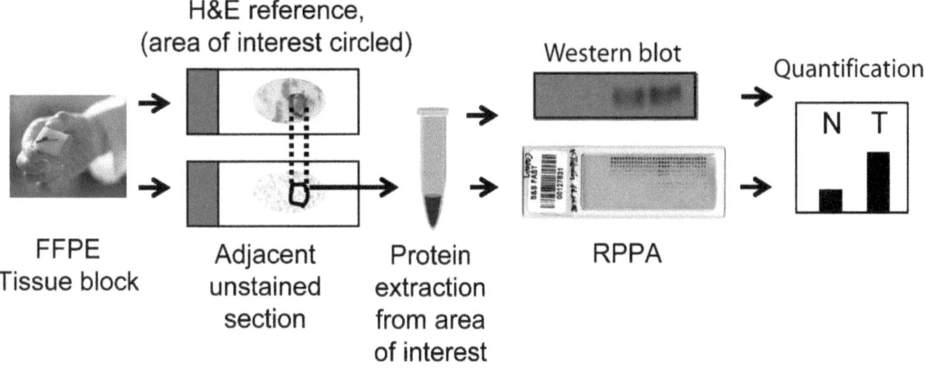

Fig. 1. Flow chart of protein lysate microarrays from FFPE tissues. 10 μm sections were cut from paraffin-embedded tissues. The tumour area was determined using an H&E stained reference. Following, the tumour area was dissected and protein lysates were prepared and subsequently analysed by RPPA. The antibody-specificity was tested by western blot. Finally, protein expression was quantified.

3. Antibody Validation Using Protein Extracts from FFPE Tissues

The most important starting point for successful detection of proteins by RPPA is the selection of antibodies with high specificity and adequate affinity. For gene transcript profiling, probes with predictable affinity and specificity can be produced. In contrast, this is not possible for antibodies (14, 15). Prior to use of an antibody in RPPA, its specificity must be confirmed by Western blot. Preferably the same materials as in the array should be utilized, e.g., protein extracts from FFPE samples. The optimal evidence for antibody specificity is a single band at the appropriate molecular weight (Fig. 2). Phosphospecific antibodies should show different signals between control and treated samples, additionally.

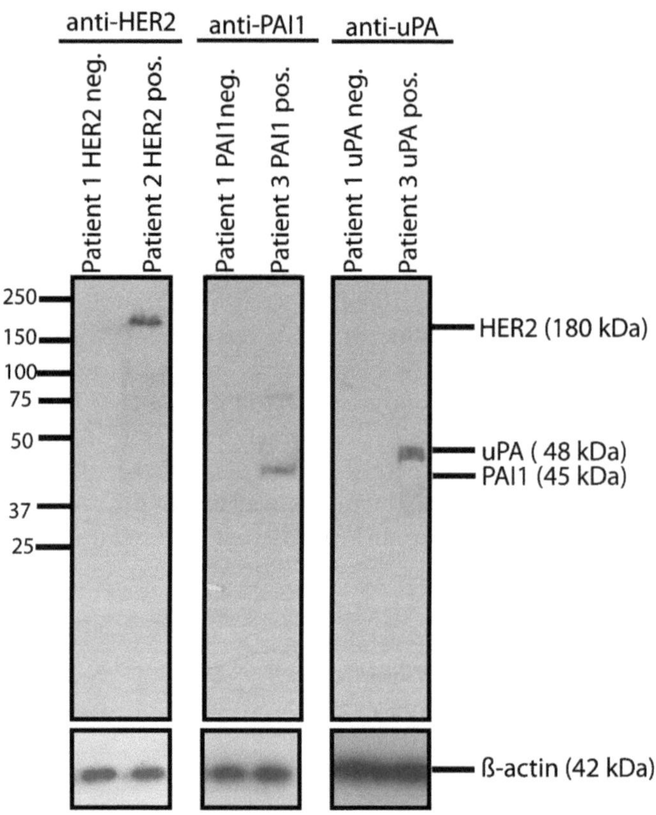

Fig. 2. Antibody validation by Western blot. HER2 and uPA/PAI-1 antibodies were tested for specificity by Western blotting (25 μg per lane) using protein lysates extracted from FFPE breast cancer tissue samples. Protein extraction from FFPE tissue samples is described in detail in Chap. 27. Extracts from HER2- and uPA/PAI1-positive tissues show bands at the predicted molecular weight. No signals were detected in HER2- and uPA/PAI1-negative tissue extracts. ß-actin was used as loading control.

4. Reverse Phase Protein Arrays

Basically, reverse phase protein arrays (RPPAs) are immobilized protein spots utilized for quantitative immunochemical detection. In general, there is no difference in the methodology whether a lysate from a cell line, frozen tissue sample, or FFPE tissue sample is applied, provided that a specific antibody against the protein of interest is available that reacts in the desired manner with the lysate applied on the array. The technology allows monitoring of changes in protein abundances over time, before and after treatment, between disease and non-disease states and between responders and non-responders (16). Usually not just one spot of the lysate of interest is put onto the slide, but a dilution curve in 3–4 replicates. Therefore, each analyte/antibody combination can be analysed in the linear dynamic range (16, 17). In addition to applying the detection antibody, one or more slides spotted in parallel are stained with a total protein detection reagent (e.g., SYPRO® Ruby staining solution). This step is necessary for normalization of the signals obtained by the antibodies to total protein. The RPPA format allows multiple samples to be analysed for expression of one protein under the same experimental conditions. So each array contains many patient samples, which are incubated with one antibody (Fig. 3). Antibody binding is measured directly; there is no need for direct labelling of patient proteins and no utilization of a two-site antibody sandwich, what reduces experimental variability (14, 18). With RPPA minimum detection levels being in the attogram (10^{-18} g) range this technology is more sensitive than an enzyme-linked immunosorbent assay (ELISA) (16, 18). Moreover, it is suited for signal transduction profiling of small numbers of cultured cells or cells isolated by laser capture microdissection from human biopsies (5, 19).

4.1. Reproducibility of Reverse Phase Protein Arrays with Lysates from FFPE Tissues

To analyse the reproducibility and variability of RPPA from FFPE tissues, we evaluated the variation between lysate preparation and variation between plate and experimental setups and array runs (intra-sample, inter-, and intra-array variations). We analysed the abundances of protein biomarkers currently used in the clinic for therapy decisions, HER2, uPA, and PAI1. Details for the clinical importance for these markers will be provided below (see Subheading 5).

First to assess the effect of array setup and preparation, we extracted proteins from seven FFPE breast cancer tissue samples and printed the protein lysates twice onto the same slide (intra-array variation). A high correlation between the expression intensities of the corresponding samples was observed. The correlation coefficients between the replicates were $r=0.99$ for HER2; $r=0.97$ for PAI1, and $r=0.99$ for uPA (Fig. 4a).

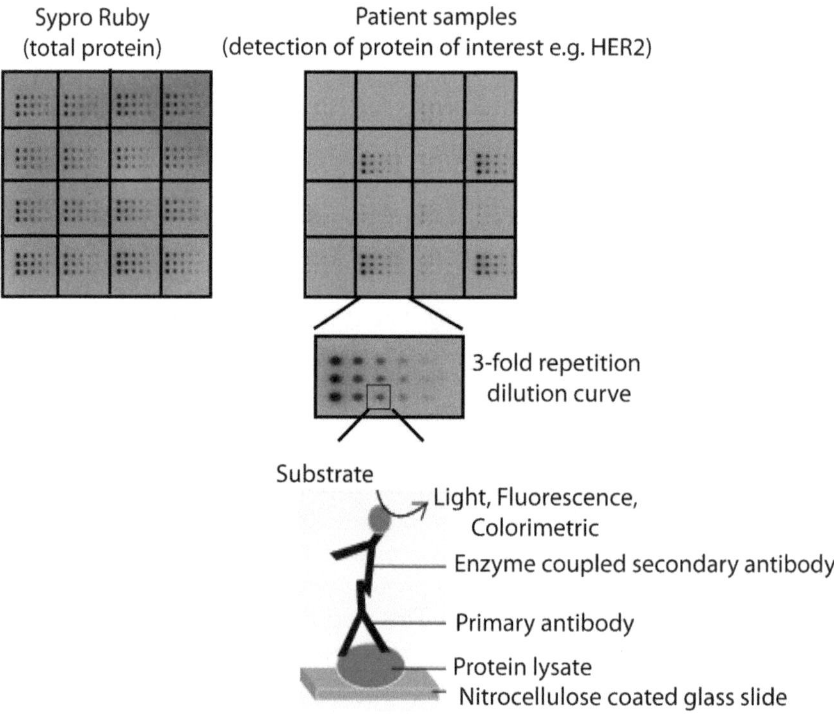

Fig. 3. Reverse phase protein microarray. Protein lysates from FFPE breast cancer tissue samples were spotted on a nitrocellulose-coated glass slide, followed by incubation with the primary antibody (*right*). Per lysate sixfold dilution curves in three replicates were spotted. Detection is carried out by using an enzyme-coupled secondary antibody together with its target substrate. For normalization of the antibody signals to total protein the slide was stained with Sypro Ruby reagent (*left*).

Second, the impact of sample preparation on RPPA variability was tested. Therefore, all seven patient samples have been extracted twice and printed onto the same slide (intra-sample variation). A high correlation was observed, indicating a high reproducibility for protein lysate production. The correlation coefficients between two independent extractions were $r=0.98$ for HER2; $r=0.80$ for PAI1, and $r=0.97$ for uPA (Fig. 4b).

Third, a particular challenge for the widespread use of RPPA in clinical settings is the variability and comparability of staining between arrays, which often hamper productive data comparisons between different hospitals or experiments. This is a highly relevant issue as multiple arrays may be required in clinical studies to consider all samples for a particular patient cohort. Ideally, a RPPA reference standard for comparing different slides detected with the same antibody should serve as a universal positive control for the staining process and antibody validation and it should also be incorporated into data analysis. Additionally, a satisfying RPPA quality reference should be renewable, reproducible in large scale, successful over a broad range of end points, stable over a long period of time, and, finally, as closely related to the test sample as possible (18). Extracts from FFPE tissues identical to the test

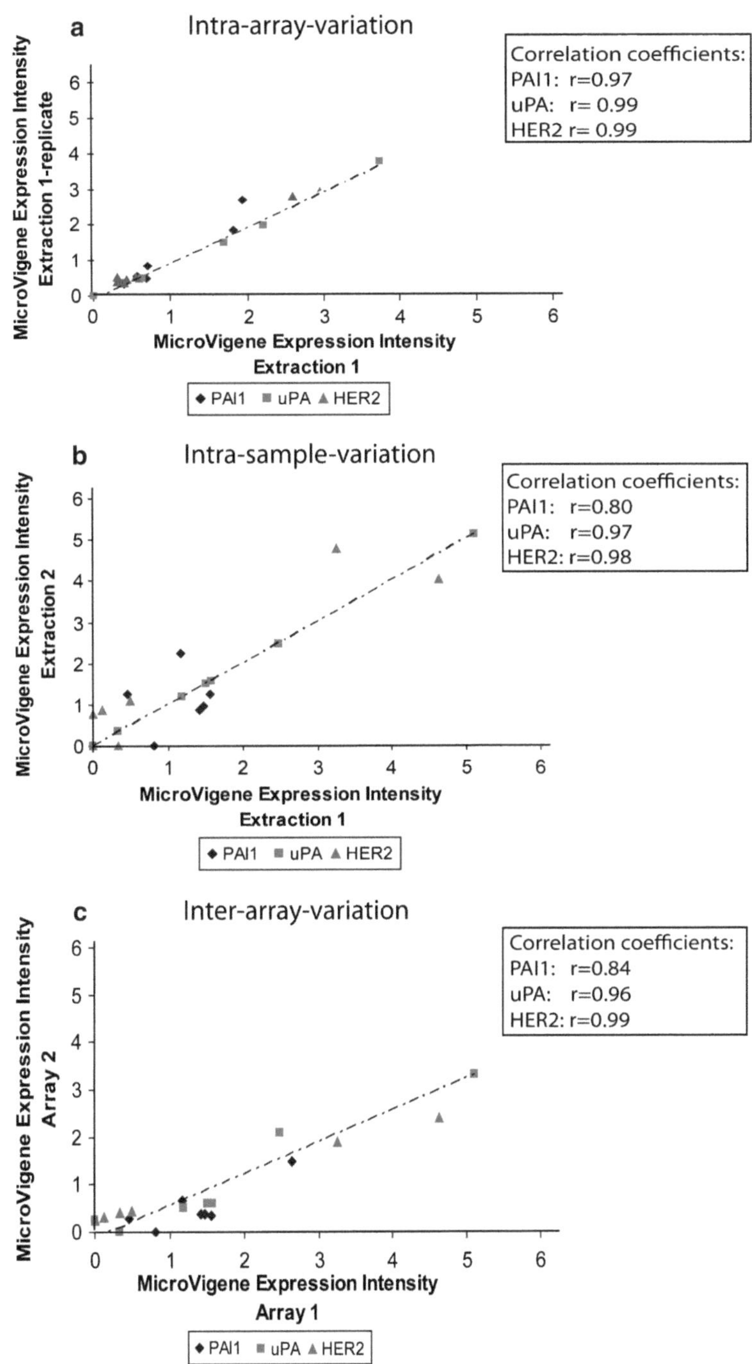

Fig. 4. Reproducibility of spotting and lysate preparation. Protein extracts from seven breast cancer patients were prepared and assayed for HER2 and uPA/PAI1 expression using RPPA technology. Each slide was incubated with HER2 and uPA/PAI1-specific antibodies to determine HER2 abundances and uPA/PAI1 levels. Total protein was determined by Sypro Ruby Protein Blot stain. Subsequently, protein expression was normalized to total protein. (**a**) After protein extraction from seven FFPE breast cancer tissue samples, the protein lysates were printed twice on a slide to assess intra-array variation. (**b**) To analyse intra-sample variation proteins of said seven patient samples were extracted twice independently and arrayed on the same slide. (**c**) The inter-array variation was determined by spotting the samples on different slides.

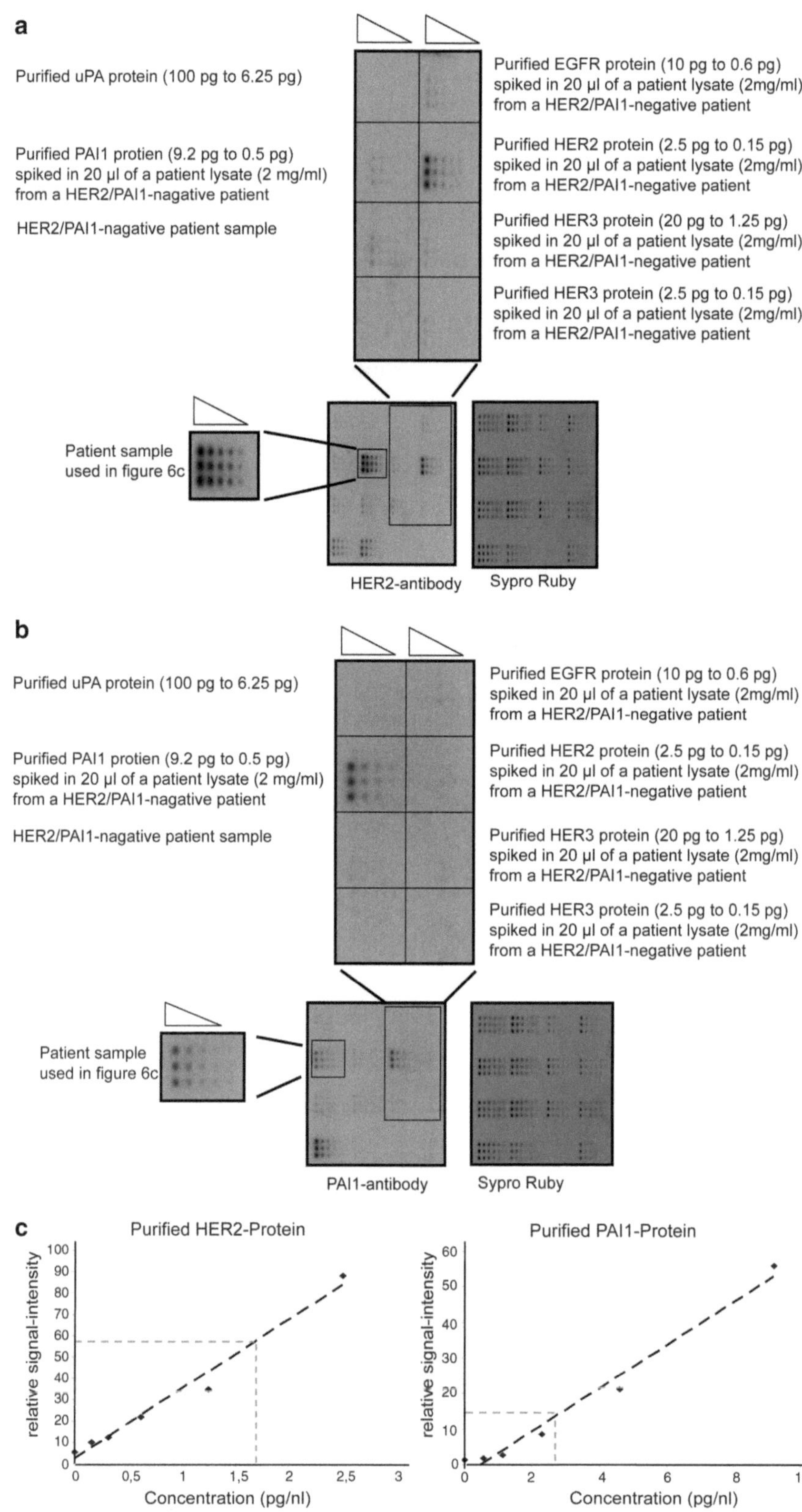
a
Purified uPA protein (100 pg to 6.25 pg)
Purified PAI1 protien (9.2 pg to 0.5 pg) spiked in 20 µl of a patient lysate (2 mg/ml) from a HER2/PAI1-nagative patient
HER2/PAI1-nagative patient sample
Purified EGFR protein (10 pg to 0.6 pg) spiked in 20 µl of a patient lysate (2mg/ml) from a HER2/PAI1-negative patient
Purified HER2 protein (2.5 pg to 0.15 pg) spiked in 20 µl of a patient lysate (2mg/ml) from a HER2/PAI1-negative patient
Purified HER3 protein (20 pg to 1.25 pg) spiked in 20 µl of a patient lysate (2mg/ml) from a HER2/PAI1-negative patient
Purified HER3 protein (2.5 pg to 0.15 pg) spiked in 20 µl of a patient lysate (2mg/ml) from a HER2/PAI1-negative patient
Patient sample used in figure 6c
HER2-antibody
Sypro Ruby
b
Purified uPA protein (100 pg to 6.25 pg)
Purified PAI1 protien (9.2 pg to 0.5 pg) spiked in 20 µl of a patient lysate (2 mg/ml) from a HER2/PAI1-nagative patient
HER2/PAI1-nagative patient sample
Purified EGFR protein (10 pg to 0.6 pg) spiked in 20 µl of a patient lysate (2mg/ml) from a HER2/PAI1-negative patient
Purified HER2 protein (2.5 pg to 0.15 pg) spiked in 20 µl of a patient lysate (2mg/ml) from a HER2/PAI1-negative patient
Purified HER3 protein (20 pg to 1.25 pg) spiked in 20 µl of a patient lysate (2mg/ml) from a HER2/PAI1-negative patient
Purified HER3 protein (2.5 pg to 0.15 pg) spiked in 20 µl of a patient lysate (2mg/ml) from a HER2/PAI1-negative patient
Patient sample used in figure 6c
PAI1-antibody
Sypro Ruby
c
Purified HER2-Protein
relative signal-intensity
100 90 80 70 60 50 40 30 20 10 0
0 0,5 1 1,5 2 2,5 3
Concentration (pg/nl)
Purified PAI1-Protein
relative signal-intensity
60 50 40 30 20 10 0
0 2 4 6 8 10
Concentration (pg/nl)

samples may be suitable candidates for reference standards but these lysates are hardly renewable and not available in quantities large enough for large-scale analysis. In contrast, extracts from various stimulated and non-stimulated cell lines can be produced in large quantities. Unfortunately, long-term reproducibility and stability of cell lines is as difficult as that of tissue extracts (18). For a first simple comparison, we spotted samples onto different slides to determine the inter-array variation. Again, we found a high correlation between two different slides with correlation coefficients of $r=0.99$ for HER2; $r=0.84$ for PAI1, and $r=0.96$ for uPA (Fig. 4c).

4.2. Quantification of Protein Expression in FFPE Samples by Reverse Phase Protein Microarrays

To determine the absolute protein concentration of HER2 and PAI1 in a patient sample we arrayed purified recombinant HER2 and PAI1 proteins of known concentration together with the patient samples on the same slide. Before spotting, 2.5–0.15 pg HER2 and 9.2–6.25 pg PAI1 were spiked in a HER2-negative (2 mg/ml) and PAI1-negative (2 mg/ml) patient lysate, respectively (Fig. 5a, b) to eliminate potential influences of protein mixtures on signal-intensity compared to purified proteins. Using HER2- and PAI1-signal-intensity–concentration curves, it was possible to determine HER2 and PAI1 concentrations in the unknown sample (1.75 pg HER2/nl; 3 pg PAI1/nl see Fig. 5c). Finally, the HER2 and PAI1 concentration was normalized to total protein. Total protein concentration was determined before arraying and was 2 mg/ml in the undiluted first spot. In our example, we used the third dilution (0.25 mg/ml total protein) for HER2 quantification and the first dilution (1 mg/ml total protein) for PAI1-quantification. Consequently, the normalized HER2 concentration in the sample was 7 pg/ng total protein. The normalized PAI1 concentration was 3 pg/ng total protein. For protein quantification, we expected that each spot contains 1 nl protein lysate, according to the manufacturer's instructions. However, the amount of protein lysate can vary dependent on humidity and buffer.

Fig. 5. Precise protein quantification by RPPA. (**a**) Purified recombinant HER2 was arrayed together with the patient samples in a dilution curve on nitrocellulose slides as protein reference. The HER2 protein (2.5 pg start concentration) was mixed with 20 μl of a HER2/PAI1-negative patient sample (2 mg/ml) before spotting to eliminate potential influences of complex protein mixtures on signal-intensity compared to purified proteins. Only a week background signal was observed in the HER2/PAI1-negative patient sample and in the patient samples spiked with EGFR, HER3, HER4, uPA, or PAI1. (**b**) As described above for HER2 the purified recombinant PAI1 protein (start concentration 9.2 pg) was mixed with a HER2/PAI1-negative patient sample (2 mg/ml). No signal was detected in the HER2/PAI1-negative patient sample and in the patient samples mixed with uPA, EGFR, HER2, HER3, or HER4. (**c**) The signal-intensity was plotted against the protein concentration generating a signal-intensity–concentration curve for HER2 and PAI1. Since the protein concentration of recombinant HER2 and PAI1 is known, the unknown HER2/PAI1-concentration in a patient sample can be determined according the standard curves. In our example, the signal-intensity in a patient sample for HER2 is 59 for PAI1 13. Hence, according to the HER2 and PAI1 standard curves the HER2 concentration in a patient sample is 1.75 pg/nl spot (a spot contains 1 nl protein lysate) and for PAI1 3 pg/nl spot. Prior to spotting, total protein concentration was determined by Bradford assay (2 mg/ml in the undiluted first spot). For HER2 quantification we used the third dilution (0.25 mg/ml total protein) and for PAI1 quantification the first dilution (1 mg/ml total protein). After normalization the HER2 concentration in the patient sample is 7 pg/ng total protein and the PAI1 concentration is 3 pg/ng total protein.

Therefore, in further studies the exact volume of protein lysate in each spot has to be defined. Nevertheless, the RPPA technology may offer an attractive method to measure protein expression more precisely than IHC. Therefore, RPPA is expected to bear advantages for patients and therapy selections. However, the correlation between protein abundances and histology is lost.

5. Protein Biomarkers in Cancer Tissues

Molecular characterization of tumour tissues becomes more and more important for individualized cancer therapy. For the selection of patients who will benefit from such individualized treatments it is crucial to precisely quantify protein biomarkers in diseased tissues. In fact, there are already several therapies available based on protein expression profiles. Two examples are discussed here.

The receptor tyrosine kinase HER2, for example is a therapy target and over-expressed in up to 30% of human breast cancer patients as a consequence of gene amplification and transcriptional activation (20–23). Signalling pathways downstream of this receptor are thought to play an important role in initiation and progression of HER2 positive breast cancers. HER2 targeted therapy using the drug Herceptin (antibody trastuzumab reacting with the extracellular domain of HER2) has proven valuable in many of HER2 positive breast cancer cases. Currently, the HER2 status is determined by IHC. For equivocal cases, fluorescence in situ hybridization (FISH) is used to determine HER2 gene amplification. As mentioned above, protein quantification using IHC is difficult and depends at least in part on the observer.

Urokinase-type plasminogen activator (uPA) and its inhibitor PAI1 (plasminogen activator inhibitor type 1) became important prognostic markers in breast cancer. The uPA/PAI system is a key factor for migration and proliferation of cancer cells. uPA converts plasminogen into active plasmin, thereby controlling matrix degradation. This activation step is regarded to be an important trigger of cell migration (24) and invasion under physiological and pathological conditions (e.g., cancer metastasis) (25). PAI1 inhibits proteolytic function of uPA (26) but was also shown to regulate adhesion and migration of cells independently of this function (27). Due to their relevance for migration and invasion uPA and PAI1 both became interesting targets for the development of new prognostic cancer markers. Node-negative breast cancer patients bearing only low levels of both factors in primary tumours have a very low risk of recurrence; thus, these patients may be spared adjuvant chemotherapy. The extraordinarily high clinical relevance of both markers has been shown at LOE-1 (the highest level of evidence) and they are used for therapy guidance in a few hospitals. Currently, the proteins are detected and quantified by an immunometric assay (ELISA).

Unfortunately, this method is only applicable to fresh or frozen tissues which are rarely available in most hospitals, thus hindering widespread utilization of this diagnostic tool. As stated at the beginning of this manuscript, FFPE tissues are the main source of clinical tissues. Quantitation of uPA and PAI1 has not been performed reliably using FFPE. Thus, improvements in protein extraction of FFPE samples and applying RPPA technology as described above could overcome the hindrance of uPA and PAI1 diagnostics due to their requirement of fresh or frozen tissue and make it available to patients all over the world in the near future.

At the moment our group is establishing RPPAs suitable for the detection of various protein biomarkers extracted from FFPE cancer tissues, including HER2, uPA, and PAI1. Even absolute protein quantifications are possible (Fig. 5a–c) and will be considered in future studies.

6. Challenges for the Application of RPPA in Clinical Settings

Recently, many studies revealed RPPA technology to be very promising for signalling pathway profiling of frozen human tissues and cell lines, thus producing basic information for the development of new therapeutics and patient selection. As many cancers exhibit great differences in cell signalling, tissue behaviour and susceptibility to chemotherapeutic drugs the most efficient way for therapy selection is probably to provide a tumour-specific network portrait prior to treatment. RPPA is suitable to analyse multiple signalling pathways simultaneously and to characterize interconnecting protein pathways as well as different phosphorylation levels in tissues and cells as described elsewhere in this book. Thus, RPPA is a promising tool for the analysis of cell and tissue physiology for research purposes.

For integration of this technology into the routine clinical setting, there are – however – several challenges that must be solved, including effects of handling, fixation, and storage of surgical tissues.

It has been shown previously that the RPPA methodology is highly reliable and gives reproducible results. In recent years, a lot of effort has been invested into the optimization of analysing methods such as RPPA or nucleic-based diagnostics. However, it turned out that a major limitation of these methods results from the strong dependency to the quality of collected biospecimens. The pre-analytical procedures including tissue collection, handling, fixation, and storage can have a significant impact on analytical data sets (Fig. 6). Currently, there are no quality-assurance guidelines available and protocols for collecting biospecimens, including collecting frozen tissue samples, are not standardized between hospitals. In the last years, the research community started to investigate pre-analytical variations more intensively and (28, 29)

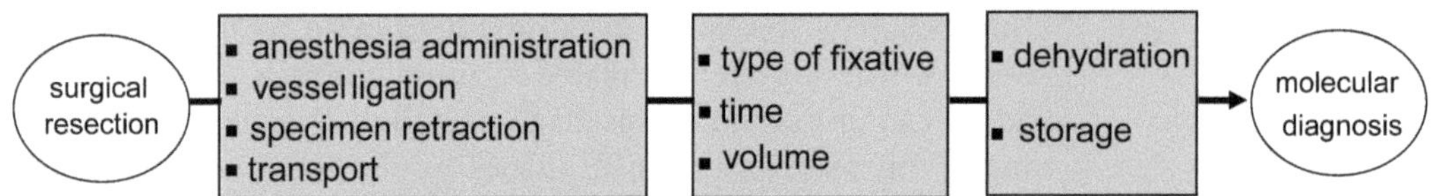

Fig. 6. Pre-analytical variations affecting quality and quantity of biomolecules in clinical tissues. These issues have to be solved by the scientific community to integrate analysis of biomarker profiles into the routine clinical setting.

recent studies could show that the time of vessel ligation until the specimen is received by the pathologist impacts the quality and quantity of molecular markers and affects phosphoprotein profiles (30–32). This period of "warm ischemia" is hard to control as it is dependent on the organ and surgical approach. However, documentation of the time of vessel ligation, anaesthesia administration, and specimen retraction from the body is recommended. Furthermore, efforts to shorten the ischemic period in combination with cooled transport to the pathology will improve the quality of biomolecules. After transportation to the pathology laboratory, tissue or organ fixation is another critical pre-analytical step. As over-fixation leads to decreased antigenicity of biomarkers in immunohistochemical stainings and insufficient fixation accelerates degradation of biomolecules (33), specimens should be dissected appropriately to ensure complete fixation in an adequate volume of neutral-buffered formalin. Following the process of dehydration, which is mainly automated, tissue samples are embedded in low-melting paraffin. This step does not further affect tissue integrity; however, it is highly dependent on complete specimen dehydration as paraffin needs to impregnate the entire sample. After the tissue sample has been processed correctly, a proper storage of the paraffin block is finalizing the whole procedure. For the storage of FFPE tissue blocks, it is recommended to have a temperature- and humidity-controlled environment to avoid further degradation. Additionally, sections for molecular approaches should be cut temporally close to the analysis to avoid loss of antigenicity possibly due to oxidation and hydration effects (34). Using FFPE tissues for diagnostic – and especially molecular – approaches one has to keep in mind that the tissue samples had undergone a long processing time. A lot of effort is now done to define the parameters affecting biomarker expression and stability in order to integrate this knowledge into diagnostic and therapeutic decisions.

Although challenging, standardization of the pre-analytical phase in hospitals and the introduction of guidelines will improve molecular characterization of tissues enormously, it is essential to relate biomarker profiles to the pathologic state of the patient and not to tissue processing.

Acknowledgments

This study is supported by the German Federal Ministry for Education and Research (BMBF), grant no 01GR0805 to Karl-Friedrich Becker. The authors wish to thank Kai Tran, Kerstin Schragner, and Christa Schott for excellent technical assistance.

References

1. Becker, K. F., Schott, C., Hipp, S., Metzger, V., Porschewski, P., Beck, R., Nahrig, J., Becker, I., and Hofler, H. (2007) Quantitative protein analysis from formalin-fixed tissues: implications for translational clinical research and nanoscale molecular diagnosis, *J Pathol 211*, 370–378.
2. Nirmalan, N. J., Harnden, P., Selby, P. J., and Banks, R. E. (2008) Mining the archival formalin-fixed paraffin-embedded tissue proteome: opportunities and challenges, *Mol Biosyst 4*, 712–720.
3. Lim, M. S., and Elenitoba-Johnson, K. S. (2004) Proteomics in pathology research, *Lab Invest 84*, 1227–1244.
4. Liotta, L., and Petricoin, E. (2000) Molecular profiling of human cancer, *Nat Rev Genet 1*, 48–56.
5. Paweletz, C. P., Charboneau, L., Bichsel, V. E., Simone, N. L., Chen, T., Gillespie, J. W., Emmert-Buck, M. R., Roth, M. J., Petricoin, I. E., and Liotta, L. A. (2001) Reverse phase protein microarrays which capture disease progression show activation of pro-survival pathways at the cancer invasion front, *Oncogene 20*, 1981–1989.
6. Addis, M. F., Tanca, A., Pagnozzi, D., Crobu, S., Fanciulli, G., Cossu-Rocca, P., and Uzzau, S. (2009) Generation of high-quality protein extracts from formalin-fixed, paraffin-embedded tissues, *Proteomics 9*, 3815–3823.
7. Becker, K. F., Mack, H., Schott, C., Hipp, S., Rappl, A., Piontek, G., and Höfler, H. . (2008) Extraction of phosphorylated proteins from formalin-fixed cancer cells and tissues, *TOPATJ 2*, 44–52.
8. Becker, K. F., Schott, C., Becker, I., and Höfler, H. (2008) Guided protein extraction from formalin–fixed tissues for quantitative multiplex analysis avoids detrimental effects of histological stains, *Proteomics Clin Appl 2*, 737–743.
9. Chu, W. S., Liang, Q., Liu, J., Wei, M. Q., Winters, M., Liotta, L., Sandberg, G., and Gong, M. (2005) A nondestructive molecule extraction method allowing morphological and molecular analyses using a single tissue section, *Lab Invest 85*, 1416–1428.
10. Chung, J. Y., Lee, S. J., Kris, Y., Braunschweig, T., Traicoff, J. L., and Hewitt, S. M. (2008) A well-based reverse-phase protein array applicable to extracts from formalin-fixed paraffin-embedded tissue, *Proteomics Clin. Appl 2*, 1539–1547.
11. Ikeda, K., Monden, T., Kanoh, T., Tsujie, M., Izawa, H., Haba, A., Ohnishi, T., Sekimoto, M., Tomita, N., Shiozaki, H., and Monden, M. (1998) Extraction and analysis of diagnostically useful proteins from formalin-fixed, paraffin-embedded tissue sections, *J Histochem Cytochem 46*, 397–403.
12. Nirmalan, N. J., Harnden, P., Selby, P. J., and Banks, R. E. (2009) Development and validation of a novel protein extraction methodology for quantitation of protein expression in formalin-fixed paraffin-embedded tissues using western blotting, *J Pathol 217*, 497–506.
13. Shi, S. R., Liu, C., Balgley, B. M., Lee, C., and Taylor, C. R. (2006) Protein extraction from formalin-fixed, paraffin-embedded tissue sections: quality evaluation by mass spectrometry, *J Histochem Cytochem 54*, 739–743.
14. Espina, V., Mehta, A. I., Winters, M. E., Calvert, V., Wulfkuhle, J., Petricoin, E. F., 3rd, and Liotta, L. A. (2003) Protein microarrays: molecular profiling technologies for clinical specimens, *Proteomics 3*, 2091–2100.
15. Templin, M. F., Stoll, D., Schrenk, M., Traub, P. C., Vohringer, C. F., and Joos, T. O. (2002) Protein microarray technology, *Trends Biotechnol 20*, 160–166.
16. Wulfkuhle, J. D., Edmiston, K. H., Liotta, L. A., and Petricoin, E. F., 3rd. (2006) Technology insight: pharmacoproteomics for cancer–promises

of patient-tailored medicine using protein microarrays, *Nat Clin Pract Oncol 3*, 256–268.

17. Liotta, L. A., Espina, V., Mehta, A. I., Calvert, V., Rosenblatt, K., Geho, D., Munson, P. J., Young, L., Wulfkuhle, J., and Petricoin, E. F., 3rd. (2003) Protein microarrays: meeting analytical challenges for clinical applications, *Cancer Cell 3*, 317–325.
18. Sheehan, K. M., Calvert, V. S., Kay, E. W., Lu, Y., Fishman, D., Espina, V., Aquino, J., Speer, R., Araujo, R., Mills, G. B., Liotta, L. A., Petricoin, E. F., 3rd, and Wulfkuhle, J. D. (2005) Use of reverse phase protein microarrays and reference standard development for molecular network analysis of metastatic ovarian carcinoma, *Mol Cell Proteomics 4*, 346–355.
19. Grubb, R. L., Calvert, V. S., Wulkuhle, J. D., Paweletz, C. P., Linehan, W. M., Phillips, J. L., Chuaqui, R., Valasco, A., Gillespie, J., Emmert-Buck, M., Liotta, L. A., and Petricoin, E. F. (2003) Signal pathway profiling of prostate cancer using reverse phase protein arrays, *Proteomics 3*, 2142–2146.
20. Emens, L. A. (2005) Trastuzumab: targeted therapy for the management of HER-2/neu-overexpressing metastatic breast cancer, *Am J Ther 12*, 243–253.
21. Piccart, M., Lohrisch, C., Di Leo, A., and Larsimont, D. (2001) The predictive value of HER2 in breast cancer, *Oncology 61 Suppl 2*, 73–82.
22. Slamon, D. J., Clark, G. M., Wong, S. G., Levin, W. J., Ullrich, A., and McGuire, W. L. (1987) Human breast cancer: correlation of relapse and survival with amplification of the HER-2/neu oncogene, *Science 235*, 177–182.
23. Slamon, D. J., Godolphin, W., Jones, L. A., Holt, J. A., Wong, S. G., Keith, D. E., Levin, W. J., Stuart, S. G., Udove, J., Ullrich, A., and et al. (1989) Studies of the HER-2/neu proto-oncogene in human breast and ovarian cancer, *Science 244*, 707–712.
24. Strand, K., Murray, J., Aziz, S., Ishida, A., Rahman, S., Patel, Y., Cardona, C., Hammond, W. P., Savidge, G., and Wijelath, E. S. (2000) Induction of the urokinase plasminogen activator system by oncostatin M promotes endothelial migration, *J Cell Biochem 79*, 239–248.
25. McMahon, B., and Kwaan, H. C. (2008) The plasminogen activator system and cancer, *Pathophysiol Haemost Thromb 36*, 184–194.
26. Cubellis, M. V., Wun, T. C., and Blasi, F. (1990) Receptor-mediated internalization and degradation of urokinase is caused by its specific inhibitor PAI-1, *EMBO J 9*, 1079–1085.
27. Germer, M., Kanse, S. M., Kirkegaard, T., Kjoller, L., Felding-Habermann, B., Goodman, S., and Preissner, K. T. (1998) Kinetic analysis of integrin-dependent cell adhesion on vitronectin--the inhibitory potential of plasminogen activator inhibitor-1 and RGD peptides, *Eur J Biochem 253*, 669–674.
28. Hewitt, S. M., Lewis, F. A., Cao, Y., Conrad, R. C., Cronin, M., Danenberg, K. D., Goralski, T. J., Langmore, J. P., Raja, R. G., Williams, P. M., Palma, J. F., and Warrington, J. A. (2008) Tissue handling and specimen preparation in surgical pathology: issues concerning the recovery of nucleic acids from formalin-fixed, paraffin-embedded tissue, *Arch Pathol Lab Med 132*, 1929–1935.
29. Leyland-Jones, B. R., Ambrosone, C. B., Bartlett, J., Ellis, M. J., Enos, R. A., Raji, A., Pins, M. R., Zujewski, J. A., Hewitt, S. M., Forbes, J. F., Abramovitz, M., Braga, S., Cardoso, F., Harbeck, N., Denkert, C., and Jewell, S. D. (2008) Recommendations for collection and handling of specimens from group breast cancer clinical trials, *J Clin Oncol 26*, 5638–5644.
30. Khoury, T., Sait, S., Hwang, H., Chandrasekhar, R., Wilding, G., Tan, D., and Kulkarni, S. (2009) Delay to formalin fixation effect on breast biomarkers, *Mod Pathol.*
31. Espina, V., Edmiston, K. H., Heiby, M., Pierobon, M., Sciro, M., Merritt, B., Banks, S., Deng, J., VanMeter, A. J., Geho, D. H., Pastore, L., Sennesh, J., Petricoin, E. F., 3rd, and Liotta, L. A. (2008) A portrait of tissue phosphoprotein stability in the clinical tissue procurement process, *Mol Cell Proteomics 7*, 1998–2018.
32. Spruessel, A., Steimann, G., Jung, M., Lee, S. A., Carr, T., Fentz, A. K., Spangenberg, J., Zornig, C., Juhl, H. H., and David, K. A. (2004) Tissue ischemia time affects gene and protein expression patterns within minutes following surgical tumor excision, *Biotechniques 36*, 1030–1037.
33. Goldstein, N. S., Hewitt, S. M., Taylor, C. R., Yaziji, H., and Hicks, D. G. (2007) Recommendations for improved standardization of immunohistochemistry, *Appl Immunohistochem Mol Morphol 15*, 124–133.
34. Fergenbaum, J. H., Garcia-Closas, M., Hewitt, S. M., Lissowska, J., Sakoda, L. C., and Sherman, M. E. (2004) Loss of antigenicity in stored sections of breast cancer tissue microarrays, *Cancer Epidemiol Biomarkers Prev 13*, 667–672.

Chapter 9

Producing Reverse Phase Protein Microarrays from Formalin-Fixed Tissues

Claudia Wolff, Christina Schott, Katharina Malinowsky, Daniela Berg, and Karl-Friedrich Becker

Abstract

In most hospitals around the world FFPE (formalin fixed, paraffin embedded) tissues have been used for diagnosis and have subsequently been archived since decades. This has lead to a sizeable pool of this kind of tissues. Till quite recently it was not possible to use this congeries of samples for protein analysis, but now several groups described successful protein extraction from FFPE tissues. In this chapter, we describe a protein extraction protocol established in our laboratory combined with the use of reverse phase protein microarray.

Key words: Reverse phase protein microarray, Formalin-fixed paraffin-embedded tissue, Protein extraction, Chemiluminescence

1. Introduction

The use of formalin as a fixative has been standard in the clinical routine for decades and it still is. Formalin fixes the tissue samples by inducing crosslinks of proteins and nucleic acids and keeps the tissue in an excellent condition (1). Admittedly, immunohistochemistry (IHC) was the only method to analyse protein expression in formalin-fixed and paraffin-embedded (FFPE) tissues until recently. The advantage of IHC is the possibility to determine the localization of the protein inside the cell; however, quantification of protein expression is difficult and at least in part dependent on the observer. But with the analysis of subtle quantitative changes on molecular level coming more and more to the fore, more sensitive methods, such as immunoblotting, enzyme-linked immunosorbent

Ulrike Korf (ed.), *Protein Microarrays: Methods and Protocols*, Methods in Molecular Biology, vol. 785,
DOI 10.1007/978-1-61779-286-1_9, © Springer Science+Business Media, LLC 2011

assay (ELISA), and protein microarrays, become the methods of choice for many research issues. Unfortunately, it was not possible to extract proteins from FFPE tissue for a long time, rendering these tissues non-accessible for the methods mentioned above. But the situation has changed: During the last few years several groups have described successful protein extraction from FFPE tissues (1–7). Their efficiency is comparable to that seen from frozen tissues, and the extracted proteins are suitable to applications such as Western blot or protein microarray analysis (8). In this chapter, we describe a method for the extraction of proteins from fixed tissues and their application to reverse phase protein microarray (RPMA). There are a variety of spotting devices which may be used to generate RPMAs (as outlined elsewhere in this book). In this chapter, we describe how to use a hand-held spotting device.

2. Materials

2.1. Deparaffinization of and Protein Extraction from Slide-Mounted FFPE Sections

2.1.1. Reagents

1. 100% (v/v) ethanol or isopropanol.
2. 70% (v/v) ethanol.
3. 96% (v/v) ethanol.
4. Qproteome FFPE Tissue Kit buffer (Qiagen, Hilden, Germany); (see Note 1).
5. Xylene.

2.1.2. Equipment

1. Collection tube sealing clip (e.g., from Qiagen, Hilden, Germany).
2. Glass slides.
3. Microtome.
4. Needles or blades.
5. Parafilm.
6. Staining dishes with proper slide holders.
7. Thermomixer.
8. Water bath.

2.2. Reverse Phase Protein Microarray Spotted with a Hand-Held Microarray System

2.2.1. Reagents

1. Extraction buffer (same as in Subheading 2.1.1).
2. 100% (v/v) ethanol or isopropanol.
3. 70% (v/v) ethanol.
4. Pin conditioner (e.g., from Whatman, Maidstone, UK).

2.2.2. Equipment

1. Blotting paper (e.g., chromatography paper, Whatman, Maidstone, UK).
2. Centrifuge with plate-adapters.
3. Microtiter plates (V-bottom).
4. Nitrocellulose slides (e.g., FAST™ slides, Whatman, Maidstone, UK).
5. Oven.
6. Sealing foils for microtiter plates (e.g., Rotilabo-Verschlußfilm, Roth, Karlsruhe, Deutschland).
7. Spotting device (e.g., MicroCaster™ Hand-Held Microarrayer System, Whatman, Maidstone, UK).
8. Vortex mixer.

2.3. Chemiluminescence Protein Detection on RPMA

2.3.1. Reagents

1. Blocking buffer matched to antibodies: e.g., 5% milk powder (MP) in TBST or 3% BSA in TBST, 0.5% casein in TBST.
2. Detection reagents for chemiluminescence (e.g., ECL plus and/or ECL Advanced, GE Healthcare, Buckinghamshire, UK).
3. Peroxidase blocking reagent (e.g., from DAKO, Glostrup, Denmark).
4. Primary antibodies against proteins of interest.
5. Reagents for film developing (e.g., from adefo-chemie, Dietzenbach, Germany).
6. Secondary antibody, species matched to primary antibody.
7. Sypro Ruby Protein Fixative Solution: 7% v/v acetic acid and 10% v/v methanol in deionized water. Store at room temperature.
8. Sypro Ruby Protein Blot Stain (e.g., Molecular Probes, Eugene, USA).
9. Tris-buffered saline with Tween (TBST buffer): 20 mM Tris pH 7.4, 140 mM NaCl, 1% (v/v) Tween in deionized water. Store at room temperature.

2.3.2. Equipment

1. Autoradiography cassettes.
2. Cool room (4°C) with shaker.
3. Darkroom (with red-light).
4. Film developing machine (e.g., Tabletop processor SRX-101A, Konica-Minolta, Tokyo, Japan).
5. Films (e.g., Amersham Hyperfilm ECL, GE Healthcare, Buckinghamshire, UK).
6. High resolution flat-bed scanner.
7. Shaker.

8. Sheet protector foils (clear and as thin as possible) (e.g., table waste bags, Hartenstein, Wuerzburg, Germany).
9. Slide incubation chambers (e.g., quadriPERM, Sigma–Aldrich, Steinheim, Germany).

3. Methods

3.1. Deparaffinization of and Protein Extraction from Slide-Mounted FFPE Sections

In this chapter, protein extraction from slide-mounted FFPE sections is described. Before starting with the extraction itself, some preliminary steps have to be carried out: cutting the FFPE tissue blocks, removal of excessive paraffin, and transferring the tissue area of interest into the extraction buffer. Subsequently, the proteins are extracted from the tissue in two heating steps (see Fig. 1).

3.1.1. Preliminary Work

1. Prepare a HE-stained slide (see Note 2).
2. The area of interest (e.g., tumour area) should be marked by a pathologist (see Note 3).

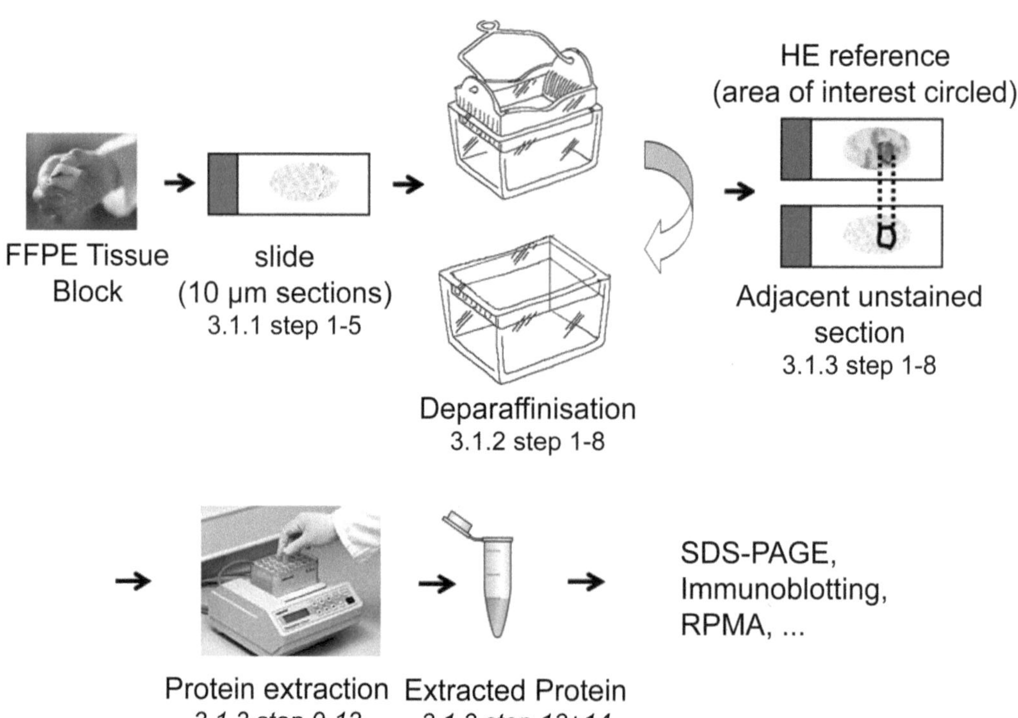

Fig. 1. Overview of the course of protein extraction from slide-mounted FFPE sections. After cutting of 10-µm sections of the FFPE tissue block, the paraffin has to be removed, because it would disturb the protein extraction. For analysis of a particular area of the slide, this area has to be marked on a HE-stained reference slide. An adjacent unstained section is used for protein extraction according to the protocol described in this chapter. The extracted proteins are compatible to subsequent methods, including SDS-PAGE, immunoblotting, or RPMA.

3. After you settled on the number of sections that are needed for 100 μl of extraction buffer, the corresponding number of 10-μm thick sections should be cut, using a microtome (see Notes 4 and 5).
4. The slides should be incubated at 50°C (for 4 h for up to 16 h).
5. Storage: At room temperature; the shorter the better, but max. 7 days.

3.1.2. Deparaffinization

1. Place the slides in a slide holder (see Note 6).
2. Prepare the alcohol row (see Note 7):

   Three staining dishes with xylene

   Two staining dishes with 100% isopropanol

   Two staining dishes with 96% ethanol

   Two staining dishes with 70% ethanol.
3. Transfer the slide to a staining dish containing fresh xylene. The slide should be completely covered. Incubate for 10 min at room temperature (15–25°C). Xylene washes should be performed in a fume hood.
4. Repeat step 1 twice, using fresh xylene each time (see Note 8).
5. Transfer the slide to a staining dish containing fresh 100% isopropanol for 10 min at room temperature (15–25°C). Repeat this step using fresh 100% ethanol.
6. Transfer the slide to a staining dish containing fresh 96% ethanol and incubate for 10 min. Repeat this step using fresh 96% ethanol.
7. Transfer the slide to a staining dish containing fresh 70% ethanol and incubate for 10 min. Repeat this step using fresh 70% ethanol.
8. Now all paraffin is removed and the slide can be transferred to a staining dish containing fresh distilled water and immerse for 30 s (slides may stay in water for up to 3 h, if protein extraction is not possible after deparaffinization).

3.1.3. Protein Extraction

1. Remove the first slide from water dish and eliminate excess water by tapping the slide carefully on a paper towel. Do not touch the section with the paper towel. Ensure that sections do not dry out.
2. Place the corresponding HE-stained slide behind the deparaffinized slide. The sections on both slides should be arranged in the same position.
3. Excise area of interest on the deparaffinized slide according to the marked HE-stained slide. This is done best with a needle or a blade. Transfer the procured cells to a 1.5-ml collection

tube containing one volume (preferably 100 μl) of extraction buffer on ice (see Note 9).

4. Repeat steps 1–3 for the amount of slides you decided to use with the volume of extraction buffer.
5. Mix by vortexing.
6. Repeat steps 1–5 for each FFPE sample you want to analyse.
7. Mix by vortexing. Seal the collection tube with a parafilm and with a collection tube sealing clip additionally.
8. Incubate on ice for another 15 min and mix by vortexing.
9. Incubate the tube in a water bath at 100°C for 20 min.
10. Incubate the tube at 80°C for 2 h with agitation at 750 rpm, using a Thermomixer (see Note 10).
11. After incubation, place the tube at 4°C for 1 min and remove parafilm plus the collection tube sealing clip.
12. Centrifuge at 14,000 × g at 4°C for 15 min. Transfer the supernatant containing the extracted proteins to a new 1.5-ml collection tube.
13. If desired, the protein concentration can be determined (see Notes 11 and 12).
14. The extracted proteins can be stored at –20°C for up to 1 week. For long-term storage, aliquot the extracted proteins and store at –80°C. Avoid repeated freeze–thaw cycles.

3.2. Reverse Phase Protein Microarrays Spotted with a Hand-Held Microarray System

In this chapter, we describe how to utilize the protein samples extracted with the method presented in Subheading 3.1 for RPMA using a hand-held microarray system (for overview see Fig. 2).

3.2.1. Preparing Lysates

1. Thaw protein extracts on ice and mix them by vortexing.
2. To avoid saturation and unspecific artefacts, dilute the lysates in the extraction buffer (undiluted, 1:2, 1:4, 1:8, and 1:16); the final volume for spotting should be more than 5 μl in each well.
3. Arrange the lysates on a 96-well plate on ice, in a special order (always skipping one well, placing the undiluted sample in the first well, the next dilution-step in the third, fifth,...) each dilution-series ending with one well of buffer (negative control) (for pipetting scheme see Note 13). This special arrangement of the lysates is due to the fixed distance between the spotting-pins of the handheld microarraying system.
4. Seal the plate with a foil.
5. Centrifuge the plate at 4°C and about 1,000 rcf for 1–2 min (to get rid of air bubbles).
6. Keep the plate on ice until use.

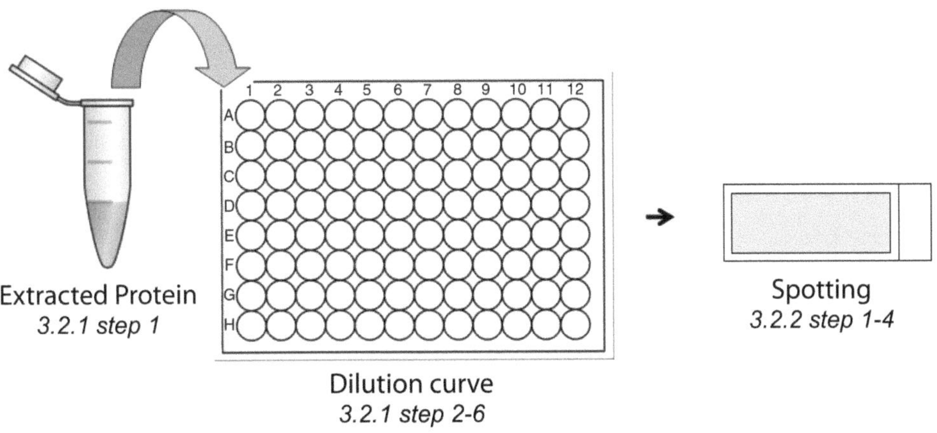

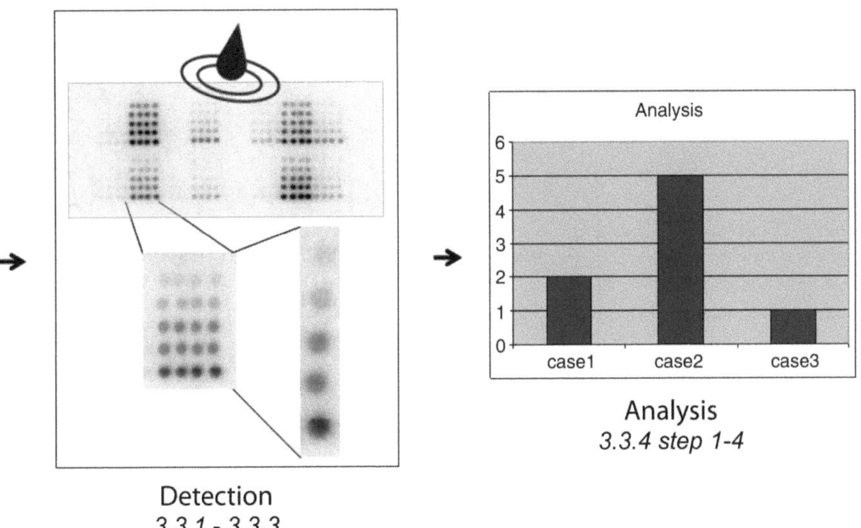

Fig. 2. Overview of the preparation and detection of reverse phase protein microarrays. The desired dilution curve is prepared in a 96-well plate. From this plate, the lysates can be spotted on nitrocellulose slides. Subsequently, SYPRO-Ruby staining and chemiluminescence protein detection are performed. The obtained results may then be quantified.

3.2.2. Spotting with a Handheld Microarraying System

As a first step the replicator pins have to be coated with surfactant:

1. Prepare a six-well plate for coating (plate A) following the scheme shown below (see Fig. 3, top).
2. Prepare a six-well plate for washing (plate B) following the scheme shown below (see Fig. 3, bottom).
3. Dip replicator pins into 5-ml pin conditioner (diluted 1:5 in A. dest.).
4. "Blot" onto blotting paper.
5. Repeat steps 3 and 4 one time.

Plate A

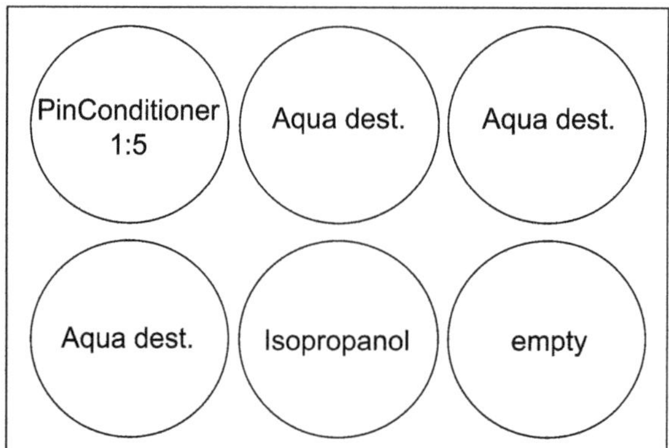

Plate B

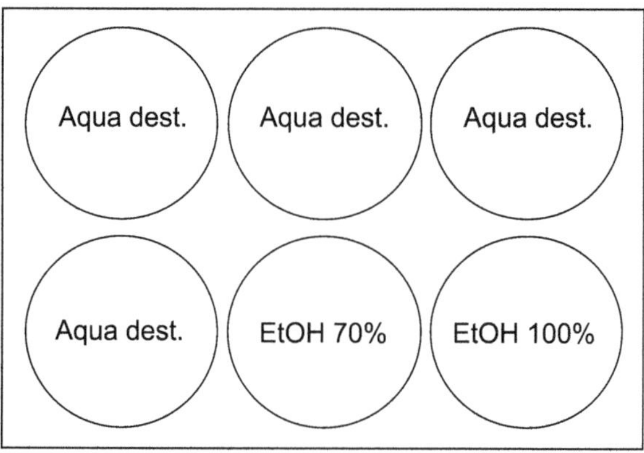

Fig. 3. Filling schema of six-well coating (plate A) and washing (plate B) plate. Volume for pin conditioner: 5 ml; Volume for A. dest. and alcohol: 7 ml each.

6. Let pins air-dry, or use hot air dryer to dry pins and slots.
7. Dip replicator pins into first A. dest. reservoir.
8. "Blot" onto blotting paper.
9. Repeat through remaining A. dest. reservoirs.
10. Dip replicator pins into isopropanol.
11. "Blot" onto blotting paper.
12. Repeat steps 10 and 11 one time.
13. Air-dry or use hot air dryer to dry pins.
14. Replicator pins are now ready for use.

Before the microarraying system can be used it has to be assembled in the right order:

15. Remove the array tool from the indexing unit (see Fig. 4). Remove the indexing needles from the indexing unit. Disassemble the indexing unit by removing the indexing deck from the base unit.

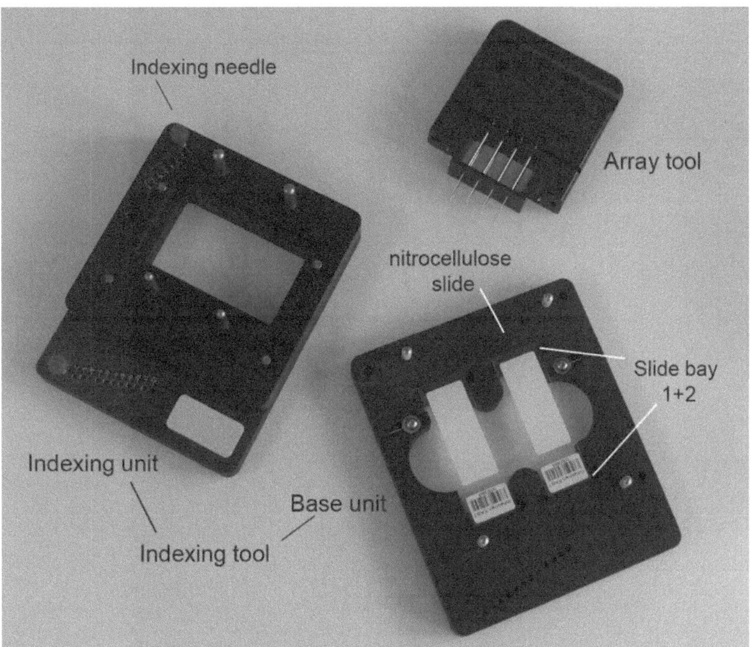

Fig. 4. The handheld microarraying system. The arraying system consists of the array tool, the indexing deck, and the base unit. Additionally, nitrocellulose slides are needed. Two slides can be kept in the correct position in the two slide bays of the base unit. During the spotting the array tool is adjusted in the correct position by the indexing needles on the indexing unit.

16. Place slides into the base unit (for the right orientation see Fig. 4).
17. Put the indexing deck back onto the base unit: Align the slots on the underside of the indexing deck with the four pins in the base unit.
18. Return indexing needles to starting position:
19. Vertical Indexing Set (Eight Holes): Place the vertical indexing pin into the first (back) alignment hole.
20. Horizontal Indexing Set (Twelve Holes): Place the horizontal indexing pin into the first (left) alignment hole.

    Steps 21–36 describe how to perform the actual arraying:

21. The array tool must be held in the proper orientation to the indexing unit.
22. The guiding pins on the indexing plate and the corresponding holes on the bottom side of the array tool are of different size (large on the back and small on the front) (see Fig. 5).
23. Dip the pins of the array tool into the appropriate wells of a microplate (start with columns 1 and 2, rows A–D; see Fig. 6).
24. Raise the pins out of the liquid slowly (see Note 14).
25. Align the guide holes on the array tool to the guide pins of slide bay 1 (Fig. 5) on the indexing unit. Lower the array tool

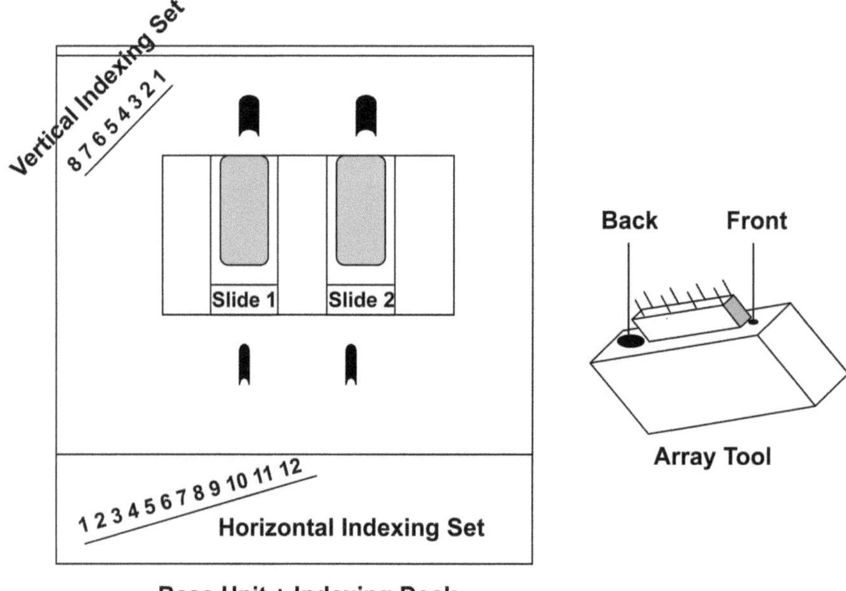

Fig. 5. Orientation of the indexing unit (base unit + indexing deck) and the array tool. The indexing unit should be placed as shown above. The horizontal indexing set is then seen on the *lower* part of the unit whereas the vertical indexing unit is seen in the *upper left corner*. Before you start spotting make sure that the array tool is orientated correctly with the small guide hole in the front.

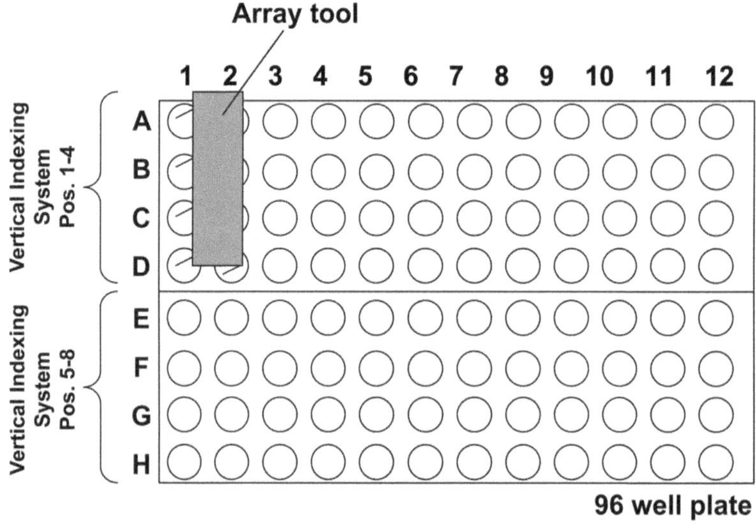

Fig. 6. Placing the pins of the array tool into the appropriate wells of a 96-well plate. The array tool has to be dipped into the plate as shown in this figure. *Rows A–D* have to be with the indexing needle in positions 1–4 of the vertical indexing system. *Rows E–H* have with the indexing needle in positions 5–8 of the vertical indexing system. After the *columns 1* and *2* are spotted in all eight vertical positions (replicates) the array tool is moved to *columns 3* and *4*, starting with *rows A–D* again.

onto the guide pins until it is resting on the spring-loaded standoff pins of the array tool.

26. Press firmly and quickly down on the top of the array tool. The motion should be a rapid and crisp to deposit the liquid onto the slide.
27. Remove the array tool from the indexing unit.
28. Move the vertical indexing needle to the next position.
29. Repeat steps 21–28 three times for four replications (dip still in columns 1 and 2, rows A–D).
30. Before dipping in the next set of wells, wash the pins in the series of water and ethanol baths as described for Plate B above (see Fig. 3).
31. Repeat steps 21–28 dipping into columns 1 and 2, rows E–H of the source plate (vertical indexing needle should start from position 5).
32. Repeat steps 21–28 again three times for four replications (ending with the vertical indexing needle in position 8).
33. Before spotting from columns 3 and 4 the horizontal indexing pin has to be moved two positions to the right, but you cannot move two positions at one time, so move one position twice.
34. Move the vertical indexing pin back to its original starting position (position 1).
35. Repeat steps 21–34 for columns 3 and 4.
36. Repeat steps 21–34 for columns 5 and 6, etc. (by following this protocol you get a slide design as shown in Note 15).

After the arraying is finished the pins have to be washed and the slides removed:

37. After spotting, wash the pins again in Plate B.
38. Then the pins need to be sonificated briefly.
39. Coat the pins again (step 3–14).
40. Disassemble indexing unit as explained in step 15 to remove slides from base unit.
41. The spotted slides can be stored at 4°C for at least 2 months.

3.3. Chemiluminescence Protein Detection on RPMA

3.3.1. Estimation of Total Protein with Sypro Ruby Staining

One of the spotted slides is stained with Sypro Ruby for normalization (see Note 16).

1. Take slide from 4°C storage and pre-wet it briefly in TBST buffer.
2. After discarding the TBST buffer, incubate the slide in 7% acetic acid and 10% methanol for 15 min in a staining dish. Gentle agitation is needed.
3. Wash in four changes of deionized water for 5 min each.

4. Incubate the slide in Sypro Ruby stain reagent for 10–15 min in a staining dish with gentle agitation.
5. Wash the slide 4–6 times for 1 min each in deionized water.
6. Visualize the staining on an imaging system, e.g., Eagle Eye (Stratagene, La Jolla, CA).
7. Save file as .tif.
8. Analyse with freeware like ScionImage (Scion Corporation, Frederick, Maryland) or use commercial software packages, e.g., MicroVigene (VigeneTech, Carlisle, Massachusetts) according to the company's instructions.

3.3.2. Incubating Slides with Antibodies

1. Take slides from 4°C storage and pre-wet them shortly in TBST buffer by gently shaking in an incubation chamber on a shaker or rocker (see Note 17).
2. Discard the TBST buffer and pour peroxidase blocking reagent on the slides completely covering it and shake it gently at room temperature for 1 h.
3. Discard the peroxidase blocking reagent and wash three times with TBST buffer for 2 min each.
4. Incubate the slides in the blocking solution suitable for the antibody of choice (see Table 1 and Note 18).
5. Shake gently at room temperature for at least 1 h.
6. Incubate the slides with the primary antibody at 4°C over night (16 h), while gently shaking (see Table 1 and Notes 19 and 20).

3.3.3. Signal Detection

1. On the next day, discard the antibody solution and wash three times for 10 min each in TBST buffer at room temperature.
2. Incubate the slide with the secondary antibody for 1 h at room temperature shaking gently (see Table 2 and Note 20).

Table 1
Suggested conditions for incubation with primary antibodies

Antibody against (provider)	Blocking	Dilution prim. Ab	Dilution solution
E-Cadherin, #610182 (BD Biosciences)	5% MP in TBST	1:5,000	5% MP in TBST
EGFR, #2232 (New England Biolabs)	5% MP in TBST	1:2,000	5% MP in TBST
β-Actin, #A1978 (Sigma-Aldrich)	5% MP in TBST	1:10,000	5% MP in TBST
Her2, #A0485 (Dako)	5% MP in TBST	1:500	TBST
ER α #EI629R06SG (DCS)	5% MP in TBST	1:25	TBST

The antibodies presented here are examples; numerous competitive reagents are available from other commercial sources. *MP* milk powder, *TBST* Tris-buffered saline with Tween 20

Table 2
Suggested conditions for incubation with secondary antibodies

Primary Ab	Secondary Ab (provider)	Dilution sec. Ab	Dilution solution	Detection
E-Cadherin	Anti-Mouse # A931 (GE Healthcare)	1:10,000	5% MP in TBST	ECLplus
EGFR	Anti-Rabbit # 7074 (New England Biolabs)	1:2,000	5% MP in TBST	Mixture ECLplus/ ECLadvanced 2:1
β-Actin	Anti-Mouse # A931 (GE Healthcare)	1:10,000	5% MP in TBST	ECLplus
Her2	Anti-Rabbit # 7074 (New England Biolabs)	1:2,000	5% MP in TBST	ECLplus
ER α	Anti-Mouse # A931 (GE Healthcare)	1:5,000	5% MP in TBST	Mixture ECLplus/ ECLadvanced 2:1

The antibodies indicated here are examples; numerous competitive reagents are available from other commercial sources (see Note 21)

3. Discard the secondary antibody and wash three times with TBST buffer for 10 min each, while shaking at room temperature.
4. Take the slides out of the incubation chamber and remove excess TBST buffer from the slides but be careful not to let them dry completely!
5. Place the slides on a glass plate and pour the detection reagent directly on top of the slides (app. 500 μl per slide); Incubate for 5 min at room temperature.
6. Remove the detection reagent, cover the slides with protector foil and put it in an autoradiography cassette.
7. Apply films for different exposure times in the darkroom. The optimal exposure period depends on various factors, e.g., the antibodies but 1 min should be a good starting time.
8. Develop the films either by hand or using a developing machine.

3.3.4. Analysis

1. Scan the slides individually on a scanner with at least 600 dpi.
2. Save files as .tif.
3. Analyse with freeware like ScionImage (Scion Corporation, Frederick, Maryland) or use commercial software packages, e.g., MicroVigene (VigeneTech, Carlisle, Massachusetts) according to the company's instructions.
4. Normalize to total protein (Sypro Ruby detection, see Subheading 3.3.1).

4. Notes

1. In this protocol, we describe how to extract proteins from FFPE tissue employing the commercially available Qproteome FFPE Tissue Kit (Qiagen, Hilden, Germany). But during the last few years several protocols for protein extraction from FFPE tissue have been established. These are shown in brief in Table 3.
2. This protocol describes how to use a particular area of a tissue slide (e.g., tumour area). If you are interested in extracting proteins from the whole slide, Subheading 3.1.1, steps 1 and 2 and Subheading 3.1.3, steps 2 and 3 may be skipped.
3. The size of this area will be important to determine the number of necessary slides per volume of extraction buffer. It emerged that the buffer volume should be around 100 µl, to obtain highest protein yields. The amount of tissue which can be extracted is depending on various factors, like tissue type and cell density. But to give an approximate value: For two slides with an area of about 0.5 cm in diameter use 100 µl.
4. We recommend using 10-µm sections. Paraffin may be more difficult to remove from thicker sections. Although using double amount of 5-µm sections may increase the protein yield we would recommend this only for very small areas like, e.g., biopsies because this will be more time and material consuming.
5. If you want to mark your slides, the pen has to be xylene and ethanol resistant (you may use a pencil).
6. For large amounts of slides it could save time and reagents to place the slides crisscross in the holder (see Fig. 7). Additionally, you may mount two sections of one block on the same slide to save time.
7. The alcohol series has to be renewed after five cycles to guarantee correct concentrations of the reagents.
8. If processing samples containing large amounts of paraffin, repeat the xylene treatment two more times.
9. Histological stains can decrease the yield of the extracted proteins. The decrease depends on the kind of dye used and on the staining-time. When a 10 second haematoxylin (Mayer) staining protocol is used, for example the protein yield drops to about 50% compared to unstained tissue. You get out even less by using Fast Red and many stainings did not work at all: Methyl blue, haematoxylin (Gil) to name some (9).
10. For some tissues it could be better to skip this step.
11. Which quantification method is best, depends on the used buffer. Before use check the compatibility of the chosen method. For Qproteome FFPE Tissue Kit buffer protein yield can be

Table 3
Summary of different extraction protocols

Extraction buffer	Protocol	Advantages	Disadvantages	References
Qproteome FFPE Tissue Kit (Qiagen, Hilden, Germany)	100°C 20 min, 80°C 120 min, −20°C for storage	No differences in protein yield and abundances between fresh frozen and FFPE tissues Analysis of phospho-specific proteins Extraction of proteins up to 190 kDa High yield	Time consuming extraction protocol	Becker et al. 2007, 2008 (1, 8, 9)
1× AgR buffer (pH 9.9) + 1% NaN_3 + 1% SDS + 10% glycerol + protease inhibitor	15 min 115°C, 10-15 psi, −80°C for storage	High yield Deparaffinization not necessary Fast extraction protocol Analysis of phospho-specific proteins Very sensitive	Smearing in FFPE samples	Chung et al. 2008 (4)
Laemmli buffer	20 min 105°C, −20°C for storage	Cheap (not commercial) Fast extraction protocol	Lower yield	Nirmalan et al. 2009 (6)
RIPA + 2% SDS	100°C 20 min, 60°C 120 min, −80°C for storage	Cheap (not commercial) High yield	Only proteins up to120 kDa were analysed Time consuming extraction protocol	Ikeda et al. 1998 (5)
NDME (BioQuick Inc., Silverspring, MD, USA)	100°C 20 min, −80°C for storage	Parallel extraction of protein and nucleic acids Short extraction protocol Extraction of proteins up to 188 kDa	Lower yield	Chu et al. 2005 (3)
20 mM Tris–HCl pH 7 or 9 + 2% SDS	100°C 20 min, 60°C 120 min, −80°C for storage	Cheap (not commercial) High yield	Time consuming extraction protocol	Shi et al. 2006 (7)
20 mM Tris–HCl pH 8.8 + 2% SDS + 200 mM DTT	100°C 20 min, 80°C 120 min, −80°C for storage	Cheap (not commercial) High yield	Time consuming extraction protocol	Addis et al. 2009 (2)

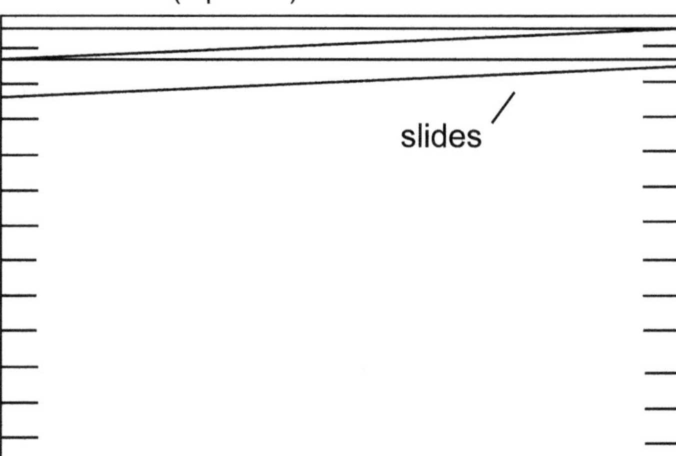

Fig. 7. Crisscross assembly of the slides in a slide holder. This arrangement can help you to save time and reagents during the deparaffinization steps.

measured by the Lowry (e.g., Protein Assay Kit, Bio-Rad) or BCA method (e.g., Micro BCA Protein Assay Kit, Pierce).

12. The following reasons may account for low protein yields:
    (a) Poor quality of starting material. Samples that were fixed for over 24 h or stored for very long periods may allow only incomplete extraction of protein
    (b) Too little starting material. Increase the amount of starting material.
    (c) Insufficient deparaffinization or too much paraffin in sample. If you are processing samples containing large amounts of paraffin, repeat the xylene treatment an additional two times. Paraffin may be more difficult to remove from thicker sections (we recommend using 10-μm sections).
13. Pipetting schema (see Fig. 8).
14. The speed at which the pins are raised out of the liquid is important. Removing the pins very fast will result in larger hanging drops on the tips of the pins, which could result in overlapping spots. It is also important to remove the pins from the centre of the wells and not near the walls, as this can affect drop size as well.
15. Slide design (see Fig. 9).
16. This slide is needed during analysis for normalization of the antibody signals. Theoretically, the most adequate way of normalization would be to stain each slide with Sypro Ruby before applying the first antibody and normalize to intern total protein in every case. There are two practical reasons anyway for using only one or a few of the spotted slides. First some antibodies do

sample1 undiluted	sample2 undiluted	sample1 1:2	sample2 1:2	sample1 1:4	sample2 1:4	sample1 1:8	sample2 1:8	sample1 1:16	sample2 1:16	buffer	buffer
sample5 undiluted	sample6 undiluted	etc.									
sample3 undiluted	sample4 undiluted	sample3 1:2	sample4 1:2	sample3 1:4	sample4 1:4	sample3 1:8	sample4 1:8	sample3 1:16	sample4 1:16	buffer	buffer
sample7 undiluted	sample8 undiluted	etc.									

Fig. 8. Pipetting schema for hand spotting device.

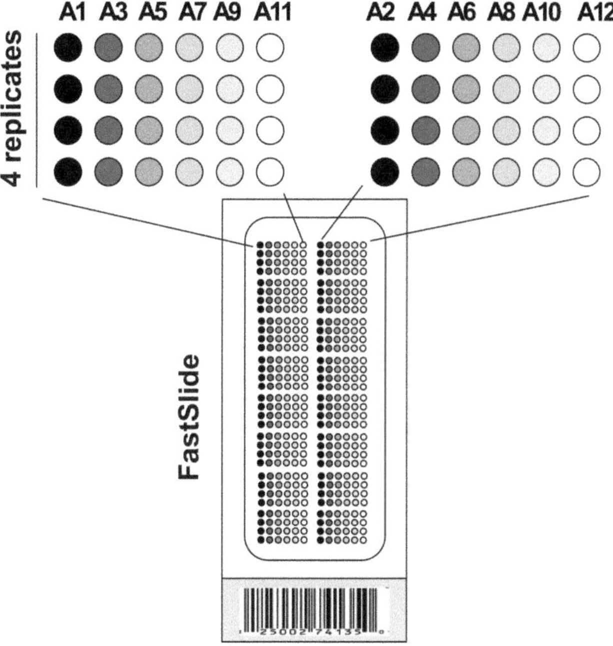

Fig. 9. Slide design. By following the protocol described in Subheading 3.2.2 you get a slide design as depicted in this figure.

not give proper signals when used on a previously Sypro Ruby stained slide and second if you have large numbers of slides the staining will get really expensive soon.

17. As an incubation chamber Heraeus Quadriperm can be used. Using this unit only 3 ml of antibody dilution and peroxidise blocking reagent are needed.

18. In general 5% milk powder in TBST buffer will do, but the blocking reagent can also be 5% BSA in TBST buffer, a mixture of both or casein-based depending on the suggestion of the antibody company and your own experience.
19. Shorter incubation times at room temperature may also be possible depending on the workflow.
20. The dilution as well as the reagent to dilute in depends on the suggestion of the antibody company and on your own experience.
21. Mixing ECLplus and ECLadvance combines advantages of both solutions: stronger and long lasting signals.

References

1. Becker, K. F., Schott, C., Hipp, S., Metzger, V., Porschewski, P., Beck, R., Nahrig, J., Becker, I. and Hofler, H. (2007) Quantitative protein analysis from formalin-fixed tissues: implications for translational clinical research and nanoscale molecular diagnosis. *J Pathol* **211,** 370–8.
2. Addis, M. F., Tanca, A., Pagnozzi, D., Crobu, S., Fanciulli, G., Cossu-Rocca, P., and Uzzau, S. (2009) Generation of high-quality protein extracts from formalin-fixed, paraffin-embedded tissues. *Proteomics* **9**, 3815–23.
3. Chu, W.S., Liang, Q., Liu, J., Wei, MQ., Winters, M., Liotta, L., Sandberg, G., and Gong, M. (2005) A nondestructive molecule extraction method allowing morphological and molecular analyses using a single tissue section. *Lab Invest* **85,** 1416–28.
4. Chung, J., Lee, S. J., Kris, Y., Braunschweig, T., Traicoff, J. L., and Hewitt, S. M. (2008) A well-based reverse-phase protein array applicable to extracts from formalin-fixed paraffin-embedded tissue. *Proteomics Clin App* **2**, 1539–47.
5. Ikeda, K., Monden, T., Kanoh, T., Tsujie, M., Izawa, H., Haba, A., Ohnishi, T., Sekimoto, M., Tomita, N., Shiozaki, H., and Monden, M. (1998) Extraction and analysis of diagnostically useful proteins from formalin-fixed, paraffin-embedded tissue sections. *J Histochem Cytochem* 46, 397–403.
6. Nirmalan, N. J., Harnden, P., Selby, P. J., and Banks, R. E. (2009) Development and validation of a novel protein extraction methodology for quantitation of protein expression in formalin-fixed paraffin-embedded tissues using western blotting. *J Pathol* 217, 497–506.
7. Shi, S. R., Liu, C., Balgley, B. M., Lee, C., and Taylor, C. R. (2006) Protein extraction from formalin-fixed, paraffin-embedded tissue sections: quality evaluation by mass spectrometry. *J Histochem Cytochem* 54, 739–43.
8. Becker, K. F., Mack, H., Schott, C., Hipp, S., Rappl, A., Piontek, G., and Höfler, H. (2008) Extraction of phosphorylated proteins from formalin-fixed cancer cells and tissues. *TOPATJ* 2, 44–52.
9. Becker, K. F., Schott, C., Becker, I., and Höfler, H. (2008) Guided protein extraction from formalin-fixed tissues for quantitative multiplex analysis avoids detrimental effects of histological stains. *Proteomics Clin Appl.* **2**, 737–43.

Chapter 10

Use of Reverse Phase Protein Microarrays to Study Protein Expression in Leukemia: Technical and Methodological Lessons Learned

Steven M. Kornblau and Kevin R. Coombes

Abstract

Leukemias are well suited to proteomic profiling by RPPA due to the ready accessibility of blasts from the blood or marrow. In this review, we review methodological and procedural issues that affect the quality of RPPA data. We recommend contact printers that minimize sample quantities and evaporation and maximize sample per slide. The impact of sample selection and handling is reviewed as well. Protein is best prepared fresh on the date of acquisition as cryopreservation changes protein expression levels in some diseases. Rapid processing is also required to avoid changes in phosphorylation over time. Sample source, blood vs. marrow does not seem to affect results as long as leukemic blast enrichment procedures are utilized. The choice of the correct "normal" control is important for comparing diseased to "normal" expression. Various means of normalizing the data are discussed.

Key words: Proteomics, Leukemia, RPPA, Reverse phase protein array

1. Introduction

The leukemias, because they are a "liquid" tumor with readily available malignant cells, are probably the easiest malignancies to study by RPPA. We have generated four different RPPAs from primary patient samples from patients with leukemia. Our first generation AML array had 550 patients' samples on it and was probed with 51 different antibodies (1). Our second generation array had 719 AML samples from 511 patients and 360 ALL samples (AML719/ALL360) (2). The third array contains matched bulk, CD34+ and CD34+ CD38− stem cell enriched samples from AML and CML cases (manuscript in progress). A fourth array has 260 CD34+ and CD34+ CD38− samples from cases with myelodysplasia and 285 freshly prepared ALL samples. We also utilize a

Ulrike Korf (ed.), *Protein Microarrays: Methods and Protocols*, Methods in Molecular Biology, vol. 785,
DOI 10.1007/978-1-61779-286-1_10,

cell line array with 186 different cell lines for validation purposes. Another group has also used RPPA to study B-cell ALL in an abstract presented at the 2008 ASCO meeting (3). The differences in the construction of these arrays represent an evolution in our understanding of how to study protein expression in leukemia. This review discusses technical and methodological issues related to the successful development of leukemia RPPAs. Additional methodological details can be found in the publications from the Kornblau laboratory (1, 4, 5).

2. Technical Aspects of Reverse Phase Protein Microarray Production

2.1. What Printer to Use?

Printers either use contact-based printing methods or inkjet like technologies to spray protein onto the slide. We extensively evaluated both with respect to amount of sample required, printing time, consistency of dot size, reproducibility, etc. While a full comparison of the various available machines is beyond the scope of this chapter a few features guided our decision. We knew that we were going to print 150–300 slides from our samples. Most printers keep all the trays open (lid off) during this process which can lead to problems with evaporation. This makes speed an important consideration. With some machines we noted that samples got sticky toward the end of a long run and did not print as well. Many machines include humidification methods, but with some this led to condensation problems within the chamber and on the slide, and in others this required frequent filling of water chambers. The inkjet technology was markedly slower than direct pin contact methods so we choose to use direct pin-based printing. We evaluated various pin diameter sizes before settling on 175-μm pins. We observed issues with dot size and reproducibility with pin sizes below 135 μm. Test data with larger pin sizes did not differ from that with the 175-μm pins, but obviously consumed more samples and permitted fewer samples per slide. We selected the Aushon 2470 arrayer which had an advantage in that it only uncovers a single tray at a time and it was clearly the fastest in our comparisons.

2.2. Slide Material

We have observed marked variation in technical features of the various commercially available nitrocellulose slides. We have observed uneven membrane thickness so that some portion is higher than others, enough so to interfere with the printer pins. We have seen issues where the machine used to roll the nitrocellulose out onto the slide in the production process was uneven resulting in a rippled effect across the slide (like the middle sheet in corrugated cardboard). There is marked batch to batch variation from the same manufacturer. Some of the membranes have greater autoflourescence than others which is a concern if fluorescent dye based analysis is planned. Others give different degrees of background staining with the DAB precipitate. We currently use ONCYTE® nitrocellulose coated film. Slides from Grace Bio-Labs (Bend Oregon, catalog # 305180). The bottom line is that each batch must be carefully checked for quality using a test print, before using them to print arrays with scarce clinical material.

2.3. Clinical Material

2.3.1. Sample Quality

Leukemic blasts can be obtained in sufficient numbers from the marrow of nearly all cases and from the blood (or leukapheresis if performed) from the majority of cases of the most common leukemias AML, ALL, CML, CLL, MDS, and MPDs. Samples from granulocytic sarcomas, lymph nodes, or isolated CNS disease are theoretically analyzable, but separation from surrounding stroma or obtaining sufficient numbers of cells presents special difficulties. Blood is collected in heparinized (green top) tubes and marrow into 3 ml of heparinized solution in 15-ml tubes (RPMI + 5% BSM + Pen Strep). The material is stored on ice or in a refrigerator until transported to the laboratory. The effect of delay in processing or temperature on protein concentration is variable. There is a growing literature (collated by the Office of Biorepositories and Biospecimen Research (OBBR) https://brd.nci.nih.gov/BRN/search.seam) documenting that variability in handling results in changes in the characteristics of the samples and that this leads to varied results in profiling arrays and biomarker studies. While mRNA and DNA have been analyzed many times there are few reports assessing changes in protein expression or phosphorylation. Most protein-based studies have analyzed paraffin-fixed material studied by immunohistochemistry; but this methodology has been shown to affect phosphorylation levels (6, 7). Maintaining tissue at room temperature compared to 4°C or 1°C prior to tissue preparation resulted in loss of phosphoprotein signal in cadaveric brain (7). Thus, sample handling prior to processing and methodology both can confound protein-based data. Since phosphorylation status is key to recognizing pathway activation, the ability to accurately ascertain this is crucial to generating consistent results. The OBBR bibliography lists no studies of protein derived from normal or malignant blood or bone marrow cells. We have conducted experiments on this processing material upon arrival, or at 24 or 48 h and we have observed that RB phosphorylation is stable, but that ERK2 phosphorylation is not. As discussed below we know that delay in processing has marked effects on protein expression in ALL. To minimize this possibility we transport material to the laboratory five times per day and rapidly process it upon arrival. There are commercially available materials that purport to stabilize cells but we have not evaluated them.

2.3.2. Enrichment of Leukemic Cells

Protein expression from tumor cells, the surrounding stroma, adjacent normal tissue, and any infiltrated cells (T and B lymphocyte infiltrates) are different. For protein expression studies the quality of the results is obviously dependent on the purity of the material studied: so purification of the malignant material from all other cells is therefore crucial to obtain a representative analysis of protein expression in the tumor. This is a major issue for studies of solid tumors but not for the leukemias. Surface markers are well defined enabling purification or enrichment of leukemic blasts from contaminating non-leukemic cells. Contaminating RBC and neutrophils are easily removed by ficoll separation. For myeloid leukemias contaminating B and T cells are removed by CD3 and CD19 depletion using magnetic antibody cell sorting. Our laboratory has successfully used the Vario, MACS, AutoMacs, and RoboSep systems. We utilize a customized mix of antibodies and find that we can use less antibody (25 μL/2.5 × 10^7 cell) than the manufacturer specifies. For the CD3/CD19 depletion we use the positive selection column and the leukemia enriched population is in the flow through. For AML and MDS this material may contain non-malignant monocytes, but we usually are processing material within 2 h after collection and the phenotype is usually unknown at that time so that we cannot deplete monocytic markers as this would remove leukemic blasts in FABM4 or M5 cases. If cryopreserved cells are used and phenotype is known, monocytes could be depleted from non-monocytic leukemias. The CD3/CD19 collection can also be processed and can serve as a control. We have used this "bulk" leukemia enriched fraction for the AML550 and AML719 arrays. For ALL purification is generally less of an issue as the percentage of blasts in the marrow and blood is usually much higher at diagnosis than in AML (The average percentage of marrow and blood blasts at MDACC since 2000 were ALL BM 78%, PB 47% and AML BM 51%, PB 27%). If phenotype is known then B or T cells can be removed. Monocytes can be removed from ALL samples as it is rare for ALL cells to express CD14. For myeloma we utilize a positive selection using anti-CD138 beads. Yields are low, but sufficient for use with RPPA.

A leukemia stem cell phenotype, minimally defined as lineage−, CD34+, CD38−, has been defined for AML (8). Protein expression may differ between the "bulk" leukemic cell and an AML stem cell raising the question of whether it would be superior to study leukemic stem cells instead of bulk cells. The critical issue is whether sufficient numbers of stem cells can be produced for use in RPPA. To print a sample set of about 200 slides once (without replicate), with five serial dilutions, requires ~300,000 cells as we utilize 30 μL of protein lysate at a concentration of 10,000 cells/μL. We have generated AML stem cells, defined as CD34+ CD38− by sequential CD34 positive selection and CD38 depletion of the ficolled material. The median yields of CD34+ cells was ~20% of the starting material and the median yield of CD34+/CD38− stem cells was about 1%. This percentage is much higher than the

Table 1
Yield of CD34+ and CD34+ CD39– "stem cells" from 93 consecutive AML samples

			34+		CD34+ CD38–	
	# Cells sorted	# Samples	# with cells	Median yield	# with cells	Median yield
BM	<2^7	8	8	$9.20E+^{5}$	5	$3.93E+^{5}$
	<3^7	10	10	$1.18E+^{6}$	6	$2.00E+^{5}$
	<4^7	16	16	$8.15E+^{5}$	9	$2.00E+^{5}$
	<7^7	9	9	$1.56E+^{6}$	6	$3.00E+^{5}$
	>7^7	5	5	$6.85E+^{6}$	5	$6.00E+^{5}$
PB	<2^7	4	4	$2.00E+^{6}$	2	$5.50E+^{5}$
	<3^7	11	11	$7.36E+^{5}$	5	$2.70E+^{5}$
	<4^7	3	3	$8.80E+^{5}$	2	$2.00E+^{5}$
	<7^7	17	17	$1.63E+^{6}$	14	$4.00E+^{5}$
	>7^7	9	9	$7.94E+^{6}$	6	$1.00E+^{6}$

1/1,000 to 1/10,000 frequency of normal stem cells. For the stem cell array we started with a median of 4×10^7 ficolled cells, we generated a median of 2×10^5 stem cells. The yields of CD34+ and stem cells from a larger consecutive series of samples are shown in Table 1. For MDS the markers to select for are the same as for AML, but the percentage of blasts in the BM and PB is by definition much lower than in AML (<20% vs. >20%). We have used a similar schema for MDS, generating CD34+ and CD34+/CD38– stem cells. As expected yields are lower as we are typically starting with less material and that material has a lower percentage of leukemic blasts (Table 2). We can successfully generate a CD34+ fraction from all AML and MDS samples, but we were only successful in generating stem cells in 60/97 fresh AML and 97/237 cryopreserved and 32/68 fresh MDS samples. Consequently, studying protein expression in only stem cells creates a bias in sample selection as not all samples produce a stem cell fraction. One alternative is to use fewer dilutions. Since we use five serial 1:2 dilutions omitting the highest cell concentration reduces the sample need by half. For CML we also have utilized CD34+/CD38– selection to generate CML SC. This debate is important as analysis of our stem cell array probed with 112 antibodies revealed highly significant difference in expression between stem cells and bulk cells for 70/112 antibodies (manuscript in preparation). A potential compromise is that most of these changes are also detected when CD34+ cells are compared to bulk cells, suggesting that CD34+ cells might serve as a partial surrogate for stem cells. Our initial study suggests that it would be superior to study CD34+ or stem cells.

Table 2
Yield of CD34+ and CD34+ CD38− stem cells from fresh and cryopreserved blood and marrow samples from patients with MDS

				CD34+			CD34+ CD38−		
		Starting # of cells	#	# with CD34+	Median yield	Med% yield	# with CD34+ CD38−	Median yield	Med% yield
Cryopreserved	Marrow	$<5\times10^6$	8	8	$1.70E+^5$	8.5	1	$3.00E+^5$	18.9
		$0.5–1\times10^7$	50	50	$1.75E+^5$	3.1	9	$2.00E+^5$	6.0
		$1–3\times10^7$	73	73	$4.30E+^5$	2.9	32	$2.90E+^5$	2.5
		$>3\times10^7$	22	22	$1.20E+^6$	3.4	15	$4.00E+^5$	1.0
	Blood	$<5\times10^6$	0						
		$0.5–1\times10^7$	30	30	$1.20E+^5$	2.0	4	$2.00E+^5$	6.7
		$1–3\times10^7$	30	30	$4.20E+^5$	2.5	20	$2.00E+^5$	1.5
		$>3\times10^7$	24	24	$7.20E+^5$	2.5	15	$4.20E+^5$	0.8
Fresh	Marrow	$<5\times10^6$	5	5	$3.80E+^5$	7.6	1	$2.00E+^5$	5.0
		$0.5–1\times10^7$	10	10	$2.80E+^5$	4.4	2	$2.00E+^5$	11.0
		$1–3\times10^7$	32	32	$6.40E+^5$	4.1	14	$2.50E+^5$	1.3
		$>3\times10^7$	14	14	$2.20E+^6$	4.8	10	$4.80E+^5$	0.7
	Blood	$<1\times10^6$	0		$1.08E+^6$	10.8		$3.20E+^5$	0.0
		$1–5\times10^6$	1	1	$2.40E+^6$	4.8	0	$8.00E+^5$	0.7
		$0.5–1\times10^7$	3	3	$2.05E+^6$	2.4	2	$4.15E+^5$	0.1
		$<5\times10^7$	3	3	$1.40E+^5$	2.9	3	$3.40E+^5$	5.0

We are currently collecting a larger collection of matched samples to study this in a larger cohort.

2.3.3. Impact of Patient Pretreatment on Sample Quality

Most of our work has utilized diagnostic samples, collected prior to the patients receiving any chemotherapy, and samples obtained at relapse. Many patients with high WBC may have received hydrea prior to referral to us. We prefer to collect a new sample after a couple of days of washout if possible, but if therapy is to be initiated and the WBC count is still high it is unlikely that limited exposure to hydroxyurea will have large effects on protein expression. So such samples could be utilized with caution. We do not have enough paired naïve vs. on hydroxyurea samples to test this. We also try to collect relapse samples. Since many relapses are discovered unexpectedly during routine BM examinations the percentage of blasts is often much lower than at diagnosis and PB involvement is much less frequent. To increase our capture of relapse specimens we collect all follow-up marrows but for financial and workload efficiency we only process those with >5% blasts. This requires keeping the sample refrigerated for a longer period of time (hours) until the differential is known. As a consequence far more of our relapse samples are BM derived. If samples are obtained shortly after the initiation of therapy it might be possible to collect cells surviving to that point by removing "doomed" annexin V positive cells. The yield is likely to be very low beyond 1–2 days after the start of chemotherapy. At day 14 from the initiation of therapy over 85% of bone marrows on AML patients have too few cells to count making it impractical to try and collect sufficient cells at that time point. Samples could be collected from primary refractory patients from day 21, 28, or 35 BM. We have not collected remission samples on patients to try and compare expression in their normal cells vs. their leukemic cells. The percentage of leukemic blasts among CD34+ CD38− cells would be variable and unless a leukemia-specific phenotype has been identified there would not be a way to separate malignant from normal blast. Furthermore, these cells are likely to be present at very low frequencies of 0.1–0.01% of cells, making recovery of sufficient cells for RPPA (or any assay) very difficult.

2.3.4. Comparison of Protein Profiles from Blood and Bone Marrow Samples

We have examined whether the source, blood vs. marrow affects protein expression patterns. In the AML719/ALL 360 array there are 140 AML and 22 ALL samples where we have paired same day blood and marrow specimens. We have analyzed this array with 176 antibodies permitting a wide ranging comparison of expression within these two compartments. In general, for both AML and ALL the majority of the antibodies tested showed no difference between the two compartments. For AML 25 were significant at a p-value of <0.01 (a more stringent criteria was used since we were performing 176 analyses) with 10 being higher in the blood and 15 higher in the marrow. For the majority of these the fold difference was <25% so whether these statistically significant differences have biological differences are questionable. Supporting the idea that some of these are real and others spurious is the observation that four of seven proteins with compartmental differences in the AML550 array produced similar results in the AML719 analysis. So for practical purposes blood and marrow produce similar results and can probably be used interchangeably. We were somewhat surprised to see as little difference as we did. At our institution research samples are collected during diagnostic procedures to avoid performing additional bone marrows on patients (who likely would decline to participate at a much higher rate if additional procedures were required, thereby introducing a selection bias). The research sample is typically the fourth pull, after the diagnostic pulls for hematopathology, cytogenetics, and flow cytometry. Although the needle is repositioned between aspirates the degree of peripheral blood contamination is likely to increase with pull number. Since we cannot estimate the degree of dilution it is possible that there are more significant differences in expression between these compartments that are obscured by the dilution.

2.3.5. Freshly Prepared vs. Cryopreserved Samples

Protein preparations can be prepared from samples on the day they arrive in the laboratory ("fresh") or from cryopreserved ("frozen") material. We asked whether protein expression changes in cells as part of the cryopreservation process. When we prepare protein from cryopreserved cells we rapidly thaw the cells, store in 20% FCS and 10% DMSO, and then dilute them in warmed media with 20% FCS. We allow the cells to stabilize for 2 h to recover from shock and then perform a ficoll separation to remove dead cells (which sink). The viable cells are then washed in TBS, counted, and then lysed at the appropriate cell concentration in the protein lysis buffer. For AML there was no significant difference in global expression across all the antibodies tested (Fig. 1a) as the distribution histograms of Fresh and frozen preps were similar. The results were distinctly different for ALL, where the "frozen" preps had significantly lower expression levels compared to the "fresh" preps (Fig. 1b). In the AML719/ALL360 RPPA, we made a frozen protein prep for 58 AML and 12 ALL cases permitting direct comparison. For AML the majority of proteins did not show a significant change between the paired samples, whereas most of the ALL samples did show a change. For ALL this was not merely a proportional reduction in protein levels, with a sample having similar expression by rank in both fresh and frozen samples. Instead the rank order was scrambled. Based on this we determined that protein expression in ALL blasts changes as part of the cryopreservation/thawing process and determined that we could not use protein prepared from cryopreserved ALL blasts for analysis of protein expression patterns. For AML this

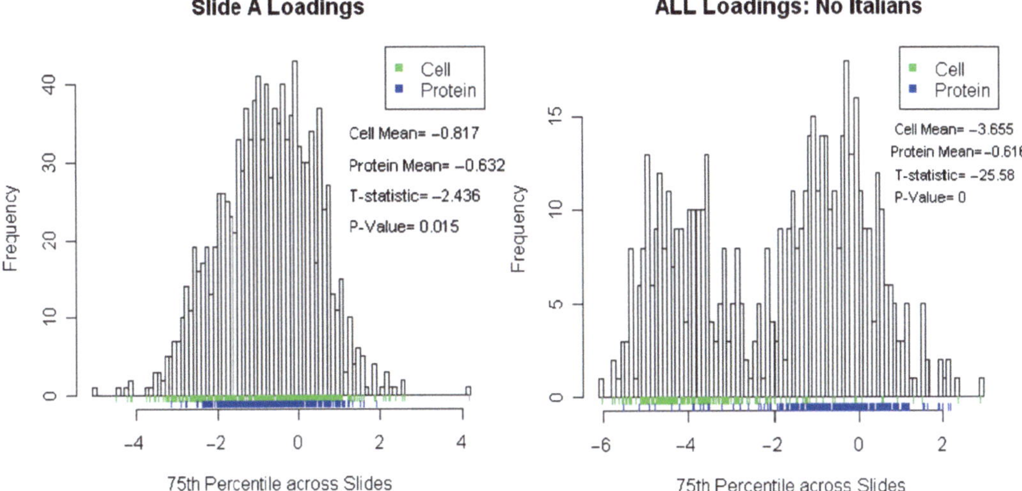

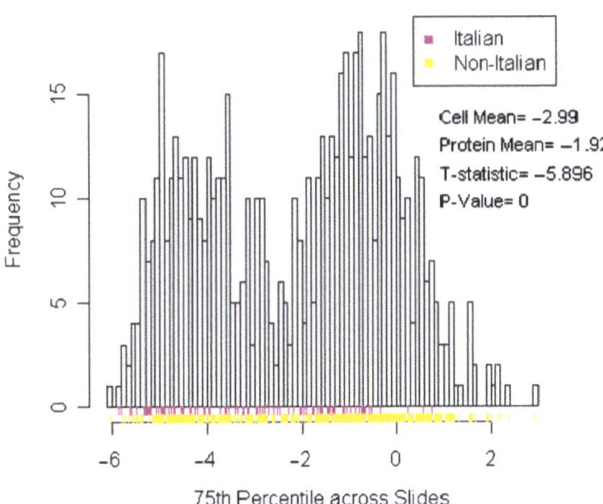

Fig. 1. Comparison or overall protein expression. These three figures all show the 75th percentile expression of 176 proteins for all cases. The *histograms* show the frequency distribution and the *colored* "*rugs*" along the bottom shows the distribution of the two conditions being compared. (**a**) Compares expression in AML protein preps made on the day of acquisition (fresh, *blue*) vs. those made from cryopreserved cells (frozen, *green*). The rugs show an equal distribution. (**b**) Shows a comparison between fresh (*blue*) and frozen (*green*) protein preps for ALL for the patients collected at MDACC. It is clear that there are two separate frequency distributions with lower levels observed in the frozen specimens. (**c**) Compares the distribution between all the samples from MDACC, both fresh and frozen, (*yellow*) and those from Italian collaborators (*pink*). The Italian samples were shipped to a central laboratory and are known to have various delays between collection and processing, but were all processed from fresh cells. It is evident that the Italian samples do not appear at the upper end of the scale and also have a wide of distribution with many samples overlapping with the frozen MDACC samples.

does not seem to be an issue. ALL cells are notoriously harder to cryopreserve and this may be another reflection of that fragility.

2.3.6. Delays in Sample Processing Affect Protein Profiles

The AML719/ALL360 array also contained 70 samples from the Italian Giemma cooperative (provided by Dr Agostino Tafuri). These samples were processed at a central laboratory, but came from across the country, so they were subject to varying time delays between collection and processing, and possibly different handling conditions (temperature). When we compared the global expression of the Italian samples to our ALL samples which were processed within 2 h, we again observed marked differences in global expression across all the tested antibodies (Fig. 1c). The yellow "rug" along the bottom shows lower expression for the Italian samples than the pink "MDACC" samples which included both fresh and frozen samples, and more closely resembles the protein prepared from cryopreserved cells. This suggests that delays in processing and possibly handling also affect protein expression in ALL.

We conclude that optimally, protein should be prepared fresh on the day of acquisition, preferably with minimal delay between collection and processing. For AML properly stored cryopreserved cells can be utilized but cryopreserved cells are not adequate for protein studies of ALL. We have not performed similar studies with CML or MDS cells, but since these are myeloid diseases they may behave more like AML than ALL.

2.3.7. Finding Appropriate Controls for Clinical Profiling

Another question is what is the appropriate control to use? Expression controls are required to verify that the antibody works as well as to provide something to compare expression to. We desire to have a positive control for each antibody as well as a negative control. It would be desirable to have a cell line known not to express each protein, but this would consume a large portion of the sample space on the array and would require foreknowledge of each epitope to be tested. Furthermore, known negatives are not available for all epitopes. As a compromise for the negative control we use protein lysis buffer (Biorad Lamelli buffer, catalog # 1610737). For the positive control we have made a protein prep from a mixture of 11 different AML cell lines. To date this has shown expression of all 190+ antibodies tested. We made a large 50 ml protein prep of this so we would have something that would be standard from array to array, thereby enabling a comparison of expression within and across different printing. We are willing to provide some of this to others using RPPA technology to facilitate comparison of results from laboratory to laboratory. We use this for both the variable slope and topographical normalization procedures described below.

In an attempt to define what can be used as a normal expression control we have included normal peripheral blood lymphocytes, Granulocyte-colony stimulating factor stimulated peripheral blood CD34+ cells collected by apheresis and bone marrow derived 34+ collected from normal (unstimulated) donors. The first two are more readily available and more economical to obtain. Unfortunately, when we compared expression between PBL,

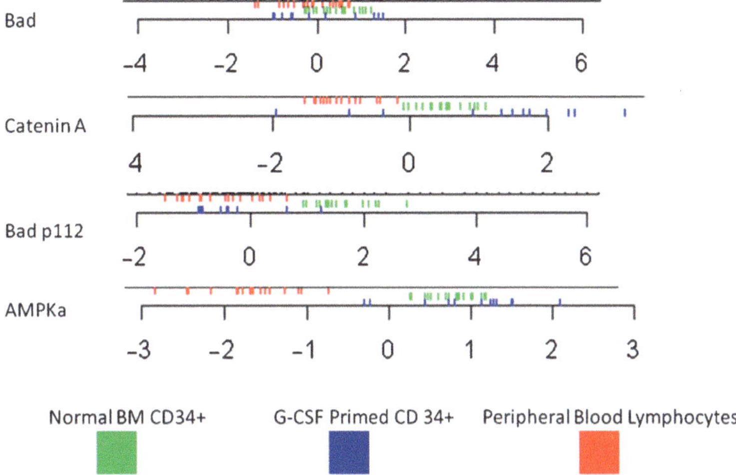

Fig. 2. Protein expression in "normal" controls varies by cell type and condition. The range of expression of normal peripheral blood lymphocytes (*red*), normal bone marrow-derived unstimulated CD34+ cells (*green*), and G-CSF mobilized CD34+ cells obtained by pheresis (*blue*) is shown for four different proteins. The numbers in *black* are in log 2 and cover the range of expression of the AML samples for each of the four proteins listed. For Bad, all three show a similar range of distribution. For Catenin-α, G-CSF primed cD34+ cells have higher expression than the unstimulated BM CD34+ cells. For Bad phosphorylated on amino acid 112 levels of G-primed CD34+ cells are similar to the normal PBLs and lower than the unstimulated CD34+ cells and for AMPkα both CD34+ samples have a similar range of expression that is higher than that of normal PBLs. This highlights the need to select an appropriate population of cells for if expression within a diseased condition is to be compared to that of normal cells.

G-primed PB-CD34+, and the BM CD34+ there were significant differences between all three potential "normals" with differences varying in both directions (higher and lower) (Fig. 2). The two CD34+ controls also showed significant differences between each other, reflecting either the consequences of G-CSF priming, or potentially compartmental differences. We have concluded that normal unprimed BM derived CD34+ cells make the best comparator to bulk and CD34+ AML and MDS leukemic cells. It would be preferable to include normal BM CD34+ CD38− stem cells as well but this is economically impractical. With normal SC comprising 1:1,000 to 1:10,000 of normal marrow mononuclear cells collecting 30,000 cells would require a starting cell number of between 6×10^7 and 6×10^8 [3×10^4 cells needed × 2 (assuming a 50% selection efficiency) × 1×10^3 or 1×10^4 population frequency]. This would require a substantial portion of a normal marrow collection from a normal donor for allogenic stem cell transplant or 6–60 vials of commercial CD34+ cells at $650 for 1×10^6 cells. For reasons of practicality we have resorted to use normal BM CD34+ cells.

3. Reverse Phase Array Data Normalization

3.1. Normalization Based on Protein Concentration

We have utilized standardization of the processing of our samples as one method of normalization and have also employed statistical measures as well. If sufficient material is available then methods to determine the concentration of protein in a sample could be utilized to determine protein concentration and sufficient buffer added to reach a consistent concentration. However, in cases where material is scarce, such as in the stem cell array it would require too much material to determine concentration. This also assumes that the concentration of protein is uniform from sample to sample which is unlikely. We have been able to deal with differences in concentration by recognizing when a protein has low levels of expression across all the tested antibodies and making adjustments using "Variable Slope" normalization techniques (this and the other algorithms mentioned below are available at http://bioinformatics.mdanderson.org/Software/OOMPA) (9).

3.2. Impact of Background Variation

Another problem is how to deal with uneven background staining. While the majority of arrays have fairly even and negligible background sometimes there is marked difference in background staining. We have also developed a technique to account for background staining differences through a technique called topographical normalization (Fig. 3). By printing the dilution series of the control cell line lysate mix described above repeatedly across the slide in every sixth column (there are eight repetitions from slide top to bottom in each column × 24 vertical blocks = 196 times on our standard 6,912 dot array) we essentially have six topographical maps showing what should be a consistent elevation across the array at six different "altitudes" (full strength, 1/2, 1/4, 1/8, 1/16th, and empty protein lysis buffer).These elevations map variation across the slide and can be used to correct for this. Topographical normalization has negligible effect on the majority of slides with minimal background variation but is very helpful on the 10–15% of slides with more significant background variation.

3.3. Using Dilution Series for Data Normalization

We utilize a dilution series consisting of several points from each sample. This allows us to build a dilution curve to assess the sensitivity of the antibody and the linearity of the detected signal. We utilize "SuperCurve" algorithms built on the curves of all samples on a slide. The SuperCurve software is freely available at http://bioinformatics.mdanderson.org/Software/OOMPA. This defines confidence intervals for the signal strength of each dilution and allows us to recognize when a misprint or large dot has occurred and to eliminate that point from the dilution curve. The signal

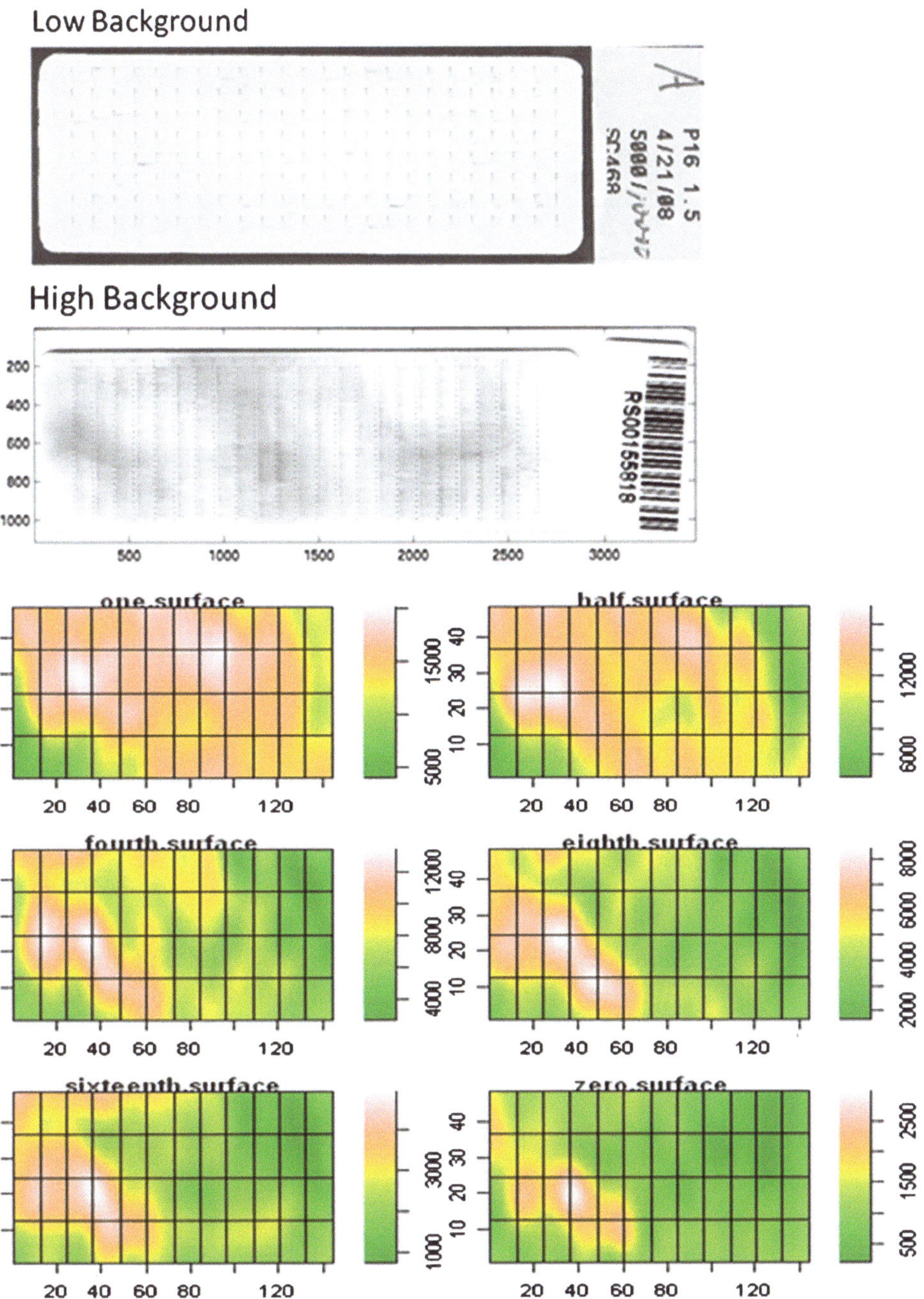

Fig. 3. Topographical normalization examples of a typical slide with low background and a slide with atypically high background and variation are shown. An example of a topographical elevation map showing variation at six altitudes (full, 1/2, 1/4th, 1/8th, 1/16th, and blank) is shown.

from each sample is then determined from the curve fitted to the dilution series. This has the advantage that a value is not based on a single dot on the slide.

4. Data Consistency and Data Reproducibility

A major question for RPPA analysis is whether these results are reproducible? The AML550 and the AML719 shared 187 samples and 47 samples but used different vials from the same protein prep and were printed over a year apart. As shown in Table 3 the majority of proteins showed highly significantly correlated expression between the two arrays using either the Pearson and Spearman rank tests. This demonstrated that protein preps stored at −80°C are stable over time. We have looked at the range of expression in samples stored for various durations of time (up to 15 years) and not observed evidence of differences. From this we conclude that the methodology is robust and reproducible within a single laboratory using the same samples. The methodology needs to have a separate laboratory, using separate samples that demonstrate similar patterns of protein expression to be fully validated.

This review did not focus on issues like antibody validation or staining techniques, which are more generic to all RPPA methodology, but these are obviously equally crucial to obtaining valid results.

Table 3
Comparisons of expression between the AML550 and AML719 arrays for 187 shared patients and 47 shared antibodies

p-Value	Pearson	Spearman	Protein (Spearman)
>0.1	1	2	P70S6K, S6RP
<0.01	1	2	MCL1* STAT3*
<0.001	3	2	Badp136, Stat1p701
<0.0001	1	4	
<0.00001	6	5	
>0.000001	1	2	
<0.000001	36	32	

For two antibodies, MCL1, and STAT3, different antibodies were used and provided poorly correlated proteins

5. Conclusions

RPPA is a valuable and facile tool for the study of protein expression in AML, but like all array methodologies obtaining accurate data requires care and attention to detail at all steps. Variability in samples selection and processing and the appropriate selection of controls is crucial. With care, repeatable results can be obtained, suggesting that RPPA could become a more generalized tool for the study of protein expression in leukemia. We are using our RPPAs to define protein expression signatures in AML and AML stem cells with the goal of developing the ability to classify the various types of AML based on the protein expression, with the idea that this can be used to match targeted therapies to the setting where the relevant pathway is crucial to the survival of leukemic cells. These are retrospective studies, but if signatures are to be used for classification and treatment selection then it is more likely that "forward phase" arrays will be required. In FPPA, numerous key antibodies, identified by RPPA, are printed on a slide, which can then be probed with protein from an individual case to provide a timely readout of protein expression and activation. Newer array technology may make protein-based classification a reality.

References

1. Kornblau SM, Tibes R, Qiu Y et al. Functional proteomic profiling of AML predicts response and survival. Blood. 2009;113: 154–164.
2. Kornblau SM, Qiu YH, Chen W et al. Proteomic Profiling of 150 Proteins in 511 Acute Myelogenous Leukemia (AML) Patient Samples Using Reverse Phase Proteins Arrays (RPPA) Reveals Recurrent Proteins Expression Signatures with Prognostic Implications [abstract]. Blood 2008;112:281–282.
3. Accordi B, Espina V, Lissandron V et al. Phosphoproteomic profiling of pediatric B-ALL patients with MLL rearrangements [abstract]. Journal of Clinical Oncology 2008;26:539S.
4. Kornblau SM, Qui YH, Tibes R, Lu Y, Mills G. Time Course Proteomic Profiling of Signal Transduction and Apoptosis Pathways in AML Survivor Cells Using Reverse Phase Protein Lysate Microarray (RPPA) Reveals Differential Effect of Time, Dose and Agent(s). [abstract]. Blood 2005; 106:355a.
5. Tibes R, Qiu YH, Lu Y et al. Reverse Phase Protein Array (RPPA): Validation of a Novel Proteomic Technology and Utility for Analysis of Primary Leukemia Specimens and Hematopoetic Stem Cells (HSC). Mol Cancer Ther 2006;5:2512–2521.
6. Baker AF, Dragovich T, Ihle NT et al. Stability of phosphoprotein as a biological marker of tumor signaling. Clin.Cancer Res. 2005;11:4338–4340.
7. Ferrer I, Santpere G, Arzberger T et al. Brain protein preservation largely depends on the postmortem storage temperature: implications for study of proteins in human neurologic diseases and management of brain banks: a BrainNet Europe Study. J.Neuropathol.Exp. Neurol. 2007;66:35–46.
8. Reya T, Morrison SJ, Clarke MF, Weissman IL. Stem cells, cancer, and cancer stem cells. Nature 2001;414:105–111.
9. Neeley ES, Kornblau SM, Coombes KR, Baggerly KA. Variable slope normalization of reverse phase protein arrays. Bioinformatics. 2009;25:1384–1389.

Part II

Antibody Microarrays

Chapter 11

Antibody Microarrays as Tools for Biomarker Discovery

Marta Sanchez-Carbayo

Abstract

The cancer biomarkers field is being enriched by molecular profiling obtained by high-throughput approaches. Targeted antibody arrays are strongly contributing to the identification of protein cancer biomarker candidates and functional proteomic analyses. Due to their versatility, novel technological and experimental design implementations are broadening the applications of antibody arrays. However, the cancer biomarker candidates delivered to date using this technology are still in their early developmental phase, requiring validation with high number of specimens focusing on specific clinical endpoints. Innovative strategies multiplexing protein measurements of protein extracts of cultured cells, tissue and body fluids using antibody arrays combined with appropriate validation approaches are enabling the discovery of cancer-associated biomarkers. This review describes these strategies and cancer biomarker candidates reported to date that may assist in the diagnosis, surveillance, prognosis, and potentially for predictive and therapeutic purposes for patients affected with solid and hematological neoplasias.

Key words: Antibody arrays, Cancer biomarker discovery

1. Antibody Arrays as Tool for Cancer Biomarker Research

Cancer can be described as a genetic disease, driven by the multistep accumulation of genetic and epigenetic factors. These molecular alterations result in uncontrolled cellular proliferation, cell cycle deregulation, decrease in cell death or apoptosis, blockage of differentiation, invasion, and metastatic spread. The particular genetic and protein expression alterations that occur as part of the crosstalk between these pathways, will in great part determine the biological behavior of the tumor including its ability to grow, recur, progress, and metastasize. The advent of high-throughput methods of molecular analysis can comprehensively survey the genetic and protein profiles characteristic of distinct tumor types and identify targets and pathways that may underlie a particular clinical behavior.

Ulrike Korf (ed.), *Protein Microarrays: Methods and Protocols*, Methods in Molecular Biology, vol. 785,
DOI 10.1007/978-1-61779-286-1_11, © Springer Science+Business Media, LLC 2011

The driving force behind oncoproteomics is the belief that certain protein signatures or patterns are associated with a particular malignancy and clinical behavior. If so, the correlation of clinical parameters with defined protein expression patterns that reflect the mutated genetic program that caused cancer, would allow tumor stratification, predict disease progression, and even define improved tailored therapeutic modalities. The technological challenges to achieve these goals are significant since the human proteome is not defined. One potential solution to finding cancer-associated protein signatures is the emerging technology of affinity proteomics. Antibody arrays represent a relatively new high-throughput technology that enables to profile multiple known proteins simultaneously. This approach addresses some of the shortcomings of traditional proteomics such as bidimensional gels or mass spectrometry and combines it with the power of high-throughput microarrays. The value of multiplexing protein measurement is being established in applications such as comprehensive proteomic surveys, studies of protein networks and pathways, validation of genomic discoveries, and the discovery and development of clinical cancer biomarkers.

2. General Guidelines for Identification and Development of Cancer Biomarkers

With the advent of high-throughput approaches, including array technologies the concept of standard unique quantitative or qualitative cancer biomarkers is being expanded to the acquisition of multiplexed information. The standard concepts describing requirements for the discovery and development of cancer biomarkers are applicable to the novel individual or protein profiles identified through antibody arrays. In order to demonstrate the clinical utility of any specific cancer biomarker candidate for any of its potential clinical uses (Fig. 1), it is necessary to show that (a) the marker can be reliably and consistently measured, (b) the marker has a combined sensitivity and specificity so that with high probability will segregate the disease status and groups under analyses, and (c) the use of the marker will improve the clinical outcome of patients by targeting diagnostic, surveillance, or therapeutic interventions.

To define appropriate experimental designs, it is important to have in mind the purpose of the use of any specific cancer biomarker candidate, meaning the endpoint of its determination. Studies can be developed to ascertain their utility for screening (who has cancer?), risk assessment (who may get cancer?), the prediction of clinical outcome (prognosis of tumor progression, overall survival?), the prediction of response to a given therapy (predictive?), or as potential targets for intervention (Fig. 1). This is a critical point in biomarker studies since experimental designs should include sample size considerations and definition of endpoints reasonable

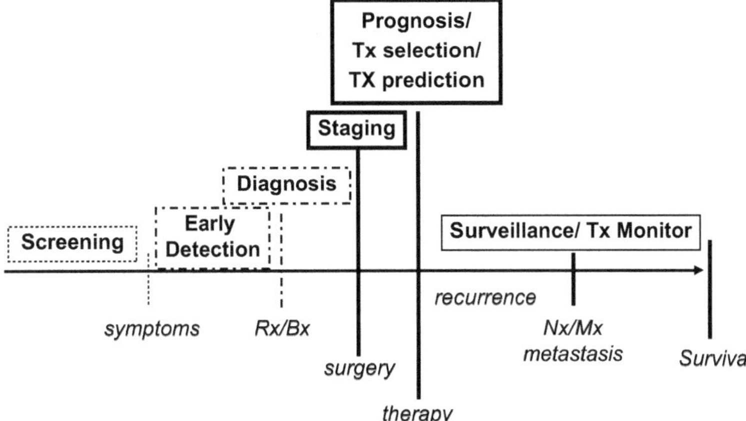

Fig. 1. Clinical utilities of cancer biomarkers along disease progression. Radiological explorations (RX); biopsies and anatomopathological explorations (Bx); therapeutical interventions (Tx); metastases in lymph nodes (Nx); metastases in distant organs (Mx).

according to the prevalence within the targeted population under study. It is important to include appropriate controls varying from healthy, benign, and premalignant conditions that may require easily be distinguishable from the cancer disease under analysis. The use of cases or controls not fully characterized may affect the identification of cancer-specific biomarker candidates using high-throughput approaches as well. This is especially critical since array technologies are measuring high number of molecular events that would require high number of individuals to enable statistical estimations. Although high-throughput platforms tend to be expensive, the experimental designs should include adequate number of samples in order to appropriately interpret the results given by these technologies. These issues may also apply to antibody arrays.

Several phases have been classically described in cancer biomarker development (Fig. 2) (1). Initially in phase I, feasibility studies are performed for assay development and evaluation of clinical prevalence. In phase II, the clinical utility for a specific endpoint is evaluated in clinical specimens. Phase III serves to confirm the clinical utility of biomarkers in independent series of cases. It is in phase IV studies where the translational research process is validated and the technology is transferred through multi-institutional evaluations leading to the incorporation of a given biomarker into clinical practice.

In the case of antibody arrays, the parallel analysis of multiple proteins in small sample volumes is being applied to measure multiple protein abundances for the discovery of biomarker candidates on biological specimens (2). Modifications to these multiplexed assays have led to profiling specific protein post-translational modifications, enzyme activities, protein cell-surface expression, or even to assist the development and characterization of antibodies (2–8). Any of these innovative variations of the technologies and applications may

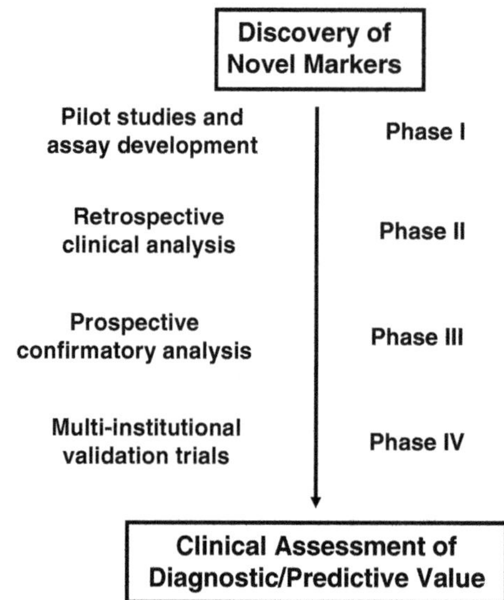

Fig. 2. Phases in cancer biomarker development.

lead to potential cancer biomarker candidates. Due to the versatility of antibody arrays, novel technological and experimental design implementations overcoming critical challenges are constantly being reported (3–5). This observation implies that phase I analyses are being conducted testing its dynamic range, reproducibility, and optimization for the newest applications. For the most accepted protocols displaying reproducibility along printing and labeling methods, such as the direct labeling, the studies reported to date may only reach the phase II or in certain cases phase III studies (1). Overall, the protein patterns obtained with antibody arrays are robust, with some variation tolerated in individual markers. Subsequent to multi-institutional validation studies, similar to any given biomarker candidate, it is expected that these patterns would be incorporated into clinical practice (6–8). Following these initial phases in technology optimization, it is becoming critical to shift the emphasis of cancer proteomics from technology development and data generation to careful study design, data organization, formatting, and mining using highly represented study-populations in order to answer clinical questions in cancer biomarker discovery research.

3. Antibody Arrays in the Scene of Proteomic Approaches

It is important to correctly classify antibody arrays in the context of other proteomic strategies that may be undertaken to investigate the cancer proteome for the discovery of cancer biomarker candidates (2–8). The terminology of untargeted and targeted proteomics refers to whether the proteins to be measured are known and considered in the experimental design (targeted) and the number of proteins that can be detected and characterized (decided at front in targeted approaches). Untargeted platforms such as two-dimensional electrophoresis (2D) and mass spectrometry are best suited for first pass comparisons of proteomes unknown at front in the experimental design to identify relatively few, novel, and known proteins that may exhibit the greatest differences in abundance. These techniques in their low- and high-resolution versions were initially considered the mainstay or standard of proteomic technologies (6–8). Targeted platforms measure and quantify known proteins of interest identified previously, and are suited for analyses of quantitative differences in abundance among known protein families and pathways. Tissue arrays and multiplexed western blots are considered targeted proteomic approaches (8). However, antibody and protein microarrays are considered the main targeted techniques used for large-scale analysis of many samples and known proteins. These two latter represent the most versatile among the proteomics techniques available to date, since antigens, peptide, complex protein solutions, or antibodies can be immobilized to capture and quantify the presence of specific either proteins or antibodies, respectively (6–8). Immobilization of proteins either as purified or phage-displayed protein versions or in format of complex protein solutions has led to tumor-associated antigen (TAAs) or reverse-phase arrays (9–12). TAAs arrays utilized on serum specimens enhance the detection of autoantibodies against TTAs, which can be utilized for cancer diagnosis and patient outcome stratification. The rationale of TAAs arrays is related to the presence in the cancer sera of antibodies which react with a unique group of autologous cellular antigens or TAAs (9, 10). Complex protein extracts can also be spotted onto membranes and probed with antibodies targeting specific proteins and pathways on the so-called reverse-phase arrays (11, 12). Overall, the versatility of targeted platforms allows controlling and estimating the reproducibility, scalability, and precise antibody and protein quantification, leading to high sensitivity and coverage. One of the major advantages of the antibody arrays approach is that it allows experimental designs to address specific hypothesis, and biological interpretation of the results obtained. However, the number of proteins amenable for these analyses depends on the availability of antibodies with high affinity and specificity to bind a target protein (6–8). Because of the little overlap between studies conducted with targeted and untargeted approaches using the same specimens, confirmation of the advantages and pitfalls of these types of high-throughput technologies for the discovery of cancer biomarker candidates remains an elusive goal. Overall, any of these proteomic strategies are impacting on the discovery of cancer-specific candidates (Table 1). In this review, these proteomic technologies have only been summarized to set up the differences with antibody arrays.

Table 1
Main characteristics of array-based proteomic techniques

Technique	Printed molecule	Pitfalls	Most frequent application
Antibody (forward-phase) arrays	Highly specific antibodies	Availability of antibodies Cross-reactivity	Protein profiling – Biomarker discovery – Signaling – Post-translational modifications
Bead-based multiplexed arrays	Antibodies coating differentially identifiable beads	Degree of multiplexing limited by number of differentially identifiable beads	Protein profiling – Cytokine – Signaling – Biomarker discovery
Reverse-phase arrays	Lysate protein extracts	Limited number of analytes analyzed even with multisectored slides Crossreactivity	Protein profiling – Biomarker discovery – Signaling – Post-translational modifications
Antigen arrays	Purified proteins and peptides	Significance of Autoabs in progression is controversial	Antibody profiling – Immune response Evaluation – Biomarker discovery

4. Antibody Array Formats

4.1. Available Antibody Array Formats

Depending on whether the antibodies are immobilized on a planar or spherical surface, antibody arrays have been classified into planar and suspension/bead formats, respectively (Fig. 3). Innovation in the immobilization surfaces and detection strategies are leading to an increasing number of planar arrays and bead-based antibody array technologies. Planar antibody arrays represent the most versatile type, as shown along the clinical applications presented for the discovery of biomarker candidates below. The main planar label-based formats comprise one-antibody and sandwich assays. One-antibody and sandwich assays present advantages and pitfalls over each other. In both formats, the target protein is always captured by one (or more) immobilized "capture" antibody in the array. In one-antibody label-based assays, the targeted proteins are detected through labeling with a tag. In sandwich assays, a second not-immobilized "detection" antibody interacts with a different epitope for a given monomeric protein enabling detection by

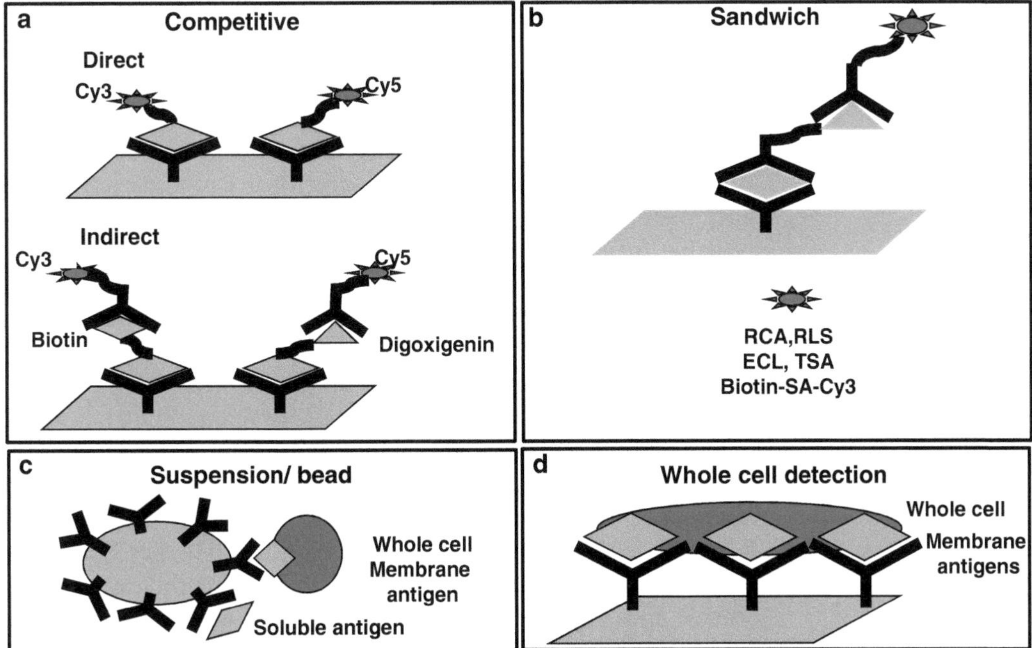

Fig. 3. Main formats of planar and suspension antibody arrays. *RCA* rolling circle amplification, *RLS* resonance light scattering, *ECL* enhanced chemiluminescence, *TSA* tyramide signal amplification, *SA* streptavidin.

forming a "sandwich" (Fig. 3). In the direct labeling, proteins are labeled with a fluorophore (including cyanines such as Cy3 or Cy5). In the indirect labeling, proteins are labeled with a tag that is later detected by a labeled antibody (7). By multiplexing with different fluorescent labels for each sample, one-antibody label-based assays may allow the incubation of more than one sample simultaneously. These assays can be designed to be competitive if the analytes belonging to the co-incubated test and reference solutions compete for binding at the antibodies. The competition in one antibody (two-color) assays is ratiometric and does not imply that the analytes are saturating the antibodies. This competition has been suggested to lead to improvements in linearity of response and dynamic range as compared to noncompetitive assays (5–8). The main disadvantage is related to the potential disruption of the analyte–antigen interaction by the label, which may also limit the detection, as well as the sensitivity and specificity.

In the sandwich label-based format, immobilized "capture" antibodies capture unlabeled proteins, which are detected by another "detection" antibody using several methods to generate the signal for detection (Fig. 3b). The use of these two "capture" and "detection" antibodies against different epitopes of a given analyte increases the specificity for the target protein to be measured as compared to label-based assays. The reduced background of these assays increases also the sensitivity. The sandwich format allows only noncompetitive assays, since only one sample can be incubated on each array (2–8). This format requires standard curves of known concentrations of analytes to achieve accurate calibration of concentrations. As compared to label-based assays, sandwich arrays are more difficult to develop in a multiplexed manner since matched pairs of antibodies and purified antigens may not be available for each target, and the potential cross-reactivity among detection antibodies increasing with additional analytes. The practical size of multiplexed sandwich assays limits to 30–50 different targets (6–8). This contrasts with one-antibody assays where only availability of antibodies and space on the substrate limit the number of targets analyzed.

Proteins in suspension can then be detected using bead/suspension arrays (Fig. 3c) (13–15). These arrays use different fluorescent beads, each coated with a different antibody and spectrally resolvable from each other (13–15). The beads are incubated with a sample to allow protein binding to the capture antibodies, and the mixture is incubated with a cocktail of detection antibodies, each corresponding to one of the capture antibodies. The detection antibodies are tagged to allow fluorescent detection. The beads are passed through a flow cytometer system, and each bead is probed by two lasers, one to read to the color, or identity of the beam and another to read the amount of detection antibody on the bead (13–15). Multiplexed bead-based flow-cytometry assays represent an active area of development. Differentially identifiable beads coated with either proteins, autoantigens, or antibodies can identify a variety of bound antibodies or proteins using a flow cytometer system (13–15). Other antibody array approaches have been developed as modifications of the one-antibody and sandwich label-based arrays. These alternate strategies allow detection of proteins on whole cells without protein isolation (Fig. 3c, d) (3, 4). Advances in instrumentation and bead chemistries are making this approach very valuable for the detection of circulating cancer cells. As another version of this concept, suspensions of cells can be incubated on antibody arrays, and the amount of cells that bound each antibody can be quantified by dark field microscopy. These arrays have the potential of characterizing multiple membrane proteins in specific cell populations or changes in cell surfaces induced by drug therapies (30).

4.2. Emerging Antibody Array Formats

Several examples can be provided to delineate recent remarkable innovations achieved to monitor specific post-translational modifications as well as to increase the limits of detection or enable the technology to profile protein extracts obtained from very few individual cells. In a first example, antibody arrays are adapted to detect differences in the content of glycans (sugars or carbohydrates) of proteins. These carbohydrate post-translational modifications on proteins are known to be important determinants of protein function in both normal and disease biology. Antibody array designs have been developed to allow efficient, multiplexed study of glycans on individual proteins from complex mixtures (16, 17). Once multiple proteins are captured using antibody microarrays, these post-translational modifications can be detected using lectins or glycan-binding antibodies (17). In pancreatic cancer, profiling of both protein and glycan variation in multiple serum samples using parallel sandwich and glycan-detection assays, has identified the cancer-associated glycan alteration on the proteins MUC1 and CEA in the serum of pancreatic cancer patients (17). These antibody arrays for glycan detection are opening a novel field of glycobiology research in the context of neoplastic diseases for discovering potential cancer biomarker candidates.

High sensitivity, in the femtomolar range, allowing protein quantification from limited sample quantities (only six cells) can be achieved by the so-called antibody "ultramicroarrays" (18). These arrays were initially tested for the detection of interleukin-6 (IL-6) and prostate-specific antigen (PSA), finding detection levels using purified proteins in the attomolar range (18). Remarkably, in the discovery of cancer biomarker candidates, this strategy should enable proteomic analysis of clinical specimens available in very limited quantities such as those collected by laser capture microdissection.

Another critical technical development that is being applied to antibody arrays for the discovery of cancer biomarker candidates is quantum dot technology. By offering remarkable photostability and brightness and low photobleaching, quantum dots allow detection of proteins in biological specimens (serum, plasma, body fluids) at pg/ml concentration, as has been shown to detect several cytokines (19). Models of quantum dot probes include conjugation of nanocrystals to antibody specific to selected markers, such as IL-10 and the use of streptavidin-coated quantum dots and biotinylated detector antibody (19). By allowing monitoring of changes in protein concentration in physiological range in body fluids, the methodology can potentially be applied to other types of planar and suspension arrays.

Another technical innovation allowing detection of protein biomarker candidates at picomolar concentrations utilizes surface plasmon resonance imaging (SPRI) measurements of RNA aptamer microarrays. The adsorption of proteins onto the RNA microarray is detected by the formation of a surface aptamer–protein–antibody complex. The SPRI response signal is then amplified using a localized precipitation reaction catalyzed by the enzyme horseradish peroxidase that is conjugated to the antibody. This enzymatically amplified SPRI methodology has initially been characterized for the detection of human thrombin at the fM concentration range. The appropriate thrombin aptamer for the sandwich assay can be identified from a microarray using several potential thrombin aptamer candidates. The SPRI method has also been optimized to detect the protein vascular endothelial growth factor (VEGF) at a biologically relevant pM concentration. This incipient technology shows a potential for increasing this sensitivity for detecting proteins in body fluids (20). In the context of cancer biomarker candidates, the sensitivity achieved for VEGF allows its measurement in the serum for selecting or monitoring antiangiogenic therapies for breast, lung, or colorectal cancer. In the same line of research, an independent study using a 17-multiplexed photoaptamer-based array has exhibited limits of detection below 10 fM for several analytes including the VEGF, IL-16, and endostatin, among others in serum samples. Since photoaptamers covalently bind to their target analytes before fluorescent signal detection, the arrays can be vigorously washed to remove background proteins, providing the potential for superior signal-to-noise ratios and lower limits of quantification in biological matrices. Interestingly, the affinity of the capture reagent can be directly correlated to the limit of detection for the analyte on the array (21).

5. Experimental Approaches for Cancer Biomarker Identification

The increasing number of strategies of antibody arrays is improving, emerging, and challenging in their applications in functional and descriptive proteomic research for the discovery of cancer biomarker candidates. Significant contributions of proteomics research using antibody arrays reported to date have derived from a wide spectra of experimental designs, varying from single experiments to comparison of relatively low or medium size datasets obtained under different conditions (e.g., normal, inflammation, cancer) and in different cells, tissues, and bodily fluids. Representative examples of these strategies are described in this section.

5.1. Cell Culture

Protein profiling studies of cultured cells using antibody arrays are allowing in-depth analyses of cancer biology. Since many of these cancer cells are derived from human tumors, they resemble human disease and may also lead to the discovery of cancer biomarker candidates if the experimental design includes validation strategies using independent sources of human clinical material. The use of antibody arrays for high-throughput profiling of cultured cells is also useful to evaluate signaling pathways including tyrosine kinases networks (22). Antibody arrays can profile enzyme activities using both protein extracts and cell culture supernatants. As compared to initial gel-based strategies utilized to assess the functional state of enzymes, they represent a convenient platform to evaluate activity-based protein profiling with high sensitivity and specificity and reducing sample consumption. While gel-based strategies basically enable protein discovery, antibody microarrays may define new patterns of expression of known proteins. The presence of phosphorylated and unphosphorylated forms of proteins can be assessed in cell cultured systems using antibody arrays if adequate antibodies are available to address the specific post-translational modifications of the target proteins under study (22, 23).

Cytokine profiles of cell lysates have also been analyzed by means of cytokine arrays and compared to those obtained on body fluids and tissue extracts (24). Commercially available cytokine arrays have been applied to conditioned media of cancer cells to dissect the cytokine secreted signature associated to the overexpression of a critical breast cancer target gene, the HER2 in breast cancer cells. This strategy revealed that the enhanced synthesis and secretion of members of the IL-8/GRO chemokine family may represent a new pathway involved in the metastatic progression and endocrine resistance of HER2-overexpressing breast carcinomas (25). This in vitro strategy not only served to identify a potentially relevant signaling pathway but also identified a cancer protein specific signature with clinical applications. Interestingly, validation analyses using sera retrospectively collected from metastatic breast cancer patients revealed that circulating levels of IL-8 and GRO cytokines were higher in HER2 positive breast cancer patients. These proteins may represent novel biomarker candidates for monitoring breast cancer responses to endocrine treatments and/or HER2-targeted therapies (25).

An independent study has screened the native cytokine expression patterns in human breast cancer cell lines associated to the expression of the estrogen receptor (ER) using cytokine arrays. ER positive cells expressed low levels of IL-8 whereas ER negative cells expressed high levels of IL-8. Such profiling served to monitor functional analyses blocking IL-8-mediated tumor cell invasion and angiogenesis using a neutralizing antibody against IL-8 as well as the exogenous overexpression of this gene, which substantially inhibited IL-8 expression. The combination of several in vitro strategies monitored by cytokine profiling using antibody arrays served to link the role of IL-8 in the development and progression of human breast cancer in association with the ER status (26, 27).

Cytokine profiles of cell supernatants of other tumor types such as the Jurkat (T-cell leukemia) and the A549 (nonsmall cell lung cancer) cells have also been monitored by means of cytokine arrays. The cytokine/chemokine response was evoked after cell stimulation with tumor necrosis factor alpha (TNFalpha), phorbol-12-myristate-13-acetate, and phytohemagglutinin. Stimulated cells showed an increase in the expression level of many of the 41-test analytes, including IL-8, TNF-alpha, and MIP-1alpha in the treated cells (28). This strategy shows the ability of antibody array analysis of cell-culture supernatants for the profiling of cellular inflammatory mediator release.

Antibody arrays can also be utilized to monitor signal transduction mediated by complex networks of interacting proteins in mammalian cells and screen cancer drugs. Microarrays created in 96-well microtiter plate formats multiplexed the measurement of amounts and modification states of signal transduction proteins in crude cell lysates. These arrays have been applied to monitor the activation, uptake, and signaling of ErbB receptor tyrosine kinases in cancer cells. This strategy was used to characterize the action of the epidermal growth factor receptor inhibitor PD153035 on cells (29). Thus, the integration of microplate and microarray methods for crude cell lysates may identify small molecules with specific inhibitory profiles against specific signaling networks. The technology is yielding comprehensive information about the mechanism of action and the efficacy of existing and novel cancer compounds in preclinical studies for the treatment of human cancer.

An interesting strategy reported on leukemias and lymphomas cells has allowed biological immunophenotyping by means of a "DotScan" antibody array, where these cells are incubated and captured based on their membrane protein expression patterns. The antibody array was initially with a set of 88 immobilized CD antibodies and later on using a higher number of 147 antibodies (30, 31). Interestingly, a high number of leukemias and lymphoma cells, as well as clinical samples were analyzed and classified by their surface protein profiles (31). The relevance of these strategies relies on the possibility of antibody arrays to capture cells based on the protein expression pattern of surface proteins. Moreover, this approach might potentially lead to a molecular classification of human blood malignancies. Thus, cell binding assays on antibody arrays might permit the rapid immunophenotyping of human living cells. The throughput of the analysis, however, is still limited due to the ability to perform parallel and quantitative detection of cells captured on the array. This limitation can be addressed using imaging techniques based on surface plasmon resonance (SPR). In addition to monitoring capture of proteins on antibody microarrays, SPR is being optimized for cell capture (32).

Protein extracts from cancer cells have been utilized to optimize and develop technological innovations of antibody arrays. The availability of cultured material allows reproducibility analyses and testing the analytical properties of a given novel innovation in antibody array technologies. Remarkably, dilution analyses varying from 100 to 4 prostatic LNCaP cells served to optimize ultramicroarrays that allow reproducible protein detection from the lysate of an average of just six cells of two known serum proteins such as secreted IL6 or PSA (18). By resembling human disease, this sensitivity improvement using cultured cells suggests the potential utility of the protein quantification of these molecules or others in human body fluids or protein extracts of laser microdissected neoplastic prostatic populations (18). Another example of technology optimization establishing limiting factors of labeling methods using protein extracts of breast cancer cells dealt with competitive binding assays under different conditions with one-color or two-color fluorescence detection methods. These analyses revealed that antibody cross-reactivity, target protein truncation and abundance, as well as the cellular compartment of origin are major factors that affect protein profiling on antibody arrays (33).

Protein extracts from murine and human lung cell lines have served to identify protein phosphatase 1 (PP1) interacting proteins (PIP) that are important in cell proliferation and cell survival by means of antibody arrays. This comparative approach identified 31 potential novel PIPs and confirmed 11 of 17 well-known PIPs included as controls, suggesting a sensitivity of at least 65%. Interestingly, validation analyzed by co-immunoprecipitation confirmed that nine of these proteins associated with PP1. By exposing these cells to nicotine, the association of PP1 with these proteins could be modulated. Thus, novel interactions with PP1 were identified and were consisting with the PP1 role at facilitating cell cycle arrest and/or apoptosis (34). The important observation is that protein profiling using carefully selected antibodies served to design functional analyses to characterize the relevance of the post-translational modifications of these proteins along cell cycle.

5.2. Tissue Specimens

To identify cancer-specific biomarker candidates, it is also feasible to characterize the protein profiles of protein extracts of tissue specimens using antibody arrays. By comparing malignant and normal counterparts, it is possible to identify differentially expressed proteins associated with disease progression. This strategy has been performed in lung cancer comparing tumor samples from patients with squamous cell lung carcinoma and normal lung tissue controls with a high number of antibodies printed on antibody arrays (35). Among the differentially expressed proteins, PEX1, MKK7, and HDAC3 up-regulated proteins were shown to correlate with a high mRNA expression obtained from paired gene microarray data. Validation analyses by immunoblot analysis revealed HDAC3 elevation in 92% of the 24 tumors analyzed. Thus, using a tumor profiling strategy, antibody microarrays served to identify lung cancer biomarker candidates (35). In line with this strategy, it is possible to characterize protein profiles of neoplastic subpopulations obtained from frozen resected tumor specimens using laser capture microdissection (36). Microdisssection is especially critical for data interpretation in heterogeneous tumors such as breast or prostatic cancer. For example, profiling of protein extracts of breast tumor versus the adjacent normal breast tissue identified a number of proteins with increased expression levels in malignant specimens such as casein kinase Ie, p53, annexin XI, CDC25C, eIF-4E, and MAP kinase 7. Decreased expressed proteins in the malignant tissue included the multifunctional regulator 14-3-3. Immunohistochemistry in paraffin-embedded normal and malignant sections derived from the same patient using antibodies against these proteins served to validate the data obtained using the antibody microarrays (36). In this exercise, protein profiling of a single neoplastic patient using a commercially available microarray served to identify molecular determinants of cancer progression in breast cancer. It seems reasonable to insist on that the clinical validation with high number of specimens on independent sets of clinical material is critical to verify the clinical significance of cancer-specific discovery analyses.

The results of protein profiling of tumor protein extracts using antibody arrays can be validated in several manners in order to confirm that potential identified biomarker candidates are cancer specific. On one hand, gene profiling of matched tumors can prove that the increased protein expression is associated with increased transcript profiles (35). At the protein level, it can also be tested that the differential expression of proteins can be detected using an independent method such as immunoblotting (35). Clinical validation of differential protein expression patterns can be confirmed by immunohistochemistry using the same antibodies that were printed on the antibody arrays on paraffin-embedded normal and malignant tissues providing high reliability on the results found by protein profiling. If tissue arrays with well-characterized independent set of tumors are available, it is possible to evaluate clinicopathological correlations of novel cancer-specific proteins with tumor stratification, disease progression, and clinical outcome (7).

The use of comprehensive gene profiling analyses using tissue material can identify tumor targets relevant of specific neoplasias for antibody arrays design. Such approach can be applied in antibody-based proteomics to generate protein-specific affinity antibodies to functionally explore the human proteome. Specific protein epitope signature tags (PrEST) can be identified and used to raise monospecific, polyclonal antibodies, and be subsequently analyzed on paraffin-embedded sections of malignant and normal tissue. Genome-based, affinity proteomics, using PrEST-induced antibodies, is an efficient way to rapidly identify a number of disease-associated protein candidates of previously both known and unknown identity (37). A descriptive and comprehensive protein atlas for tissue distribution and subcellular localization of human proteins in both normal and cancer tissues is being created (38). The subsequent antibodies generated can be used for analysis of corresponding proteins in a wide range of assay platforms, including (1) immunohistochemistry for detailed tissue profiling, (2) specific affinity reagents for various functional protein assays, and (3) capture ("pull-down") reagents for purification of specific proteins and their associated complexes for structural and biochemical analyses (38).

A critical part in proteomic research aiming at the discovery of cancer biomarker candidates deals with the optimization of sample preparation for comprehensive protein measurements.

Proteases inhibitors can be added in order to overcome accelerated protein degradation due to the presence of secreted proteases. Novel tissue sample handling approaches to enrich (>95% purity) epithelial cells from fresh human tissue samples include the use of an epithelial cell surface antibody (such as the Ber-Ep4). This purification method showed several advantages for proteomic analyses on tissue specimens since a large quantity of cells available for downstream analysis were available and it showed an increased reproducibility (39). Flow cytometry, sorting analyses, pulldowns of protein extracts, HPLC, and spectrometry techniques represent alternative approaches to enrich cell populations of interest before protein profiling using antibody arrays.

5.3. Body Fluids

The initial report applying antibody microarrays in serum cancer for the discovery of biomarker candidates was performed using direct labeling methods for prostate cancer, comparing several substrates for antibody printing (40). As part of optimization analyses, data from "reverse-labeled" experiment sets accurately predicted the agreement between antibody microarrays and enzyme-linked immunosorbent assay measurements (40). Comparison of protein profiles of patients with prostate cancer and control serum samples identified five proteins (von Willebrand Factor, immunoglobulin M, Alpha1-antichymotrypsin, Villin, and immunoglobulin G) that had significantly different levels between the prostate cancer samples and the controls. This initial study using direct labeling protocols is one of the critical analyses that led to multiple developments enabling the immediate use of high-density antibody and protein microarrays for studies dealing with the discovery of biomarker candidates (40). In this regard, antibody arrays could be considered as multiplexed ELISAs, especially for sandwich assays (Table 2). The use of amplification protocols, such as a two-color RCA method served to improve the detection of low abundant proteins. This method has also been shown to provide adequate reproducibility and accuracy for protein profiling on serum specimens and clinical applications (41–43). Sandwich assays can also measure protein abundances in body fluids using amplification detection methods such as resonance light scattering (44), enhanced chemiluminescence (45), or the tyramide signal amplification method (46) (Fig. 3, reviewed in (7)). A recent report designed antibody arrays for bladder cancer by selecting antibodies against targets differentially expressed in bladder tumors versus their respective normal urothelium identified by gene profiling (47). Serum protein profiles obtained by two independent sets of antibody arrays served to segregate bladder cancer patients from controls. Protein profiles provided predictive information by stratifying patients with bladder tumors based on their overall survival. In addition, serum proteins, such as c-met, that were top ranked at identifying bladder cancer patients were associated with pathological stage, tumor grade, and

Table 2
Main characteristics of antibody arrays compared to standard individual enzyme immunoassays (ELISA) (3–8, 13–15)

Feature	Antibody array (multiplex ELISA)[a]	ELISA (Individual ELISA)
Sample type	Cell culture supernatant, cell culture protein extract, serum, plasma, body fluids, tissue	Cell culture supernatant, [b]serum, plasma, body fluids
Volume/assay	2–5 μL	50–100 μL
Analytes/assay	100–1,000	1
Reproducibility	<10%	5%
Sensitivity	1–50 fg/mL	1–50 pg/mL
Specificity	High	High
Dynamic range	3–4 logs	2 logs
Assay time	4 h	4 h
Instrumentation	Simple	Simple

[a]Please note that antibody arrays commercially available in planar or bead versions are frequently described as multiplex ELISA kits by many suppliers, especially for cytokine and specific cell signaling measurements
[b]Although ELISAs can be used for any type of protein extract, usually their most recommended applications do not include total protein extracts of cell lysates or tissue. These samples may require dilution optimization analyses of the protein extract depending on the varying level of expression of the target protein in these specimens

survival when validated by immunohistochemistry of tissue microarrays containing bladder tumors (47). Such strategy provides experimental evidence for the use of several integrated technologies strengthening the discovery process of cancer-specific biomarker candidates.

Cytokine profiling on serum and plasma specimens represents one of the most described applications of antibody arrays technology, especially for autoimmune diseases. In neoplastic diseases, they have been evaluated to a lower extent, although the implementation of cytokine antibody arrays is increasing in many aspects of cancer research, such as the discovery of biomarker candidates, molecular mechanisms of cancer development, preclinical studies, and the effects of cancer compounds (49). Studies in clinical material and in vitro systems have revealed the potential of cytokine profiling using antibody arrays for characterizing hematological neoplasias (8–10), or in serum of patients with breast cancer (25). Cytokine profiles can support differentiation between cancer patients from control subjects and also stratify patients with leukemia based on clinical outcome. Several reports have also compared the reproducibility and differences among the several technologies available for multiplexing cytokine measurements, including not only planar antibody arrays but also bead-based technologies (13–15).

The tumor interstitial fluid (TIF) which perfuses the tumor environment has also been utilized for protein profiling using antibody arrays. Analysis of the TIF could identify factors present in the tumor microenvironment that may be associated with tumor growth and progression. TIFs collected from small pieces of freshly dissected invasive breast carcinomas have been analyzed by cytokine-specific antibody arrays. The approach provided a snapshot of more than 1,000 proteins – either secreted, shed by membrane vesicles, or externalized due to cell death – produced by the complex network of cell types that make up the tumor microenvironment. Considering that the protein composition of the TIF reflects the physiological and pathological state of the tissue, it should provide a new and potentially rich resource for the discovery of diagnostic biomarker candidates and for identifying more selective targets for therapeutic intervention (48, 49). Interestingly, labeling and hybridization methods have been optimized for multiple protein detection on CSF specimens, characterized by low protein concentrations (50). This approach would have diagnostic applications, allowing differential diagnosis of neoplastic disease from other benign conditions. Noninvasive body fluids such the saliva, sputum, or urine specimens represent potential samples for clinical application of antibody arrays. It is required to optimize labeling and hybridization protocols to the sensitivities required for such specimens. The development of diagnostic and prognostic markers using these samples may improve current clinical management of the cancer patient using noninvasive approaches.

A recent protein profiling in prostate cancer using custom-made antibody array analyses of serum specimens using RCA methods revealed that patients with prostatic tumors repressed the expression of trombospondin-1 levels as compared to benign prostatic disease (51). Although thrombospondin-1 levels did not correlate with PSA levels, they differentiated benign from malignant disease with sensitivity and specificity around 80%. Importantly, the strategy of this study included validation analyses using immunoblots, immunoprecipitation/mass spectrometry, and sandwich immunoassays. Another critical point is the comparison of the new findings using the antibody profiling with the standard of practice tumor marker in prostate cancer, PSA. This observation led to conclude that the measurement of thrombospondin-1 could be used as adjunct information to assist in a critical clinical decision to obtain a biopsy in men with suspected prostate cancer.

6. Conclusions

The parallel analysis of multiple proteins in small sample volumes is being applied to measure multiple protein abundances for the discovery of cancer biomarker candidates using antibody arrays. Application on biological specimens is serving to address disease progression, clinical subtypes and outcomes in exploratory analyses.

Modifications to antibody arrays are leading to other protein profiling strategies that may also result into novel cancer biomarker candidates such as (a) detecting specific protein post-translational modifications, (b) measurement of enzyme activities, (c) quantification of protein cell-surface expression, (d) characterizing signaling pathways, and (e) the development and characterization of antibodies including identification of binding partners to proteins derived from functional studies for drug discovery or novel epitope mapping for determining regions of proteins than bind specific antibodies.

The use of antibody array methods not only results in added benefit for cancer diagnostics and patient stratification but also provides complementary information for the characterization of the biology underlining tumorigenesis and tumor progression. Protein profiling using antibody arrays is contributing to reveal the importance of monitoring multiple cell signaling endpoints and thus, mapping specific cellular networks not only in protein extracts from cell lines but also form tissue or body fluid specimens. Changes in glycan contents, phosphorylation status or cleaved states of key signaling proteins can easily be evaluated using antibody arrays as well. It is possible to test whether one pathway might become blocked with chemotherapeutic agents. Analyses of these pathways might reveal relevant information for designing individual targeted therapies and/or combinatorial strategies directed at multiple nodes in a cell signaling cascade. This strategy might be tested to predict response to novel drug therapies using the protein extracts of the tumors or in body fluids specimens.

It is important to adequately interpret the value of any potential "biomarker candidates" identified through antibody arrays. Depending on the experimental design, many of the detected protein partners identified using "in vitro" material may be highly interesting for understanding protein interactions in cancers but have limited value for biomarker discovery. Only the adequate clinical validation using appropriate human material would allow addressing the clinical relevance of such discovery approaches. A great amount of the data presented above concern the differential expression of target proteins and/or their abundance in cancers. A priori it is clear that many of these proteins cannot be used as biomarkers in clinical routine practice because of different reasons. In some cases, because of lack of any clinical validation analyses performed with adequate human specimens to address the clinical objective under analyses. In others, because of lack of appropriate reagents and antibodies of high specificity for the protein under study compatible and with appropriate sensitivity for the dynamic range of expression of the protein in the specimens where it is going to be measured. In many of them, because of pending multi-institutional validation studies confirming the initial discovery patterns in independent sets of samples. It is also important to be aware that biomarker

Table 3
Summary of representative cancer biomarker candidates discovered by protein profiling using antibody arrays per specific tumor types

Tumor type	Biomarker candidates	Clinical utility	Reference
Pancreatic	Glycan variation of MUC1 and CEA	Diagnostics	(17)
Breast	Cytokine signaling mediated by HER2: IL-8 and GRO	Therapeutic response	(25)
	Casein kinase Ie, p53, annexin XI, CDC25C, eIF-4E and MAP kinase 7,14-3-3	Tumor progression	(36)
Lung	PEX1, MKK7, and HDAC3	Tumor progression	(35)
Blood malignancies	CD antigens	Diagnosis, classification	(30, 31)
Prostate	von Willebrand factor, immunoglobulin M, Alpha1-antichymotrypsin, Villin, and immunoglobulin G	Diagnostics	(40)
	Trombospondin-1	Diagnostics	(51)
Bladder	Protein profile, C-MET	Diagnosis, clinical outcome	(47)

candidates (individual or patterns) may be useful for diagnostic purposes and not for clinical outcome prediction and vice versa. Along the progression of the disease, some could adequate to detect or stratify early stages and not for advanced or metastatic disease, and vice versa. Thus, it is critical to emphasize that independent validation analyses should be conducted including human material adequate enough to address the clinical endpoint under study. Overall, good marker candidates need to be supported by well-documented data obtained with various experimental approaches and independent sets of experiments.

Interestingly, this review has commented on several examples of cancer biomarker candidates discovered by protein profiling of human specimens using antibody arrays, per several specific tumor types, as summarized in Table 3. In pancreatic cancer, analyses of post-translational modifications such as glycan variation in the serum of cancer patients are allowing the identification of biomarker candidates for cancer diagnostics (17). In breast cancer, the application of the cytokine array in serum specimens served to identify cytokines differentially expressed in HER2 oncogene-dependent breast carcinomas, related to increased metastatic potential and resistance to current anticancer therapies. These proteins may represent biological determinants of potential clinical relevance to monitor and predict resistance to therapeutic drugs targeting the HER2 receptor (25). Tumor progression candidates can be identified using strategies comparing neoplastic and normal counterparts, as has been reported for breast (36), and lung cancer (35). Immunophenotyping of blood malignancies has been improved by means of antibody arrays including several sets of immobilized CD antibodies that captured circulating neoplastic cells through the corresponding surface molecules (30, 31). In prostate cancer, the utility of the current tumor marker in clinical practice, PSA, has been shown to be complimented by the information provided by novel individual or protein patterns identified using antibody arrays (40, 51). In bladder cancer, antibody arrays represented comprehensive means for the identification of protein patterns specific for cancer diagnosis and clinical outcome stratification. Validation analyses with ELISA and immunohistochemistry on tissue microarrays confirmed the relevance of identified proteins for tumor progression (47). These approaches provided experimental evidence for the use of several integrated technologies strengthening the process of the discovery of cancer biomarker candidates.

7. Take Home Messages

Antibody-based microarrays represent a rapidly emerging technology for the discovery of biomarker candidates that is advancing from the first proof-of-concept studies to increasing protein profiling applications in cancer biomarker development. The increasing number, scope, and effectiveness of the formats, methods, and applications of antibody arrays are likely to markedly accelerate the characterization of cancer-specific pathways, networks, and post-translational modifications. Identifying cancer-associated protein changes may lead to the discovery of cancer-associated biomarker candidates that may assist in disease predisposition, diagnosis, prognosis, patient monitoring, and possibly for therapeutic purposes on various sample types, such as serum, plasma, and other bodily fluids; cell culture supernatants; tissue culture lysates; and resected tumor specimens. As standards do not yet exist that bridge all of these applications, the current recommended best practice for validation of results is to approach study design in an iterative process using independent sets of human clinical material and to integrate data from several measurement technologies. The main problems described in poorly delineated experimental designs include lack of uniform patient inclusion and exclusion criteria, low patient numbers, poorly supporting clinical data, absence of standardized sample preparation, and limited analytical verification providing estimations of the intra- and inter-assay reproducibility.

Several challenges and limitations remain to be improved in the design and application of antibody arrays for the appropriate interpretation of the identified cancer biomarkers obtained through protein profiling (52, 53): (1) the mechanisms by which proteins or antibodies are immobilized in substrates such as nitrocellulose are poorly understood for certain technological innovations; (2) the limited dynamic ranges of two or three orders of magnitude for certain labeling protocols can be increased; (3) achieving accuracy and reproducibility similar to clinical immunoassays at the very low pico/femtomolar detection level; (4) the immunoreactivity might be affected by the molecular protein complexity and potential protein denaturation; (5) lack of standards and calibrators for all the antibody and reagents utilized; (6) development of high-affinity and highly specific antibodies are not available for all the potential target antigens under study.

The highly increasing technical modalities of antibody arrays are requiring standardized processes for storing and retrieving data obtained from different technologies by different research groups. In this regard, it is necessary to acknowledge the multi-institutional effort of the Human Proteome Organization (HUPO) toward the standardization of critical parameters in serum or plasma proteomic analyses, including protein profiling using antibody arrays. Initial studies provided guidance on pre-analytical variables that can alter the analysis of blood-derived samples, including choice of sample type, stability during storage, use of protease inhibitors, and clinical standardization (52). As part of the HUPO approach, it is also critical to standardize statistical strategies for high confidence protein identification and data analysis. These efforts and strategies toward integrating proteomic datasets would lead toward accurate and comprehensive representation of human proteomes (53).

Thus, the most significant contribution of proteomics research using antibody arrays for the discovery of cancer biomarker candidates is expected to derive not from single experiments, but from the synthesis and comparison of large datasets obtained under different conditions (e.g., normal, inflammation, cancer) and in different tissues and organs. The technology will continue providing unique opportunities in cancer diagnostics, patient stratification, predicting clinical outcome, and therapeutic response (54–56). Continued progress in the technology will surely lead to extensions of these applications and the development of new ways of using the methods. Further innovations in the technology and in the experimental strategies will further broaden the scope of the applications and the type of information that can be gathered. In the near future, the detailed characterization of the specific protein expression profiles or protein atlases of each tumor will serve to better detect, monitor, and stratify the clinical outcome risk of each specific cancer patient so that they may be benefit of tailored interventions based on the aggressiveness of their disease.

References

1. Pepe, M.S., Etzioni, R., Feng, Z., Potter, J.D., Thompson, M.L., Thornquist, M., Winget, M., Yasui, Y. (2001) Phases of biomarker development for early detection of cancer *J Natl Cancer Inst* **93**, 1054–61.
2. Haab, B.B., Dunham, M.J., Brown, P.O. (2001) Protein microarrays for highly parallel detection and quantitation of specific proteins and antibodies in complex solutions *Genome Biol* **2**, RESEARCH 0004.
3. Kingsmore, S.F. (2006) Multiplexed protein measurement: technologies and applications of protein and antibody arrays. *Nat Rev Drug Discov* **5**, 310–321.
4. Chan, S.M., Ermann, J., Su, L., Fathman, C.G., Utz, P.J. (2004) Protein microarrays for multiplex analysis of signal transduction pathways *Nat Med* **10,** 1390–1396.
5. Angenendt, P., Glökler, J., Murphy, D., Lehrach, H., Cahill, D.J. (2002) Toward optimized antibody microarrays: a comparison of current microarray support materials *Anal Biochem* **309**,253-60.
6. Kopf, E., Zharhary, D. (2007) Antibody arrays--an emerging tool in cancer proteomics *Int J Biochem Cell Biol* **39**, 1305–1317.
7. Sanchez-Carbayo, M. (2006) Antibody arrays: technical considerations and clinical applications in cancer *Clin Chem* **52**, 1651–1659.
8. Borrebaeck, C.A., Wingren, C. (2007) High-throughput proteomics using antibody microarrays: an update *Expert Rev Mol Diagn* 7, 673–686.
9. Wang, X., Yu, J., Sreekumar, A., Varambally, S., Shen, R., Giacherio, D., Mehra, R., Montie, J.E., Pienta, K.J., Sanda, M.G., Kantoff, P.W., Rubin, M.A., Wei, J.T., Ghosh, D., Chinnaiyan, A.M. (2005) Autoantibody signatures in prostate cancer *N Engl J Med* **353, 1224–1235**.
10. Anderson, K.S., LaBaer, J. (2005) The sentinel within: exploiting the immune system for cancer biomarkers *J Proteome Res* **4**, 1123–1133.
11. Nishizuka, S., Charboneau, L., Young, L., Major, S., Reinhold, W.C., Waltham, M., Kouros-Mehr, H., Bussey, K.J., Lee, J.K., Espina, V., Munson, P.J., Petricoin, E. 3rd., Liotta, L.A., Weinstein, J.N. (2003) Proteomic profiling of the NCI-60 cancer cell lines using new high-density reverse-phase lysate microarrays *Proc Natl Acad Sci USA* **100**, 14229–14234.
12. Petricoin, E.F. 3rd., Bichsel, V.E., Calvert, V.S., Espina, V., Winters, M., Young, L., Belluco, C., Trock, B.J., Lippman, M., Fishman, D.A., Sgroi, D.C., Munson, P.J., Esserman, L.J., Liotta, L.A. (2005) Mapping molecular networks using proteomics: a vision for patient-tailored combination therapy *J Clin Oncol.* **23**, 3614–3621.
13. Lash, G.E., Scaife, P.J., Innes, B.A., Otun, H.A., Robson, S.C., Searle, R.F., Bulmer, J.N. (2006) Comparison of three multiplex cytokine analysis systems: Luminex, SearchLight and FAST Quant *J Immunol Meth* **309**, 205–208.
14. de Jager, W., Rijkers, G.T.(2006) Solid-phase and bead-based cytokine immunoassay: a comparison *Methods* **38**, 294–303.
15. Waterboer, T., Sehr, P., Pawlita, M. (2006) Suppression of non-specific binding in serological Luminex assays *J Immunol Methods* **309**, 200–204.
16. Dotan, N., Altstock, R.T., Schwarz, M., Dukler, A. (2006) Anti-glycan antibodies as biomarkers for diagnosis and prognosis *Lupus* **15**, 442–450.
17. Chen, S., LaRoche, T., Hamelinck, D., Bergsma, D., Brenner, D., Simeone, D., Brand, R.E., Haab, B.B. (2007) Multiplexed analysis of glycan variation on native proteins captured by antibody microarrays *Nat Methods* **4**, 437–444.
18. Nettikadan, S., Radke, K., Johnson, J., Xu, J., Lynch, M., Mosher, C., Henderson, E. J. (2006) Detection and quantification of protein biomarkers from fewer than 10 cells *Mol Cell Proteomics* **5,** 895–901.
19. Zajac, A., Song, D., Qian, W., Zhukov, T. (2007) Protein microarrays and quantum dot probes for early cancer detection *Colloids Surf B Biointerfaces* **58**, 309–314.
20. Li, Y., Lee, H.J., Corn, R.M. (2007) Detection of protein biomarkers using RNA aptamer microarrays and enzymatically amplified surface plasmon resonance imaging *Anal Chem* **79**, 1082–1088.
21. Bock, C., Coleman, M., Collins, B., Davis, J., Foulds, G., Gold, L., Greef, C., Heil, J., Heilig, J.S., Hicke, B., Hurst, M.N., Husar, G.M., Miller, D., Ostroff, R., Petach, H., Schneider, D., Vant-Hull, B., Waugh, S., Weiss, A., Wilcox, S.K., Zichi, D. (2004) Photoaptamer arrays applied to multiplexed proteomic analysis *Proteomics* **4**, 609–618.
22. Gembitsky, D.S., Lawlor, K., Jacovina, A., Yaneva, M., Tempst, P. (2004) A prototype antibody microarray platform to monitor changes in protein tyrosine phosphorylation *Mol Cell Proteomics* **3**, 1102–1118.
23. Ivanov, S.S., Chung, A.S., Yuan, Z.L., Guan, Y.J., Sachs, K.V., Reichner, J.S., Chin, Y.E. (2004) Antibodies immobilized as arrays to

profile protein post-translational modifications in mammalian cells *Mol Cell Proteomics* **3**, 788–795.

24. Lin, Y., Huang, R., Cao, X., Wang, S.M., Shi, Q., Huang, R.P. (2003) Detection of multiple cytokines by protein arrays from cell lysate and tissue lysate *Clin Chem Lab Med* **41,** 139–145.
25. Vazquez-Martin, A., Colomer, R., Menendez, J.A. (2007) Protein array technology to detect HER2 (erbB-2)-induced 'cytokine signature' in breast cancer *Eur J Cancer* **43**, 1117–1124.
26. Lin, Y., Huang, R., Chen, L.P., Lisoukov, H., Lu, Z.H., Li, S., Wang, C.C., Huang, R.P. (2003) Profiling of cytokine expression by biotin-labeled-based protein arrays *Proteomics* **3**, 1750–1757.
27. Lin, Y., Huang, R., Chen, L., Li, S., Shi, Q., Jordan, C., Huang, R.P. (2004) Identification of interleukin-8 as estrogen receptor-regulated factor involved in breast cancer invasion and angiogenesis by protein arrays *Int J Cancer* **109**, 507–515.
28. Garcia, B.H. 2nd., Hargrave, A., Morgan, A., Kilmer, G., Hommema, E., Nahrahari, J., Webb, B., Wiese, R. (2007) Antibody microarray analysis of inflammatory mediator release by human leukemia T-cells and human non small cell lung cancer cells *J Biomol Tech* **18**, 245–251.
29. Nielsen, U.B., Cardone, M.H., Sinskey, A.J., MacBeath, G., Sorger, P.K. (2003) Profiling receptor tyrosine kinase activation by using Ab microarrays *Proc Natl Acad Sci USA* **100**, 9330–9335.
30. Belov, L., de la Vega, O., dos Remedios, C.G., Mulligan, S.P., Christopherson, R.I. (2001) Immunophenotyping of leukemia using a cluster of differentiation antibody microarray *Cancer Res* **61**, 4483.
31. Belov, L., Mulligan, S.P., Barber, N., Woolfson, A., Scott, M., Stoner, K., Chrisp, J.S., Sewell, W.A., Bradstock, K.F., Bendall, L., Pascovici, D.S., Thomas, M., Erber, W., Huang, P., Sartor, M., Young, G.A., Wiley, J.S., Juneja, S., Wierda, W.G., Green, A.R., Keating, M.J., Christopherson, R.I. (2006) Classification of human leukemias and lymphomas using extensive immunophenotypes from an antibody microarray *Brit J Haem* **135,** 184–197.
32. Kato, K., Ishimuro, T., Arima, Y., Hirata, I., Iwata, H. (2007) High-Throughput Immunophenotyping by Surface Plasmon Resonance Imaging *Anal Chem.* 79, 8616–23.
33. Yeretssian, G., Lecocq, M., Lebon, G., Hurst, H.C., Sakanyan, V. (2005) Competition on nitrocellulose-immobilized antibody arrays: from bacterial protein binding assay to protein profiling in breast cancer cells *Mol Cell Proteomics* **4**, 605–617.
34. Flores-Delgado, G., Liu, C.W., Sposto, R., Berndt, N. (2007) A limited screen for protein interactions reveals new roles for protein phosphatase 1 in cell cycle control and apoptosis *J Proteome Res* **6**, 1165–1175.
35. Bartling, B., Hofmann, H.S., Boettger, T., Hansen, G., Burdach, S., Silber, R.E., Simm, A.T. (2005) Comparative application of antibody and gene array for expression profiling in human squamous cell lung carcinoma *Lung Cancer* **49**, 145–154.
36. Hudelist, G., Pacher-Zavisin, M., Singer, C.F., Holper, T., Kubista, E., Schreiber, M., Manavi, M., Bilban, M., Czerwenka, K. (2004) Use of high-throughput protein array for profiling of differentially expressed proteins in normal and malignant breast tissue *Breast Cancer Res Treat* **86**, 281–291.
37. Ek, S., Andréasson, U., Hober, S., Kampf, C., Pontén, F., Uhlén, M., Merz, H., Borrebaeck, C.A. (2006) From gene expression analysis to tissue microarrays - a rational approach to identify therapeutic and diagnostic targets in lymphoid malignancies *Mol Cell Proteomics* **5**, 1072–1081.
38. Uhlén, M., Björling, E., Agaton, C., Szigyarto, C.A., Amini, B., Andersen, E., Andersson, A.C., Angelidou, P., Asplund, A., Asplund, C., Berglund, L., Bergström, K., Brumer, H., Cerjan, D., Ekström, M., Elobeid, A., Eriksson, C., Fagerberg, L., Falk, R., Fall, J., Forsberg, M., Björklund, M.G., Gumbel, K., Halimi, A., Hallin, I., Hamsten, C., Hansson, M., Hedhammar, M., Hercules, G., Kampf, C., Larsson, K., Lindskog, M., Lodewyckx, W., Lund, J., Lundeberg, J., Magnusson, K., Malm, E., Nilsson, P., Odling, J., Oksvold, P., Olsson, I., Oster, E., Ottosson, J., Paavilainen, L., Persson, A., Rimini, R., Rockberg, J., Runeson, M., Sivertsson, A., Sköllermo, A., Steen, J., Stenvall, M., Sterky, F., Strömberg, S., Sundberg, M., Tegel, H., Tourle, S., Wahlund, E., Waldén, A., Wan, J., Wernérus, H., Westberg, J., Wester, K., Wrethagen, U., Xu, L.L., Hober, S., Pontén, F. (2005) A human protein atlas for normal and cancer tissues based on antibody proteomics *Mol Cell Proteomics* **4**, 1920–1932.
39. Kellner, U., Steinert, R., Seibert, V., Heim, S., Kellner, A., Schulz, H.U., Roessner, A., Krüger, S., Reymond, M. (2004) Epithelial cell preparation for proteomic and transcriptomic analysis in human pancreatic tissue *Pathol Res Pract* **200**, 155–163.
40. Miller, J.C., Zhou, H., Kwekel, J., Cavallo, R., Burke, J., Butler, E.B., the, B.S., Haab, B.B. (2003) Antibody microarray profiling of human prostate cancer sera: antibody screening and identification of potential biomarkers *Proteomics* **3**, 56–63.
41. Schweitzer, B., Roberts, S., Grimwade, B., Shao, W., Wang, M., Fu, Q., Shu, Q., Laroche, I.,

Zhou, Z., Tchernev, V.T., Christiansen, J., Velleca, M., Kingsmore, S.F. (2002) Multiplexed protein profiling on microarrays by rolling-circle amplification *Nat Biotechnol* **20**, 359–365.

42. Zhou, H., Bouwman, K., Schotanus, M., Verweij, C., Marrero, J.A., Dillon, D., Costa, J., Lizardi, P., Haab, B.B. (2004) Two-color, rolling-circle amplification on antibody microarrays for sensitive, multiplexed serum-protein measurements *Genome Biol* **5**, R28.
43. Shao, W., Zhou, Z., Laroche, I., Lu, H., Zong, Q., Patel, D.D., Kingsmore, S., Piccoli, S.P. (2003) Optimization of Rolling-Circle Amplified Protein Microarrays for Multiplexed Protein Profiling *J Biomed Biotechnol* **5**, 299–307.
44. Saviranta, P., Okon, R., Brinker, A., Warashina, M., Eppinger, J., Geierstanger, B.H. (2004) Evaluating sandwich immunoassays in microarray format in terms of the ambient analyte regime *Clin Chem* **50**, 1907–1920.
45. Huang, R., Lin, Y., Shi, Q., Flowers, L., Ramachandran, S., Horowitz, I.R., Parthasarathy, S., Huang, R.P. (2004) Enhanced protein profiling arrays with ELISA-based amplification for high-throughput molecular changes of tumor patients' plasma *Clin Cancer Res* **10**, 598–609.
46. Varnum, S.M., Woodbury, R.L., Zangar, R.C. (2004) A protein microarray ELISA for screening biological fluids *Methods Mol Biol* **264,** 161–172.
47. Sanchez-Carbayo, M., Socci, N.D., Lozano, J.J., Haab, B.B., Cordon-Cardo, C. (2006) Profiling bladder cancer using targeted antibody arrays. *Am J Pathol* **168**, 93–103.
48. Celis, J.E., Gromov, P., Cabezón, T., Moreira, J.M., Ambartsumian, N., Sandelin, K., Rank, F., Gromova, I. (2004) Proteomic characterization of the interstitial fluid perfusing the breast tumor microenvironment: a novel resource for biomarker and therapeutic target discovery *Mol Cell Proteomics* **3**, 327–344.
49. Celis, J.E., Moreira, J.M., Cabezón, T., Gromov, P., Friis, E., Rank, F., Gromova, I. (2005) Identification of extracellular and intracellular signaling components of the mammary adipose tissue and its interstitial fluid in high risk breast cancer patients: toward dissecting the molecular circuitry of epithelial-adipocyte stromal cell interactions *Mol Cell Proteomics* **4**, 492–522.
50. Romeo, M.J., Espina, V., Lowenthal, M., Espina, B.H., Petricoin, E.F. 3rd., Liotta, L.A. (2005) CSF proteome: a protein repository for potential biomarker identification Expert Rev Proteomics **2**, 57–70.
51. Shafer, M.W., Mangold, L., Partin, A.W., Haab, B.B. (2007) Antibody array profiling reveals serum TSP-1 as a marker to distinguish benign from malignant prostatic disease *Prostate* **67**, 255–367.
52. Rai, A.J., Gelfand, C.A., Haywood, B.C., Warunek, D.J., Yi, J., Schuchard, M.D., Mehigh, R.J., Cockrill, S.L., Scott, G.B., Tammen, H., Schulz-Knappe, P., Speicher, D.W., Vitzthum, F., Haab, B.B., Siest, G., Chan, D.W. (2005) HUPO Plasma Proteome Project specimen collection and handling: towards the standardization of parameters for plasma proteome samples *Proteomics* **5**, 3262–3277.
53. States, D.J., Omenn, G.S., Blackwell, T.W., Fermin, D., Eng, J., Speicher, D.W., Hanash, S.M. (2006) Challenges in deriving high-confidence protein identifications from data gathered by a HUPO plasma proteome collaborative study *Nat Biotechnol* **24**, 333–8.
54. Haab, B.B. (2005) Antibody arrays in cancer research *Mol Cell Proteomics* **4**, 377–383.
55. Borrebaeck, C.A. (2006) Antibody microarray-based oncoproteomics. *Expert Opin Biol Ther* **6**, 833–838.
56. Sanchez-Carbayo, M. (2010) Antibody array-based technologies for cancer protein profiling and functional proteomic analyses using serum and tissue specimens *Tumour Biol* **31**, 103–12.

Chapter 12

Assessment of Antibody Specificity Using Suspension Bead Arrays

Jochen M. Schwenk and Peter Nilsson

Abstract

With the increasing collection of affinity reagents, their validation in terms of functionality and binding specificity becomes a challenge. To match this growing need, miniaturized and parallelized platforms have become available to corroborate the applicability for a broad range of binder scaffolds. Among the commonly used systems, planar microarrays have been a platform of choice for a long time but alternative systems are emerging, of which one is based on color-coded beads for the creation of arrays in suspension. The latter systems offer to perform a two-dimensional multiplexing by now analyzing up to 384 samples against up to 500 analytes in a single experiment. While the analyte parameter is flexible in terms of its composition, an extended screening can be facilitated without the need to set up a microarray production facility.

Key words: Suspension bead array, Antibody specificity, Cross-reactivity

1. Introduction

Challenges of the postgenomics era demand the development of elaborate methods that meet the requirements of improved sensitivity and throughput. Within these, the use of specific binding molecules to study target proteins is one possibility when examining expression patterns, subcellular localization, biochemical function, splice variants, and posttranslational modifications of proteins.

However, before any affinity reagent is applied to a downstream analysis platform its binding properties have to be assessed. Currently, differing tools such as Western blots are chosen to evaluate the binding characteristics, while more multiplexed methods such as protein arrays allow a parallel determination of binding specificities and help to judge cross-reactivity. One such method is performed on a planar protein microarray platform used for the specificity analysis of all generated antibodies antigen fragments (1),

Ulrike Korf (ed.), *Protein Microarrays: Methods and Protocols*, Methods in Molecular Biology, vol. 785,
DOI 10.1007/978-1-61779-286-1_12,

proteins (2), or peptides (3). These systems use an arraying device to create a two-dimensional arrangement of immobilized molecules on functionalized microscope slides. Assay results are displayed with the help of reporter dyes, a biochip reader system, and subsequent image analysis. An alternative platform for a parallelized and miniaturized analysis of proteins is offered by bead-based technologies (4). One available system is built on the principle to use spectrally distinguishable beads (5). A red and an infrared dye are incorporated at different ratios into these microspheres. This creates a set of now up to 500 beads of different color code signatures and those also allow utilizing coupled beads in a series of experiments. Mixtures of these beads are used to create arrays in suspension, and their composition can be adjusted individually to address a scientific question. A flow cytometer analyzes the co-occurrence of the color code and bead bound reporter dye to display bead assigned interactions. In the context of analyzing specificity and selectivity of antibodies, beads can be coupled with different proteins (6) or peptides (7), combined and mixed, and the binding molecules are added to the arrays to obtain a binding pattern. An overview of binding patterns from a number of binders is given in a heat map in Fig. 1. In the following, the terms "antibody" and "antigens" are used to represent any type of affinity reagent and respective target proteins.

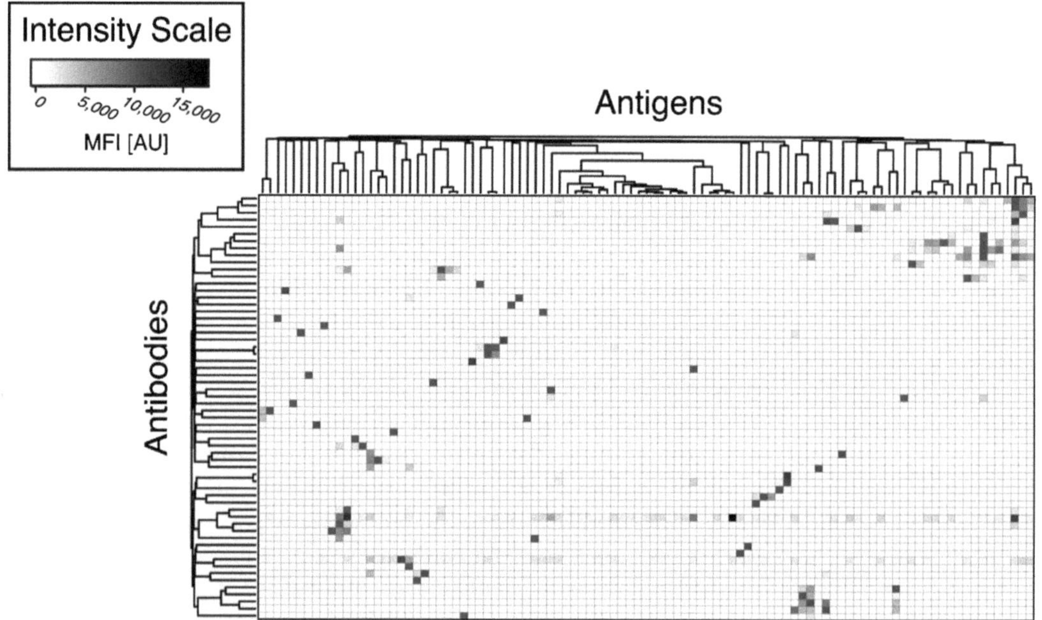

Fig. 1. Specificity analysis overview. A series of 60 antibodies was analyzed for specificity against up to 100 protein targets and the resulting intensity values were translated to a color scale and summarized as heat map. Clustering was performed based on Pearson's correlation in order to group antibodies and antigen's of similar binding and recognition pattern.

2. Materials

2.1. Bead Coupling

1. Beads: MagPlex or MicroPlex microspheres (Luminex Corp).
2. Activation buffer (1×): 100 mM monobasic sodium phosphate (Sigma), pH 6.2, store at +4°C for up to 3 months and at −20°C for long term.
3. EDC solution: Prepare aliquots of 1-ethyl-3-(3-dimethylaminopropyl) carbodiimide hydrochloride (EDC, Pierce) in screw-capped tubes and store at +4°C. Dissolve in activation buffer to 50 mg/ml directly prior usage.
4. S-NHS solution: Prepare sulfo-*N*-hydroxysuccinimide (NHS, Pierce) aliquots in screw-capped tubes and store at −20°C. Dissolve in activation buffer to 50 mg/ml directly prior usage.
5. Coupling buffer: 100 mM 2-(*N*-morpholino)ethanesulfonic acid (MES) pH 5.0, store at +4°C for up to 3 months and at −20°C for long term.
6. Wash buffer: 0.05% (v/v) Tween 20 in 1× PBS pH 7.4 (PBST).
7. Coupling validation solution: R-Phycoerythrin-modified antibodies directed against the tag of the antigen diluted in PBST (see Note 4).

2.2. Assay Procedure

1. Assay buffer (1×): 0.05% (v/v) Tween 20 in 1× PBS pH 7.4 (PBST).
2. Assay wash buffer (1×): PBST.
3. Detection solution: Dilute R-Phycoerythrin-modified anti-antibody reagent in assay buffer to 0.5 μg/ml directly before use and protect from light.

3. Methods

3.1. Bead Coupling

In the following, a method for protein coupling is described, for which magnetic and nonmagnetic beads can be utilized. The main difference between these two bead types is on how the beads are manipulated during an exchange of a surrounding liquid phase. For coupling batches not exceeding the number of positions found in common bench top microcentrifuges, we suggest to use 1.5-ml tubes or tubes with filter inserts to pellet the beads via centrifugation. For magnetic beads, magnetic forces are used to attract and to temporarily retain the particles. In both cases, the coupling of more than 24 bead IDs in parallel is preferably performed in microtiter plates. Hereby, proteins can be immobilized on nonmagnetic beads in filter bottomed microtiter plates (Millipore) with a filter pore sizes below bead diameter and vacuum devices (Millipore) accommodate these plates to remove liquid. For magnetic bead coupling in plates, dedicated plate magnets are available (LifeSept, Dexter Magnetic Technologies) to facilitate bead sedimentation and fixation. Depending on the future usage of the beads, they can be transferred to or kept in plates without filter bottom, or transferred to tubes. In the following, an example for coupling is given based on magnetic beads.

1. Prepare antigen at the desired concentration in coupling buffer. We suggest to use 10 μg or a solution of 100 μg/ml per 1×10^6 beads (see Note 1).
2. The beads are to be distributed in desired portions (e.g., 80 μl = 1×10^6 beads) into the wells of a microtiter plate and the beads are washed with 3× 100 μl activation buffer (see Note 2).
3. Prepare fresh solutions of NHS and EDC, both at 50 mg/ml in activation buffer. Calculate a use of 0.5 mg of each chemical per bead ID and coupling, and combine 10 μl NHS, 10 μl EDC, and 80 μl activation buffer for each well/ID.
4. Incubate 20 min under permanent, gentle shaking, and wash thereafter with 3× 100 μl coupling buffer.
5. Continue without interruption (see Note 3) by adding antigen solution to the activated beads and incubate for 2 h under permanent gentle shaking.
6. The beads are washed 3× with 100 μl wash buffer.
7. The beads are then recovered from the wells and transferred into microcentrifuge tubes using 3× 100 μl wash buffer. The liquid is removed and 100 μl storage buffer are added prior to the bead storage at +4°C in the dark for at least 1 h.

3.2. Bead Mixture Preparation

The yield of protein immobilized on beads can be judged after the coupling if the proteins do carry a tag to which affinity reagents are available. To allow an economic amount of beads to be utilized for the experiments and to combine equal numbers of beads in a bead mixture, the bead concentration obtained after coupling can be determined. This allows calculating the required volumes to be added in a common bead mixture solution.

1. The tubes with antigen-coupled beads are to be vortexed and sonicated for 5 min.
2. From each bead solution a 1/100 dilution is created in coupling validation solution (see Notes 4 and 5) directly in the wells of a microtiter plate.
3. The plates are incubated for 20 min and measured. Set the software to count all bead IDs over 80 s in a volume of 100 μl (see Note 6).
4. The number of counts per ID is multiplied by a correction factor of 3.3 for a 1/100 dilution to obtain a first estimation of

beads per μl storage solution. From this number the volumes of beads in storage solution can be calculated which are to be applied into the bead mixture that will match the number of experiments. The required number of beads supplied should be adjusted for each assay procedure and be based on the quantity of beads being counted by the instruments. We suggest to always obtain ≥32 counts per ID and to prepare the bead mixture for an extra 10% in terms of number of assays.

5. For the preparation of the next bead array mixtures, the number of counted beads is to be averaged from the previously measured wells. From each bead ID average, new volumes are calculated to adjust the number of beads that are to be included in the next bead array. We suggest to correct the volumes to a count 20% above the values set in the system software. In an example, where 100 beads were to be counted, the instrument had actually determined a count of 60 after the assay. A new volume of this bead ID is then calculated with theoretical count of 120, leading to a new volume of this beads ID which is 2× of the initial volume.

3.3. Assay Procedure

3.3.1. No-Wash Procedure

1. The antibodies are diluted in assay buffer and 45 μl of these solutions are added to the wells that have previously been loaded with 5 μl of bead mixture (see Note 7).
2. The mixtures are then incubated for 60 min at 23°C under permanent shaking on a microtiter plate mixer.
3. Of the detection solution 50 μl are added and incubated for another 60 min at 23°C under permanent shaking (see Note 7) and the plates are then measured with the Luminex instrumentation.
4. Select the utilized bead IDs in the system software and count at least 100 beads per ID. We suggest using the median fluorescence intensity for further data processing.

3.3.2. Wash Procedure

1. The antibodies are diluted in assay buffer and 45 μl of these solutions are added to the wells that have previously been loaded with 5 μl of bead mixture (see Note 7).
2. The antibodies are then incubated for 60 min at 23°C under permanent shaking on a microtiter plate mixer.
3. The plates are then washed 3× with 75 μl wash buffer.
4. Of the detection solution 50 μl are added and incubated for another 60 min at 23°C under permanent shaking (see Note 7).
5. The plates are then finally washed 3× with 75 μl wash buffer and 100 μl of wash buffer are added before the plates are measured with the Luminex instrumentation.
6. Select the utilized bead IDs in system software and count at least 100 beads per ID. We suggest using the median fluorescence intensity for further data processing.

4. Notes

1. Employ solutions of purified proteins and avoid other stabilizing proteins, Tris, or other amine-based buffers as they reduce the coupling efficiency.
2. At all times, try to minimize the light exposure, especially to direct sunlight, as the internal fluorescence of the beads as well as reporter fluorophores could be bleached. During incubation, protect the plates with an opaque cover or place plate into a light-tight box.
3. Do not interrupt the activation process after dissolving EDC and NHS, as theses active substances are susceptible to hydrolysis resulting in reduced activity.
4. When combining beads with solutions for counting and assay procedure, always distribute small volume bead solution (e.g., 5 μl) into the well first, then add larger volume buffer portion (e.g., 45 μl) to allow an instant distribution of the beads.
5. Other fluorescent dyes such as Alexa555, Alexa532, or Cy3 can be utilized as well but have been shown to yield lower signal intensities.
6. If you experience aggregation of beads, vortex the beads followed by a more extended sonication for a longer period of time. If aggregations persist, rub the tubes on the inner wall of the sonication bath and stay above the sonication liquid. Safety measures regarding the handling of sonication baths are to be observed!
7. The amount of the antibody and anti-antibody detection reagent depends on properties such as purity, affinity, detectability via a secondary reagent, and antigen accessibility. Therefore, both should be tested and adjusted in terms of their concentrations and incubation times to achieve a significant intensity level separation between signal and background.

Acknowledgments

We are grateful to the entire staff within the Human Protein Atlas project for their tremendous efforts. This work is supported by the PRONOVA project (VINNOVA, Swedish Governmental Agency for Innovation Systems), and by grants from the Knut and Alice Wallenberg Foundation.

References

1. Nilsson, P., Paavilainen, L., Larsson, K., Odling, J., Sundberg, M., Andersson, A. C., Kampf, C., Persson, A., Al-Khalili Szigyarto, C., Ottosson, J., Bjorling, E., Hober, S., Wernerus, H., Wester, K., Ponten, F., and Uhlen, M. (2005) Towards a human proteome atlas: high-throughput generation of mono-specific antibodies for tissue profiling, *Proteomics 5*, 4327–4337.
2. Michaud, G. A., Salcius, M., Zhou, F., Bangham, R., Bonin, J., Guo, H., Snyder, M., Predki, P. F., and Schweitzer, B. I. (2003) Analyzing antibody specificity with whole proteome microarrays, *Nat Biotechnol 21*, 1509–1512.
3. Poetz, O., Ostendorp, R., Brocks, B., Schwenk, J. M., Stoll, D., Joos, T. O., and Templin, M. F. (2005) Protein microarrays for antibody profiling: specificity and affinity determination on a chip, *Proteomics 5*, 2402–2411.
4. Templin, M. F., Stoll, D., Schwenk, J. M., Potz, O., Kramer, S., and Joos, T. O. (2003) Protein microarrays: promising tools for proteomic research, *Proteomics 3*, 2155–2166.
5. Fulton, R. J., McDade, R. L., Smith, P. L., Kienker, L. J., and Kettman, J. R., Jr. (1997) Advanced multiplexed analysis with the FlowMetrix system, *Clin Chem 43*, 1749–1756.
6. Schwenk, J. M., Lindberg, J., Sundberg, M., Uhlen, M., and Nilsson, P. (2007) Determination of Binding Specificities in Highly Multiplexed Bead-based Assays for Antibody Proteomics, *Mol Cell Proteomics 6*, 125–132.
7. Larsson, K., Eriksson, C., Schwenk, J. M., Berglund, L., Wester, K., Uhlen, M., Hober, S., and Wernerus, H. (2009) Characterization of PrEST-based antibodies towards human Cytokeratin-17, *J Immunol Methods 342*, 20–32.

Chapter 13

Quantitative Analysis of Phosphoproteins Using Microspot Immunoassays

Frauke Henjes, Frank Götschel, Anika Jöcker, and Ulrike Korf

Abstract

Protein microarrays are an ideal technology platform which allow for a robust and standardized profiling of the cellular proteome. Many cellular functions are not simply controlled by the presence of certain proteins, especially the propagation of external stimuli, which depend on transient post-translational modifications that determine whether a protein is in its active or inactive state. Thus, complex biological processes require the availability of a sound set of quantitative and time-resolved measurements to be understood. For this reason, new assay platforms which allow for the investigation of several proteins in parallel are necessary. The current best understood mode of cellular regulation occurs via phosphorylation and dephosphorylation processes, which are mediated via a large panel of kinases and phosphatases. The microspot immunoassay technique described here allows for an exact determination of several different phosphorylated proteins in parallel, as well as from small sample amounts, and is therefore an appropriate system to deepen the understanding of the complex regulatory networks implicated in health and disease.

Key words: Protein microarray, Protein quantification, Immunoassay, Phosphoprotein, Phosphoproteomics, Kinase, Phosphatase, Signal transduction, QuantProReloaded

1. Introduction

Understanding signaling networks at the systems level still remains a challenge, and requires the development of robust and accurate methods for the quantification of changes on the genomic, transcriptomic, and proteomic level, as well as for protein activation. With the invention of DNA microarrays, high-throughput screening of biological samples on the transcript level became possible. This technique was a first step toward obtaining a more global view of biological systems, and is nowadays accepted as a standard method to analyze the complete transcriptome of a cell in a single experiment. However, transcript profiling is only of limited use when the behavior of biological systems needs to be predicted.

Ulrike Korf (ed.), *Protein Microarrays: Methods and Protocols*, Methods in Molecular Biology, vol. 785, DOI 10.1007/978-1-61779-286-1_13, © Springer Science+Business Media, LLC 2011

First and most especially, the half-life of key regulators in signaling is short and tightly regulated, and as a result, the abundance of signaling proteins rarely shows a direct correlation with the expression level of the corresponding transcripts. Secondly, the fine tuning of cellular signal transduction processes is regulated via transient, but highly specific post-translational modifications of proteins, and thus, provides an additional level of regulation to control protein function. Modifications like acetylation, glycosylation, and also phosphorylation are very important means to regulate protein–protein interactions, proteolysis, and signal transduction, and most essential cellular processes, such as differentiation and proliferation, are regulated this way. Phosphorylation and dephosphorylation mediated by kinases and phosphatases are a common biological mechanism to regulate the activation of a broad spectrum of signaling pathways. A deregulation of the sophisticated balance between phospho-versus-dephospho state is considered as a hallmark for the development of serious diseases like cancer (1).

As a result, methods for high-throughput screening of proteins were required to gain insights into the tight network of biological regulation (2, 3). During the last few years, the field of protein microarrays has made tremendous progress, and this technique has become a central proteomics tool for basic research, as well as for commercial applications. The advantage of microarray formats offers a reliable option to obtain a large amount of high-quality data, while consuming only minimal amounts of sample material. Thus, they are a fast and efficient tool for systems biology-type data generation, as well as for monitoring disease-associated changes of the proteome in clinical samples.

The use of quantitative microspot immunoassays for phosphoproteomics may therefore provide a direct approach to improving our understanding of cellular responses at the pathway level (4, 5). In general, the microspot immunoassay format (Fig. 1) works like a miniaturized sandwich ELISA (enzyme-linked immunosorbent assay), and requires two antibodies directed against different epitopes of the same protein. Due to the two-antibody approach, sensitivity and specificity of the assay are increased, and the detection of low-abundance proteins, or modifications, is also feasible. However, in contrast to the traditional ELISA technique, only a small amount of capture antibodies are immobilized to a solid surface. This reduces the amount of capture agent required for the analysis, and also prevents target protein depletion (6). A second advantage of the microarray format is the possibility to multiplex the detection of different target proteins.

In greater detail, the quantification of phosphoproteins works best by capturing proteins via their phosphorylation sites whereas the detection antibody is often polyclonal, and recognizes a different epitope. Nevertheless, it might be necessary to switch the order of the antibody combination, depending on the antibody properties

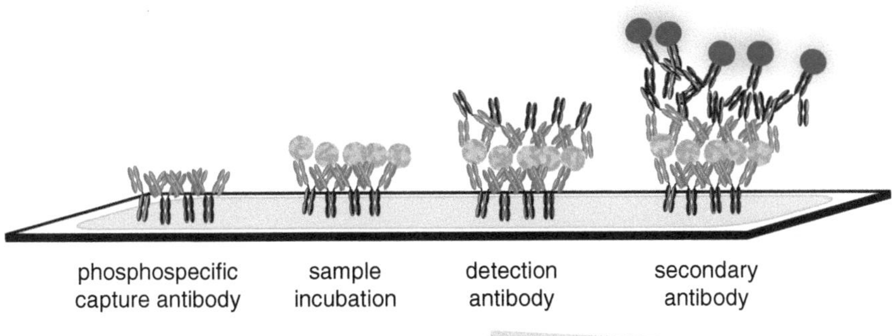

Fig. 1. Schematic presentation of sandwich antibody arrays.

and availability. The exact quantification of phosphoproteins is a special feature of the approach described in this chapter and requires that the phosphorylation rate is determined by mass spectrometry. A dilution series containing defined amounts of the different phosphorylated proteins is employed to generate phosphoprotein-specific calibration curves. Phosphoproteins are available either from commercial sources, or can be produced by in vitro phosphorylation of recombinantly expressed and purified proteins. Data processing is performed with a tailored software program (e.g., QuantProReloaded) (7). Measurements from the standard dilution curve serve as a reference to quantify the concentration of the phosphoproteins of interest in the samples.

In sum, microspot immunoassays allow a robust and standardized quantification of the cellular phosphoproteome, and are therefore, an appropriate system for examining complex regulatory interactions, thus providing a solid basis for molecular research, systems biology and clinical diagnostics (8, 9).

2. Materials

2.1. Slides, Antibodies

1. Aushon 2470 arrayer (Aushon Biosystems, Billerica, MA, USA), or comparable printer (see Note 1).
2. Nitrocellulose-coated glass slides (Oncyte Nitrocellulose Film Slides, Grace Bio-Labs, Blend, OR, USA) (see Note 2).
3. 384-Well plates with lids (Cat# AB-1056, Thermo Scientific).
4. Mouse monoclonal phosphospecific antibodies as capture antibodies (see Notes 4 and 5).
5. Rabbit antibodies for the detection step (see Notes 4 and 5).
6. Alexa 680-labeled secondary anti-rabbit antibody (Invitrogen) (see Note 12).

2.2. Cell Culture Experiments and Lysis

1. In vitro-generated samples from, e.g., dynamic measurements.
2. Cell lysis buffer: M-PER (Pierce), or a comparable lysis buffer.
3. PhosStop (phosphatase inhibitor, Roche).
4. Complete Mini (protease inhibitor, Roche).
5. Cell scraper.
6. Dry ice.

2.3. Microspot Immunoassay

1. Phosphate-buffered saline (PBS) was prepared from a 10× stock. The 10× stock was prepared as follows:
   - 1.37 M NaCl
   - 27 mM KCl
   - 18 mM KH_2PO_4
   - 100 mM Na_2HPO_4

   PBS was stored at room temperature.
2. 1% Tween 20 stock solution in PBS, stored at room temperature.
3. 10% (w/v) SDS stock solution was stored at room temperature.
4. Assay buffer to dissolve antibodies: 1% (w/v) BSA, 0.5% (w/v) NP-40, 0.02% (w/v) SDS, 50 mM Tris–HCl, pH 7.4, 150 mM NaCl, 1 mM EDTA, 5 mM NaF, 1 mM vanadate. Add one complete mini tablet per 10 ml assay buffer. Prepare freshly each day.
5. Blocking buffer to prevent unspecific binding of proteins to the nitrocellulose surface: 5% (w/v) milk, 0.5% (w/v) NP-40, 50 mM Tris–HCl, pH 7.4, 150 mM, 1 mM EDTA. Can be stored in the refrigerator.
6. Washing buffer to remove unbound sample and reagents from the wells: 0.1% (w/v) Tween-20/PBS.
7. Recombinant phosphoproteins are derived either from commercial sources, or are prepared in-house. Protein phosphorylation grade has to be calibrated by mass spectrometry for each new batch of phosphoprotein.
8. Incubation chamber for 16-pad slides, e.g., 3/16, Metecon, Mannheim, Germany. This chamber holds three slides in parallel, so that 30 samples can be processed in parallel (see Note 3).

2.4. Scanning, Data Analysis, and Data Presentation

1. Odyssey Infrared Imaging System (LI-COR Bioscience, Lincoln, Nebraska, USA) or suitable microarray scanner.
2. GenePix Pro Software:

   http://www.moleculardevices.com/pages/software/gn_genepix_pro.html
3. QuantProReloaded:

   http://code.google.com/p/quantproreloaded/

3. Methods

3.1. Printing Slides

Phosphospecific antibodies are printed using the *Aushon 2470* solid-pin tool arrayer to deliver capture agents onto nitrocellulose-coated glass slides. Sixteen identical subarrays of capture antibodies are printed onto each slide. The printhead conformation was set to 2×4 pins with 9×9-μm spacing.

1. Antibodies are diluted with PBS to a concentration matching 0.2–0.5 μg/μl.
2. Add Tween-20 to a final concentration of 0.05% (w/v).
3. 3–5 μl of antibody dilution are transferred to a 384-well plate. Using the Aushon spotter, a single extraction corresponds to wells A1, A3, A5, A7, C1, C3, C5, and C7, and so on. In each of the eight corresponding wells, the same antibody is deposited.
4. Centrifuge 384-well plate for 1 min at 700×*g* to collect dissolved capture antibodies at the bottom of the wells.
5. Place the lid on top of the 384-well plate, and load this as source plate into the elevator of the Aushon instrument.
6. Fix slides onto slide platens, and place the platens into the platen elevator.
7. Start the software, and create a suitable layout.
8. Set top offset to 6,200 mm, and left offset to 3,150 mm.
9. Set superarray replicates to 2.
10. Choose the number of replicates: five for six replicates, linear (horizontal).
11. Set humidity to 80%.
12. After checking all doors and containers, start print run.

3.2. Time-Resolved Measurement in Cell Culture Samples

1. Cells are seeded in plates, and grown to 70–80% confluence.
2. Cells are serum-starved prior to stimulation.
3. To collect time-point samples, medium is removed and replaced by ice-cold PBS to stop all ongoing enzymatic processes.
4. Subsequently, the PBS is aspirated, and ice-cold lysis buffer is added to the plate (see Note 8).
5. Cells are scraped on ice with a cell scraper, transferred to a 1.5-ml reaction tube, and placed on dry ice.
6. After collecting all samples of a time series, samples are thawed, lysed for 20 min on an end-over-end shaker at 4°C, and centrifuged for 5 min at 16,000×*g*.
7. As a last step, the total protein concentration is determined, and lysates are diluted to adjust equal protein concentration.

3.3. Microspot Immunoassay

1. Add 10% SDS to a final concentration of 1% to lysates, and stock solutions containing standard phosphoproteins (see Note 9).
2. Heat all samples for 5 min at 95°C (see Note 9).
3. Place the samples on ice, and centrifuge briefly.
4. Dilute the recombinant proteins with assay buffer to generate a six-step serial dilution of the phosphoprotein standard (see Note 7).
5. Dilute the lysates according to the protein of interest with assay buffer to a final concentration of 10–50 μg/ml.
6. Block slides with blocking buffer (see Note 10).
7. Mount slides in a suitable frame to create an incubation chamber for each subarray.
8. Pipette 150 μl washing buffer per well with a multichannel pipette.
9. Aspirate the washing buffer completely from the first six wells, and pipette 200 μl of the standard proteins into the wells (see Note 11).
10. Aspirate the washing buffer from the remaining ten wells, and add 200 μl of the diluted lysates to a different pre-assigned well (see Note 11).
11. Incubate the slides overnight at 4°C on a rocking platform.
12. Aspirate the samples from all wells, and fill 150 μl assay buffer into the wells using a multichannel pipette.
13. Wash the slides three times for 5 min on a rocking platform.
14. Aspirate the washing buffer, and remove the slides from the incubation chamber.
15. Place the slides into a dark box, and wash again for 5 min with assay buffer.
16. Dilute detection antibodies 1:400 to 1:1,500 in assay buffer, and incubate slides for 2 h at 4°C on a rocking platform.
17. Slides are washed four times, as described before.
18. Dilute secondary antibody 1:5,000 in assay buffer, and incubate for 30 min at 4°C on a rocking platform that is protected from light.
19. Wash slides four times with washing buffer, and once with ddH_2O protected from light.
20. Dry the slides for 15 min in a slide chamber lined with soft kim wipe paper.

3.4. Scanning and Data Analysis

1. Slides are scanned with the Odyssey Infrared Imaging System at a resolution of 21 μm at 700 nm.

2. The images are quantified using the GenePix Software. The results are saved as .gpr files.
3. The analysis is performed with the QuantProReloaded software tool. This software was developed for the analysis of microspot immunoassays, and fits standard curves of calibrator measurements that determine the concentrations of the different phosphoproteins.

3.5. Data Analysis and Data Presentation

1. The analysis is done via the open source software tool QuantProReloaded, which is written in R and Java. This software tool was developed for the quantification of microspot immunoassay samples. Calibration curves are fitted via the raw standard data (also called calibrator spots); based on these curves, the concentration of the target phosphoproteins is estimated within each spot.
   (a) Start the QuantProReloaded software tool.
   (b) Calculate calibrator curves using the task "Performance Analysis" which offers different options. Select a regression function which matches best the calibrator data points, and start the analysis. In the output window, the *Performance Plot* is shown (see upper-right plot in Fig. 2); this plot illustrates how accurate the concentration of the target protein within each calibrator spot can be estimated from the calibrator curve. A value of 1 indicates that the mean-estimated concentration is identical with the real concentration. The error bars indicate the standard deviations of these values. Make sure that the accuracy approximates 1 as best as possible, within the measurement range of the expected target protein concentration. Start the test with a linear regression function, and continue with nonlinear regression functions with a low number of parameters; choose the curve with the best performance for the final analysis (see Note 13).
   (c) Estimate the concentration of the phosphoproteins by starting a new "Measurement Analysis" in QuantProReloaded.
   (d) Select a certain antibody pair. Apply the linear or the nonlinear regression function identified from Subheading 3.5, step 1b, and choose "time-course analysis." Furthermore, define a directory where the result plots will be stored. Click the Submit button, and wait until the analysis is finished. For every antibody pair, a new result window will open up. This allows a direct comparison between different antibody pairs.
   (e) Control your analyses using the *Calibrator Panel* tab, to ensure that the estimated readout is within a valid calibration range.

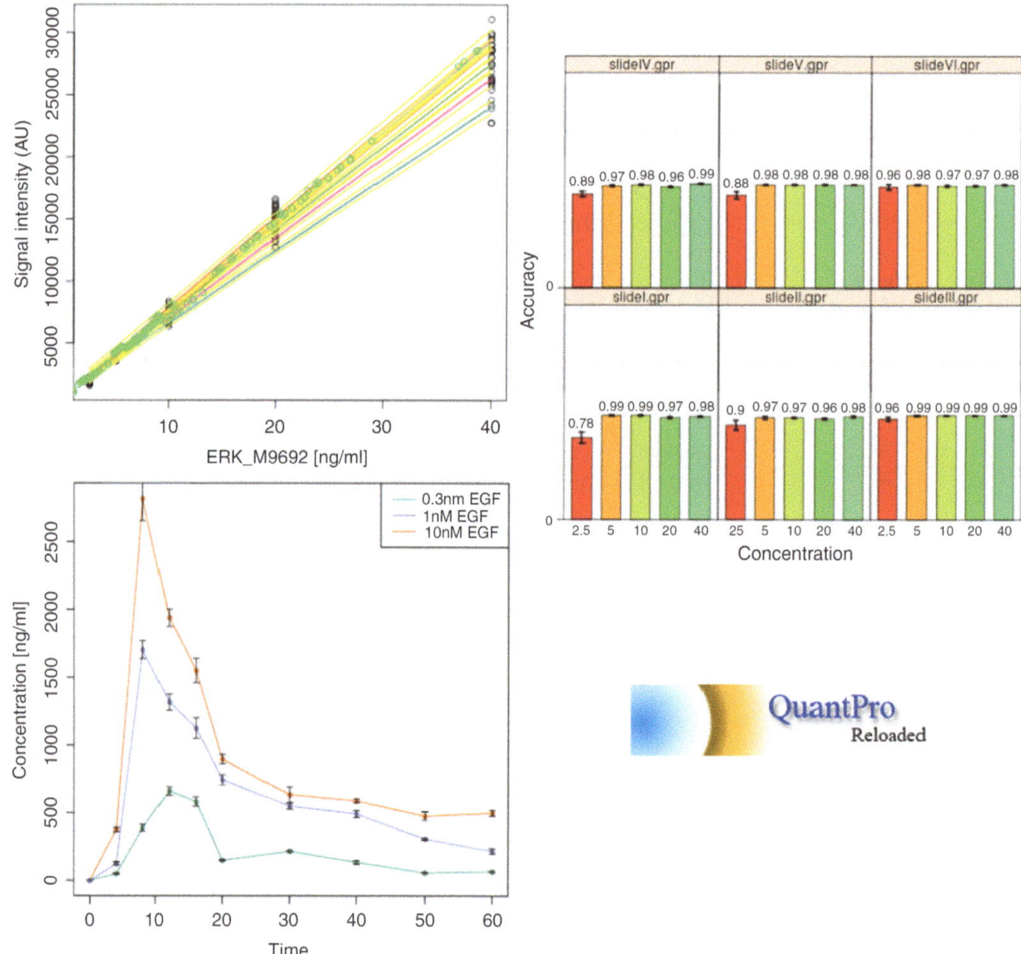

Fig. 2. Output of QuantproReloaded. *Upper-left side*: Calibrator plot. *Black points* are the experimentally measured calibrator data points used for the calculation. *Colored lines* show the calculated calibrator curves. In this case the analysis was done for each slide separately, therefore, there are six calibrator lines shown. *Green points* indicate the estimated concentration within each spot of the target phosphoproteins (diluted concentration is shown). *Lower-left side*: Time-course plot. Each line shows the median changes of phospho-ERK (target protein) under one condition at different time points. In this example case, the phospho-ERK concentration is maximal after 8 min of EGF treatment. Increased concentrations of EGF ligand lead to elevated concentration levels of phospho-ERK at all time points. *Upper-right side*: Performance plot. For each slide and each concentration, the accuracy is shown. The accuracy indicates how accurate the concentration of the protein within each calibrator spot can be estimated, given the calibrator curve. In this case, the accuracy is very high at all concentrations (>0.9), and therefore, the estimation of the target protein concentration should also be accurate.

(f) In the *Time Panel* tab, the time-course plot is visualized (see the lower-left plot in Fig. 2); this facilitates a direct comparison between the median changes of a certain target protein under different conditions, at different time points. This plot allows the identification of differences in phosphoprotein abundance for certain conditions.

4. Notes

1. Two principally different spotting techniques are available: contact and noncontact (10). Noncontact, or piezo spotters, take up a large volume of several μl, and disperse small drops of a defined volume in the pl–nl range. This method is very accurate, but a considerable volume of antibody dilution is wasted as part of the uptake step. The accuracy of a contact spotter depends on the consistency between single pins and consumes only the amount of sample transferred to slides.
2. 16-Pad slides often show a high variability of thickness between different pads, resulting in an apparently in-homogeneous dispersion of capture antibodies. Consequently, all spots corresponding to a certain pad appear as "outliers" of the calibration slopes, or in measurements. Thus, the coating of slides with a 21-mm × 71-mm nitrocellulose is more consistent, and can better be used, instead of 16-pad slides.
3. In collaboration with Metecon, a reusable 16-pad slide incubation chamber was developed, and tailored to subdivide 21-mm × 71-mm coated nitrocellulose slides into a 16-well surface. Several companies also distribute 16-pad slides incubation chambers (e.g., FAST Frame, Whatman). However, any multiwell incubation chamber must be tested to ensure that the seal is hermetic.
4. Phosphoepitopes can be detected via the capture, or the detection antibody step. In our experience, a phosphospecific capture step leads to more consistent results. Detection with a polyclonal antibody results in most cases in higher signal intensities.
5. In a sandwich approach, it is crucial to use two antibodies from different species; otherwise, the labeled secondary antibody will directly bind to the immobilized capture antibody. Another option is the use of biotinylated detection antibodies.
6. The efficiency weakly binding or highly diluted capture antibodies, or the capturing of lowly expressed proteins, can be improved by two or more depositions per feature, which doubles the amount of antibody per spot.
7. To ensure the exact quantification of phosphoproteins, the phosphorylation degree has to be determined by mass spectrometry (11). As phosphoprotein resource, different options can be explored.
   (a) Numerous proteins are available in their phosphorylated form from companies.
   (b) Proteins available in the unphosphorylated form can be phosphorylated in vitro.

   (c) Proteins can be expressed, purified, and phosphorylated in vitro using standard biochemistry tools (12).

8. Phosphatase and protease inhibitors are very important additions to the lysis buffer. When measuring the phosphorylation state at distinct time points, it is important to prevent degradation, and dephosphorylation events during sample processing (13). Additionally, the promiscuous kinase inhibitor staurosporine can be added to prevent phosphorylation.
9. Instead of boiling the proteins with SDS, the microspot immunoassay can be performed with native proteins. The best working sample preparation has to be investigated for each individual protein.
10. Some antibodies reveal considerable background fluorescence when used in the capture step. Instead of milk, blocking buffers designed to reduce fluorescence (Odyssey Blocking Buffer, Rockland) can be tested to optimize the background of a single antibody.
11. The most crucial step for the subsequent quantitative readout is the incubation with standard protein and samples. Thus, it is important to remove the washing buffer completely from the wells by careful aspiration. Small volumes of washing buffer leftovers can affect the measurement, since the measurements are performed in a small volume of only 100–200 μl.
12. A large variety of different fluorescently labeled secondary antibodies are commercially available. It is necessary to test the cross-reactivity between secondary antibodies, and capture antibodies. Occasionally, fluorescent dyes cross-react unspecifically with antibodies; this is probably due to the charged groups of the dye.
13. If the variance of the replicates increases with the concentration log, the values before the calibration curve are fitted (use parameter "log values" in QuantProReloaded).

Acknowledgments

The authors would like to acknowledge S. Schumacher, M. Wosch, C. Becki, and C. Schmidt for their excellent technical support in the development of technically robust protocols. This work was supported by the German Federal Ministry for Education and Science in the framework of the Program for Medical Genome Research (grants 01GS0890 and 01GS0864), the Program for Medical Systems Biology (grant 0315396B), as well as the Helmholtz Systems Biology Initiative (SBCancer).

References

1. Haab, B. B. (2005) Antibody arrays in cancer research, *Mol Cell Proteomics 4*, 377–383.
2. Stoevesandt, O., Taussig, M. J., and He, M. (2009), Protein microarrays: high-throughput tools for proteomics, *Expert Rev Proteomics 6*, 145–157.
3. Wingren, C., and Borrebaeck, C. A. (2009), Antibody-based microarrays, *Methods Mol Biol 509*, 57–84.
4. Korf, U., Henjes, F., Schmidt, C., Tresch, A., Mannsperger, H., Lobke, C., Beissbarth, T., and Poustka, A. (2008), Antibody microarrays as an experimental platform for the analysis of signal transduction networks, *Adv Biochem Eng Biotechnol 110*, 153–175.
5. VanMeter, A., Signore, M., Pierobon, M., Espina, V., Liotta, L. A., and Petricoin, E. F., 3 rd. (2007), Reverse-phase protein microarrays: application to biomarker discovery and translational medicine, *Expert Rev Mol Diagn 7*, 625–633.
6. Ekins, R. J. (1989), Pharmaceutical & biomedical analysis, *2*, 155–168.
7. Joecker, A., Sonntag, J., Henjes, F., Goetschel, F., Tresch, A., Beissbarth, T., Wiemann, S., and Korf, U. (2010), QuantProReloaded: Quantitative analysis of Microspot Immunoassays, *Bioinformatics 26*, 2480–1.
8. Hartmann, M., Roeraade, J., Stoll, D., Templin, M. F., and Joos, T. O. (2009), Protein microarrays for diagnostic assays, *Anal Bioanal Chem 393*, 1407–1416.
9. Yu, X., Schneiderhan-Marra, N., and Joos, T. O. (2010) Protein microarrays for personalized medicine, *Clin Chem 56*, 376–387.
10. Barbulovic-Nad, I., Lucente, M., Sun, Y., Zhang, M., Wheeler, A. R., and Bussmann, M. (2006), Bio-microarray fabrication techniques, *Crit Rev Biotechnol 26*, 237–259.
11. Korf, U., Derdak, S., Tresch, A., Henjes, F., Schumacher, S., Schmidt, C., Hahn, B., Lehmann, W. D., Poustka, A., Beissbarth, T., and Klingmuller, U. (2008), Quantitative protein microarrays for time-resolved measurements of protein phosphorylation, *Proteomics 8*, 4603–4612.
12. Korf, U., Kohl, T., van der Zandt, H., Zahn, R., Schleeger, S., Ueberle, B., Wandschneider, S., Bechtel, S., Schnolzer, M., Ottleben, H., Wiemann, S., and Poustka, A. (2005), Large-scale protein expression for proteome research, *Proteomics 5*, 3571–3580.
13. Thingholm, T. E., Jensen, O. N., and Larsen, M. R. (2009), Analytical strategies for phosphoproteomics, *Proteomics 9*, 1451–1468.

Chapter 14

Robust Protein Profiling with Complex Antibody Microarrays in a Dual-Colour Mode

Christoph Schröder, Mohamed S.S. Alhamdani, Kurt Fellenberg, Andrea Bauer, Anette Jacob, and Jörg D. Hoheisel

Abstract

Antibody microarrays are a multiplexing technique for the analyses of hundreds of different analytes in parallel from small sample volumes of few microlitres only. With sensitivities in the picomolar to femtomolar range, they are gaining importance in proteomic analyses. These sensitivities can be obtained for complex protein samples without any pre-fractionation or signal amplification. Also, no expensive or elaborate protein depletion steps are needed. As with custom DNA-microarrays, the implementation of a dual-colour assay adds to assay robustness and reproducibility and was therefore a focus of our technical implementation. In order to perform antibody microarray experiments for large sets of samples and analytes in a robust manner, it was essential to optimise the experimental layout, the protein extraction, labelling and incubation as well as data processing steps. Here, we present our current protocol, which is used for the simultaneous analysis of the abundance of more than 800 proteins in plasma, urine, and tissue samples.

Key words: Antibody microarray, Proteomic profiling, Dual-colour analysis, direct sample labelling, Plasma profiling, Multiplexed immunoassay

1. Introduction

Antibody microarrays represent a relatively new technology in proteomics, which facilitates the analyses of hundreds of analytes in a parallel manner. Only small sample volumes are required (1–3). In the last decade, they have gained importance due to their advantageous combination of multiplexing capacity and very high sensitivities of low femtomolar range, even without signal amplification (4–6). Therefore, even for complex protein samples no expensive and elaborate protein pre-fractionation or protein depletion steps are needed (7). As with DNA microarrays, assay robustness and reproducibility could be improved dramatically by the

Ulrike Korf (ed.), *Protein Microarrays: Methods and Protocols*, Methods in Molecular Biology, vol. 785, DOI 10.1007/978-1-61779-286-1_14,

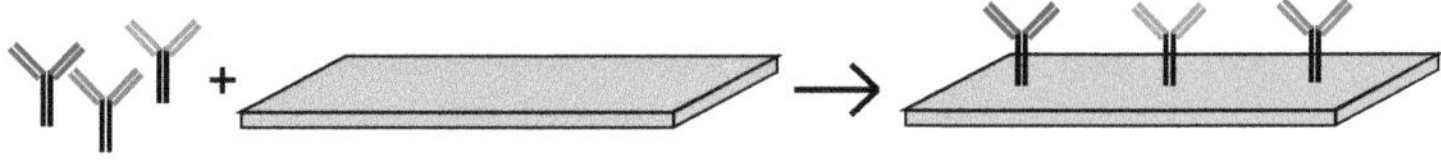

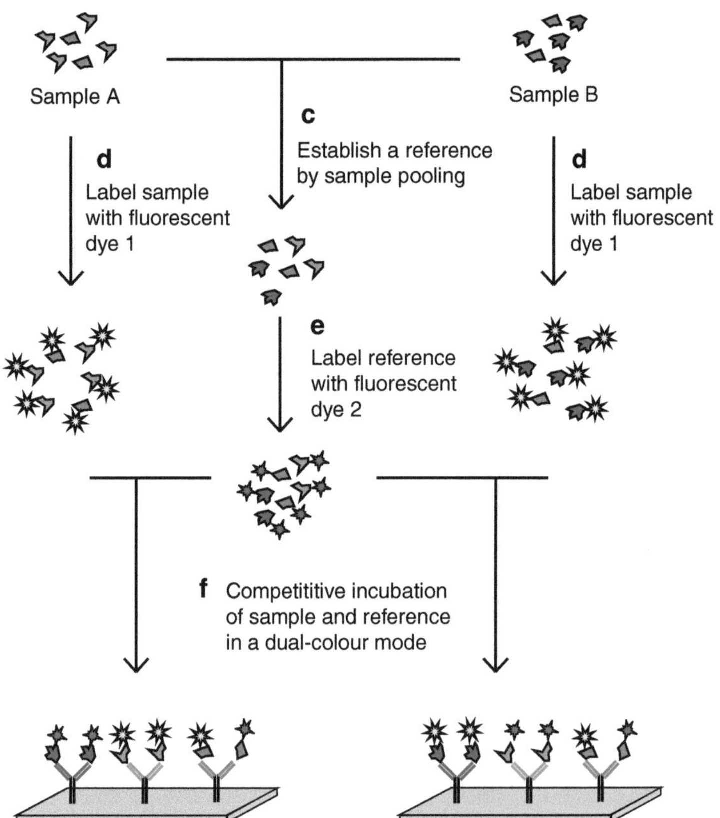

Fig. 1. Schematic representation of an antibody microarray experiment in a dual-colour mode and with a reference-based design.

implementation of a dual-colour assay (7, 8), which is therefore a standard element of our current protocol.

Antibody microarray experiments comprise five major steps: array production (Subheading 3.1), protein extraction (Subheading 3.2), sample labelling (Subheading 3.3), incubation (Subheading 3.4), and finally image acquisition and data analysis (Subheading 3.5). For array production (Fig. 1a), a set of different antibodies is immobilised at distinct locations on a planar surface. Protein samples are extracted from different sources such as plasma, serum, urine, tissue, or cell culture (Fig. 1b). Subsequently, there are different experimental design options for dual-colour assays. For a direct comparison, two

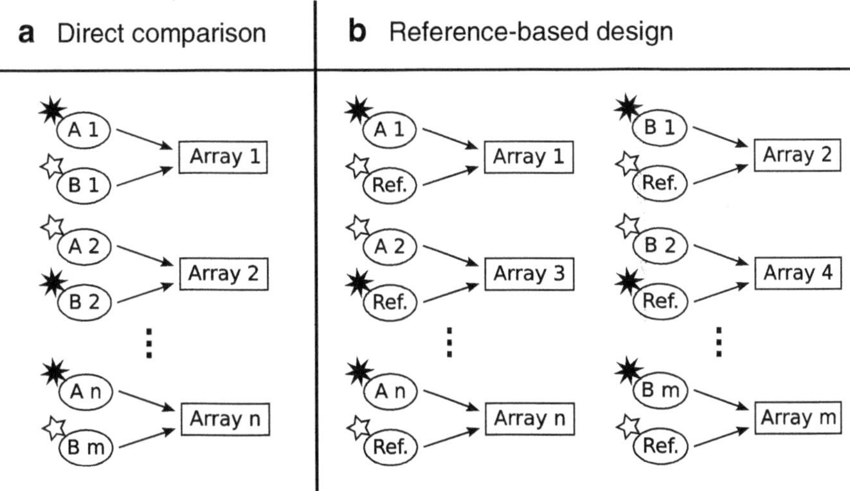

Fig. 2. Experimental layouts for microarray experiments. A scheme is shown for two different biological factors (**a**, normal; **b**, cancer) with biological replicates (1, 2, n, m). Samples can either be compared directly on the same array (**a**) or indirectly via a common reference (**b**).

samples or sample types are labelled with different fluorescent dyes and incubated competitively on the same array (Fig. 2a). Another option is a reference-based design (Fig. 2b). Herein, each sample is labelled with the same fluorescent dye (Fig. 1d) and incubated competitively with a common reference (Fig. 1f), which is labelled with a second dye (Fig. 1e). Such a reference sample can be established by pooling all samples (Fig. 1c) or a certain subset of samples represented in the study. The reference should be available in sufficient quantity for repeated incubation with all individual samples and encompass all sample types analysed in the study. A direct comparison is favourable for smaller studies in which a small number of parameters is analysed. The reference-based design facilitates the supplemental analyses of the impact of additional parameters such as gender, age, and the presence of a certain medication or disease without changes in the experimental design.

After incubation, slides are scanned and resulting fluorescence images transformed into signal intensities at the two colour channels using a software for spot recognition. The ratios of the two colour channels are used for the identification of differences. As with DNA-microarrays, most technical variation effects are eliminated by considering the ratios, leading to reproducible data (7). Besides sensitivity, a good reproducibility and consequently assay robustness are an essential prerequisite for proteomic profiling studies recording expression differences. In order to achieve high performance, we optimised the experimental layout, array production (4, 5, 9), protein extraction (10), labelling (4, 5) and sample incubation conditions (6) as well as data processing steps. Here, we present our current protocols which we use for a robust analysis of

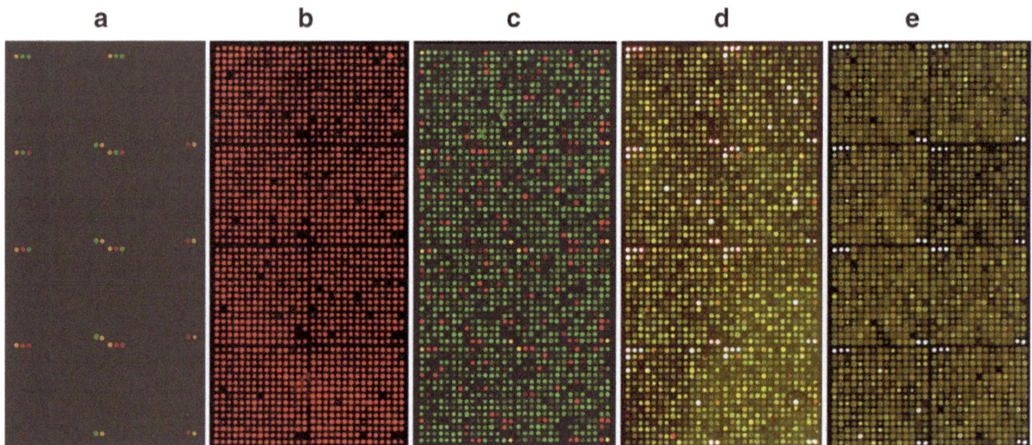

Fig. 3. Quality control of an antibody microarray consisting of some 1,800 features. Antibodies have been spotted in duplicates. (**a**) Two-colour positional controls facilitate easy tracking of the grid and verification of spot segmentation. (**b**) Sypro Ruby staining acts as a quality control measure of spotting. Negative controls do not show any immobilised protein. Panel (**c**) shows an antibody microarray incubated with 5 nM each of secondary fluorescently labelled antibodies against rabbit IgG (*green*) and mouse IgG (*red*). The majority of antibodies on the microarray were produced in rabbits. Panels (**d**) and (**e**) present antibody microarrays that were incubated with two human plasma (**d**) or cell culture samples (**e**) following the protocols presented here.

the abundance of more than 800 proteins in plasma, urine, and tissue samples (7, 10). Applying these protocols, we could demonstrate a high quality of the array production (Fig. 3). In a profiling study on urine samples, for example, pancreatic cancer patients and healthy controls could be differentiated on the basis of the protein patterns obtained (Fig. 4).

2. Materials

The following stock solutions and buffers are used in more than one of the methods. Stock solutions were prepared in ultrapure water, unless stated otherwise in the text.

1. 20% Triton X-100 stock solution (w/v); filter for sterilisation. *Attention: Triton X-100 is irritant and dangerous for the environment.*
2. 20% Tween-20 stock solution (w/v); filter for sterilisation.
3. 1 M Sodium bicarbonate buffer (pH 9.0); autoclave for sterilisation; store in aliquots at −20°C until use.
4. 5% Sodium azide stock solution (w/v).
5. 100× Halt protease and phosphatase inhibitor cocktail (Thermo Scientific, Bonn, Germany).
6. 10× PBS stock solution.

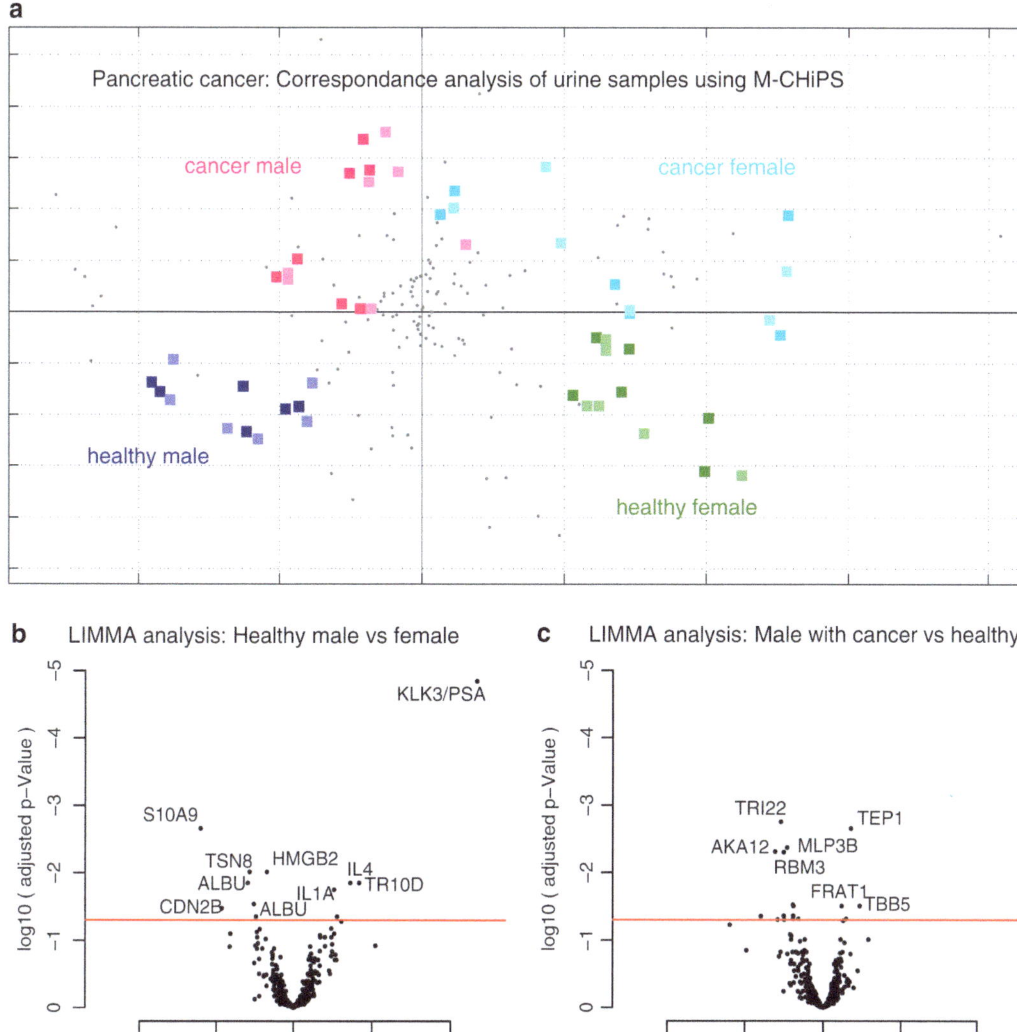

Fig. 4. The protocols presented here were applied for a profiling of urine samples from patients with pancreatic adenocarcinoma and healthy controls. (**a**) Correspondence analysis with M-CHiPS (15–17) resulted in a biplot of differentially abundant proteins and the samples. Samples are depicted as squares that are coloured according to disease-state and gender; black spots represent differentially expressed proteins. Samples located in the same direction from the centroid of the plot exhibit a similar expression pattern. The smaller the distance between two samples the higher is the concordance of their expression profiles. Proteins were found, which are particularly associated with the different sample groups. This association is indicated in the correspondence analysis plot by localization in the same direction off the centroid as the respective sample type. (**b**–**c**) Volcano plots summarise the results of LIMMA analyses (12). Log-fold changes and adjusted *p*-values are shown for gender-specific comparisons of the healthy (**b**) as well as disease-specific comparisons of the male subgroup (**c**). The *red line* marks a significance level of $p=0.05$. This research was originally published in Molecular and Cellular Proteomics (7), © the American Society for Biochemistry and Molecular Biology.

7. Washing buffer A: 0.01% sodium azide (w/v), 0.05% Tween-20 (w/v), 0.05% Triton X-100 (w/v) in 1× PBS.
8. Washing buffer B: 0.5× PBS.

2.1. Antibody Microarray Production

1. Microarraying robot: several commercial models are available. Protocols have been established using the contact printers MicroGrid 2 (Genomic Solutions, Ann Arbor, USA) and SDDC-2 (ESI, Toronto, Canada) as well as the contact-free piezo spotter NP-2 (GeSiM, Großerkmannsdorf, Germany).
2. Centrifuge suitable for 384-well plates.
3. Epoxy-coated slides (Nexterion E, Schott, Jena, Germany).
4. Poly or monoclonal antibodies, affinity-purified in PBS with a concentration of 2 mg/mL (see Note 1 and Subheading 3.1.1).
5. Sypro Ruby protein blot stain (no. S4942, Sigma-Aldrich Corp., St. Louis, USA).
6. 10% Dextran (no. 31394, Sigma-Aldrich Corp., St. Louis, USA) stock solution (w/v) in H_2O; store at 4°C until use.
7. 10% Trehalose stock solution (w/v).
8. 1% Igepal CA-630 (no. I3021, Sigma-Aldrich Corp., St. Louis, USA) stock solution (v/v) in H_2O.
9. 100 mM Sodium borate buffer (pH 9.0).
10. 2× Spotting buffer: mix 2 mL 100 mM sodium borate buffer, pH 9.0, 20 μL 5% sodium azide, 0.5 mL 10% dextran stock solution, 10 μL 1% Igepal stock solution and adjust the volume to 10 mL with H_2O; filter for sterilisation. Prepare aliquots and store at −20°C until use.
11. Washing buffer C: 10% (v/v) methanol, 70% (v/v) acetic acid in H_2O.

2.2. Protein Extraction

1. Cooled centrifuge.
2. Small size porcelain mortar and pestle.
3. Cell scrapers.
4. 2–5 mL syringes with 25-gauge needles.
5. Liquid nitrogen.
6. Dulbecco's phosphate-buffered saline (DPBS).
7. 250 U/μL of benzonase.
8. 60 mM Magnesium chloride ($MgCl_2 \cdot 6H_2O$) stock solution.
9. 100 mM EDTA (pH 8.5) stock solution.
10. 200 mM Phenylmethanesulfonyl fluoride (PMSF) stock solution in iso-propanol; store in small aliquots at −20°C for up to 6 months.
11. 10% Nonidet P-40 substitute (NP-40S) in H_2O (w/v).
12. 10% Cholic acid sodium salt; filter for sterilisation (w/v).

13. 5% Amidosulfobetaine-14 (ASB-14) (w/v); filter for sterilisation and store up to 6 months at 4°C. ASB-14 may precipitate upon cooling, but can be solubilised again by bringing to room temperature while vortexing.
14. 2.5% *n*-dodecyl-β-d-maltoside (12-Malt) (w/v) (GenaXXon Bioscience GmbH, Ulm, Germany); filter for sterilisation and store up to 6 months at 4°C.
15. Glycerol (BioUltra grade).
16. Working solution of the extraction/labelling buffer: prepare from stock solutions by mixing 2.0 mL glycerol, 500 μL EDTA, 1.0 mL of each of carbonate buffer, NP-40S, cholate, 12-Malt, and ASB-14, 167 μL magnesium chloride, 50 μL PMSF, 4.0 μL benzonase, and 100 μL of proteases and phosphatases inhibitors cocktail. Complete the volume to 10 mL with H_2O, and keep on ice until use. The working solution is stable for 1 week at 4°C. However, PMSF loses its activity 30 min after dilution in the working buffer, and hence, always needs to be added directly before cell lysis.

2.3. Labelling of Protein Samples

1. *N*-Hydroxysuccinimide (NHS)-ester of fluorescent dyes (e.g., Dy549-NHS, DY649-NHS, Dyomics, Jena, Germany).
2. Pierce Zeba Spin Desalting columns (Thermo Fisher Scientific, Waltham, USA).
3. Hydroxylamine (Sigma-Aldrich Corp., St. Louis, USA). *Attention: hydroxylamine is harmful and dangerous for the environment.*

2.4. Sample Incubation

1. Advalytix Slidebooster (Olympus Life Science Research, Munich, Germany).
2. Homemade Plexiglas incubation chambers, which have a slightly larger inner dimension than the spotting area and can be reversibly attached to the slides by double-sided adhesive tape. As an alternative, LifterSlips or Gene Frames (Thermo Fisher Scientific, Waltham, USA) can be used (see Note 2).
3. Slide racks and containers (no. 2285.1, Carl Roth GmbH, Karlsruhe, Germany).
4. Blocking buffer: 5% non-fat dry milk (Biorad, Munich, Germany), 0.01% sodium azide, 0.05% (w/v) Tween-20 in 1× PBS. Mix well for at least 30 min on a magnetic stirrer in order to allow the milk powder to dissolve completely. Store at 4°C and use within a few days.

2.5. Scanning and Data Analysis

1. Microarray scanner: several commercial models are available. The protocols described here have been established using the ScanArray 4000XL (Perkin Elmer, Waltham, USA).
2. Software for image segmentation and spot recognition of the signals obtained on the microarrays: e.g., GenePix Pro

(Molecular Devices, Sunnyvale, USA), Mapix (Innopsys, Carbonne, France), or TIGR SpotFinder (11).

3. Software for data analysis: several open-source packages as well as commercial programmes are available (see Note 3).We used the LIMMA-package (12) of R-Bioconductor and M-CHiPS.

3. Methods

According to the experimental process, the protocols below are divided into five sections: antibody microarray production (Subheading 3.1), protein extraction (Subheading 3.2), sample labelling (Subheading 3.3), incubation of the samples on the microarrays (Subheading 3.4), and finally image acquisition and data analysis (Subheading 3.5).

3.1. Production of Antibody Microarrays

3.1.1. Pre-processing and Storage of Antibodies

Perform all subsequent steps at 4°C or on ice.

1. Purify antibodies that are delivered in a crude formulation like ascites fluid, whole antiserum, or in the presence of stabilisers such as bovine serum albumin (BSA) or gelatine. Use Protein A or G columns (e.g., Pierce Nab Protein G Spin Kit, Thermo Fisher Scientific, Waltham, USA) depending on the exact host used for antibody production and antibody isotype. Follow the instructions of the respective user manual.
2. For antibodies formulated in another buffer system or with the addition of glycerol, exchange buffer to PBS by dialysis (Pierce Slide-A-Lyzer, Thermo Fisher Scientific, Waltham, USA).
3. If necessary, adjust antibody concentration to 2 mg/mL (see Note 1) by filtration (Microcon YM-100, Millipore, Schwalbach, Germany) or dialysis (Pierce Slide-A-Lyzer and Pierce Slide-A-Lyzer concentrating solution; Thermo Fisher Scientific, Waltham, USA).
4. Prepare 5 μL aliquots (see Note 4) of antibody solution to avoid additional freeze–thaw cycles and store antibody aliquots at −20°C until use.

3.1.2. Preparation of Antibody Spotting Microtiter Plates

Prepare the spotting microtiter plate(s) directly prior to microarray spotting. Handle all tubes and plates on ice.

1. Thaw antibodies on ice.
2. Mix 5 μL antibody with 5 μL 2× spotting buffer (see Note 4) in order to have a final spotting concentration of 1 mg/mL.
3. Transfer the mix to the appropriate wells of the spotting microtiter plate. Pipette carefully in order to prevent any air bubbles.

4. Include positional controls: add 0.25 mg fluorescently labelled protein, 0.75 mg BSA, and 1× spotting buffer in a volume of 10 μL to wells of the spotting microtiter plate(s). Positional controls (Fig. 3a) facilitate an easy identification of slide orientation and grid recognition as well as spot segmentation during image processing.
5. Include negative controls: add 5 μL 2× spotting buffer and 5 μL PBS to some wells of the spotting microtiter plate(s).
6. Mix liquid in the microtiter plate wells thoroughly using a plate vortexer.
7. Centrifuge plate(s) at 1,000 × *g* for 2 min and keep at 4°C covered with a lid until array spotting.

3.1.3. Spotting Process

Prior to the final array production, a test-spotting run should be performed in order to assess the performance of the existing infrastructure. The test system should mimic the final production run with regard to the number of antibodies, complexity of the arrays as well as duration of spotting. BSA or immunoglobulins can be used for the test spotting (see Note 5). The quality of array production can be assessed as described in Subheading 3.1.4.

1. Program the robot and fill the washing buffer reservoirs and the air humidifier. If the robot has a cooling system, set temperature to 10°C.
2. For pin spotters, clean the pin tool and the pins thoroughly (see Note 6). Follow the manual of the pin manufacturer.
3. Start a pre-spotting by delivering at least 1,000 spots of 1× spotting buffer containing 1.0 mg/mL BSA. Make sure that all pins or piezo needles are performing well.
4. Place the slides in the robot using powder-free nitrile gloves. Pay attention not to touch the slides on the surface.
5. Place the spotting microtiter plates in the robot. If the robot has no cooling device, allow the plates to adapt to room temperature beforehand.
6. Allow the relative humidity in the robot to reach a value of about 50%.
7. Start the spotting process.
8. After spotting is finished, keep slides for two more hours within the robot at 50% humidity.
9. Leave the slides overnight at 4°C in the dark.
10. Perform quality control analysis (Subheading 3.1.4) for a part of the slides picked randomly from different positions within the microarraying robot.
11. Store slides dry at 4°C (e.g., in an exsiccator).

3.1.4. Quality Control

For quality control, immobilised antibodies can be stained after microarray production by a fluorescent dye such as Sypro Ruby (Fig. 3b). It is essential to perform such staining prior to blocking the slide surface:

1. Wash slides 4× 5 min on a shaker in washing buffer A.
2. Cover slides for 1 h with the ready-to-use Sypro Ruby staining solution.
3. Wash slides 4× 5 min in washing buffer C.
4. Wash slides 2× 5 min with H_2O.
5. Dry each slide individually by pointing a sharp stream of air to each spotting area. Always keep slides wet beforehand and do not allow remaining droplets to move into the spotting area.
6. Scan the slides with a microarray fluorescence scanner recording the emission at 610 nm using excitation at 280 or 450 nm.

As an additional control, antibody microarrays can be incubated with 10 nM fluorescently labelled secondary antibodies (Fig. 3c) in blocking buffer for 2 h to control protein immobilisation and functionality. See Notes 7–10 for how to solve the most common problems faced in array production.

3.2. Protein Extraction

Sample handling has a major effect on the protein quality and composition. Therefore, treat all specimens within an experiment series in a uniform manner. Make sure that this happens also prior to their arrival in the microarray laboratory, e.g., during sampling. Typical factors that can affect quality are the kind of columns used for plasma/serum preparation, the period that a blood or tissue sample remains at room temperature prior to freezing, the time of adding protease inhibitors, and the number of freeze–thaw cycles. In general, all protein samples should be thawed and handled on ice in order to minimise degradation by proteases. Aliquot samples as soon as possible and avoid repeated freeze–thaw cycles.

While plasma or serum is prepared using standard procedures, the extraction of proteins from tissue or cell culture should be performed according to the protocols given below in order to facilitate an effective extraction, which is compatible with the subsequent label reaction and analyses on antibody microarrays (10).

Attention: Human samples can potentially be infectious and should therefore be handled as biohazard.

3.2.1. Tissue Samples

1. Prepare from stock solutions a working solution of the extraction/labelling buffer.
2. Mince tissues with a scalpel or a scissor to small pieces of 2–3 mm^3.

3. Immerse the mortar and pestle in liquid nitrogen for a minute until bubbling ceases. *Attention: Liquid nitrogen is extremely hazardous and may cause severe burns. Care should be taken to wear protective gear during handling* (see Note 11).
4. Snap-freeze the minced tissues in liquid nitrogen and transfer them immediately to the mortar. Layer the tissue with few millilitres of liquid nitrogen and pulverise with the pestle until the tissue has become a fine powder.
5. Transfer the powder into a pre-weighed microfuge tube and add 10 μL of the extraction/labelling buffer for each 1 mg of tissue (e.g., 500 μL buffer to 50 mg of tissue).
6. Vortex vigorously to disperse the sample in the buffer.
7. Keep it on ice for 20 min with occasional vortexing.
8. Pipette the sample up and down ten times with a small syringe and 25-gauge needle.
9. Centrifuge at 20,000 × *g* at 4°C for 20 min.
10. Aspirate the supernatant with a fine needle.
11. Label samples and store at –20°C.

3.2.2. Cell Culture

1. All buffers must be cooled and procedures must be carried out on ice (see Note 12).
2. Prepare from stock solutions a working solution of the extraction/labelling buffer.
3. Completely remove culture medium from the vessel (see Note 13).
4. Wash cells three times with ice-cold DPBS.
5. Remove completely the wash buffer and leave the vessel for 1 min in an upright position to drain minute amounts of remaining buffer.
6. Add the minimal volume of the extraction/labelling buffer, which is sufficient to cover the entire surface of the culture vessel (see Note 14).
7. Keep flask in flat position on ice or in a refrigerator for 20 min. Inspection of cells under the microscope may give a good indication for the lysis efficiency.
8. Collect cells with cell scraper and transfer them to a 2-mL Eppendorf tube.
9. Pipette the sample up and down ten times with a small syringe and 25-gauge needle. This operation ensures shredding of chromosomal DNA and facilitates the effect of benzonase.
10. Centrifuge at 20,000 × *g* at 4°C for 20 min.
11. Aspirate the supernatant with a fine needle.
12. Label samples and store them at –20°C.

3.3. Labelling of Protein Samples

3.3.1. Label Reaction

Fluorescent labels are covalently attached to the amino groups of the proteins using NHS-ester chemistry. For competitive two-colour assays with a common reference, all samples are labelled by a NHS-fluorescent dye (e.g., Dy-649). In addition, a reference sample is labelled with a second NHS-fluorescent dye (e.g., Dy-549). In order to have sufficient volume for all incubations in the study, the reference sample is usually labelled in multiple reactions and then mixed. For each label reaction

1. Thaw protein samples on ice.
2. Measure the protein concentration by BCA assay; the protein concentration should be at least 5 mg/mL for blood samples and not less than 1.0 mg/mL for proteins extracted from tissue or cell culture (see Note 15).
3. Label blood samples in a final concentration of 4 mg/mL with 400 μM NHS-ester of a fluorescent dye in 1% Triton X-100 and 100 mM sodium bicarbonate in a final volume of 250 μL (see Notes 16 and 17). For labelling of tissue and cell culture samples, use a protein concentration of 1 mg/mL and a dye concentration of 200 μM. After protein extraction according to Subheadings 3.2.1 or 3.2.2, proteins can be labelled directly without the addition of carbonate or detergent, since these are already present in the extraction/labelling buffer.
4. Incubate reaction tubes on a shaker (200 rpm) at 4°C, protected from light.
5. To stop the reaction, add to each reaction tube hydroxylamine to a final concentration of 1 M and incubate for 30 min at 4°C.
6. Use Zeba Desalt columns (Pierce) in order to remove unreacted dye and exchange buffer to PBS according to the protocol provided by the manufacturer.
7. Add 1× protease and phosphatase inhibitors cocktail.
8. Prepare aliquots and store them protected from light at –20°C.

3.4. Sample Incubation

1. Wash slides three times in washing buffer A by quickly moving the rack up and down.
2. Incubate slides in blocking buffer (see Note 18) for at least 3 h at room temperature with shaking at 150 rpm.
3. Wash slides 4× 5 min in washing buffer A.
4. Wash slides 2× 5 min in washing buffer B.
5. Dry each slide individually by aiming a sharp stream of air to each spotting area. Always keep slides wet beforehand and do not allow remaining droplets to move into the spotting area.
6. Attach home-made Plexiglas incubation chambers to the slide using a double-adhesive tape (see Note 2).

7. Place the slides on a slidebooster instrument (see Note 19).
8. Prepare incubation buffer by adding Tween-20 to a final concentration of 1% (w/v) to the blocking buffer.
9. Dilute labelled blood samples and the common reference 1:20 each with incubation buffer (e.g., mix 30 μL sample A, 30 μL reference, and 540 μL incubation buffer). Instead of 1:20 use a ratio of 1:100 for tissue and cell culture samples. Transfer the incubation mix into the incubation chambers.
10. Cover incubation chambers, start mixing, and incubate overnight. Keep incubation time consistent for all samples.
11. Remove the incubation mix and wash each array 4× 5 min with washing buffer A under mixing conditions. Add the washing buffer quickly in order to keep the slide surface wet.
12. Stop the slidebooster and place slides in a container filled with washing buffer A.
13. Detach incubation chambers. Take care to remove all remaining residues of adhesives and keep the array surface wet during removal.
14. Immediately place arrays in a slide rack in a container filled with washing buffer A and wash for 3× 5 min.
15. Wash 2× 5 min in washing buffer B.
16. Dry each slide individually by aiming a sharp air stream at the spotting area. Always keep slides wet beforehand and do not allow remaining droplets to move into the spotting area.
17. Store slides protected from light until scanning.

3.5. Scanning and Data Analysis

3.5.1. Scanning the Microarrays

Detect signal intensities in a microarray scanner. Beforehand, adjust the scanner settings of the photomultiplier tube (PMT) and the laser power (LP) in order to obtain visible signals for most spots, with only a small number of saturated spots for abundant proteins (see Note 20). For two-colour incubations, adjust scanner settings additionally in a way that the signal intensity distributions for both dyes match each other. Keep scanner settings fixed for all arrays within an experimental series.

3.5.2. Spot Recognition

Convert recorded image files into signal intensities by a software for semi-automatic spot recognition as well as signal quantification such as GenePix Pro, Mapix, or the freeware TIGR Spotfinder. Adjust the size of the spots and if possible flag the spots according to their quality.

3.5.3. Analysis of Differential Expression

There are several freeware and commercial analysis platforms available (see Note 3). For detailed information how to use M-CHiPS for correspondence analysis (Fig. 4a) refer to the online manuals (http://www.mchips.org). Here, we describe the procedure for the identification of differential proteins using the LIMMA package within R-Bioconductor (Fig. 4b, c). More detailed information can be found in the user's guide available with the package (http://www.bioconductor.org/packages/release/bioc/html/limma.html).

1. Build up a "Targets.txt" file. This file contains the filenames of the results files after spot segmentation. In two additional columns, a definition of the sample types used for incubation in the two colour channels is provided. For an analysis of the factors cancer/healthy and male/female the following sample types would be defined in the two columns: "cancer_male", "cancer_female", "healthy_male", "healthy_female", or "reference".
2. Load the limma library.

   ```
   library(limma)
   ```
3. Import the Targets.txt file.

   ```
   targets <- readTargets()
   ```
4. Import the mean of the signal intensities and the median of the background intensities.

   ```
   RG<-read.maimages(targets$FileName,source="genepix",
       columns=list(R="F649 Mean",G="F549 Mean",Rb="B649
       Median",Gb="B549 Median"))
   ```
5. Subtract the local background using the normexp method.

   ```
   RGb <-backgroundCorrect(RG,method="normexp",
       offset=50)
   ```
6. Log-transform the data and normalise the two colour channels using loess normalisation (see Note 21) in one go.

   ```
   MA<-normalizeWithinArrays(RGb,method="loess",
       iterations=10)
   ```
7. Assess the quality of your incubations and the efficiency of normalisation by inspecting signal intensity distributions and MA-plots prior (a) and after normalisation (b). For the MA-plots inspect each array by incrementing `i` from 1 to the number of arrays by `i=1; i=2` …

   (a) `plotMA(RGb(,i))` and `plotDensities(RGb)`

   (b) `plotMA(MA(,i))` and `plotDensities(MA)`
8. Build up a design matrix defining the experimental layout for the linear models as well as a contrasts matrix defining the type of comparisons to be performed.

   ```
   design <- modelMatrix(targets, ref="Reference")
   contrasts.matrix <- makeContrasts
   (cancer=(cancer_male + cancer_female)-(healthy_
       male + healthy_female),
   ```

```
gender=(cancer_male + healthy_male)-(cancer_female +
    healthy_female),
levels=design)
```

9. Apply the linear models and search for differential expression.

```
fit <- lmFit(MA,design)
fit2 <- contrasts.fit(fit, contrasts.matrix)
fit2 <- eBayes(fit2)
```

List proteins with differential abundance for the disease (a) and gender-specific (b) comparison.

(a) `topTable(fit2, coef=1, adjust="BH")`

(b) `topTable(fit2, coef=2, adjust="BH")`

4. Notes

1. *Antibodies*: All antibodies that are working with high specificity and sensitivity in Western and ELISA assays can be used for antibody microarray experiments. Western blotting is the current method of choice to assess the specificity of antibodies. Antibodies should be purified and formulated in PBS without addition of stabilisers such as BSA, gelatine, or glycerol, which are negatively affecting the spotting process. It is beneficial to immobilise the antibodies in a comparably high concentration of 1 mg/mL after addition of the spotting buffer. However, it is possible to immobilise antibodies at lower concentrations, if sensitivity is of lower importance than antibody consumption.
2. *Incubation chamber*: Also LifterSlips or Gene Frames (Thermo Fisher Scientific, Waltham, USA) can be used to keep the incubation volume. However, higher sensitivity will be obtained with the increased volumes that were made possible by an adapted, homemade incubation chamber.
3. *Software for data analysis*: Many very versatile tools for the normalisation, filtering, and statistical testing are available within the Bioconductor package (13) for R. Most important functions are also integrated in the online analysis platform Expression Profiler (14). Also the freeware TIGR MultiExperiment Viewer (11) allows an application of many different analysis algorithms to the data. We used the LIMMA package (12) in R-Bioconductor and M-CHiPS (15–17). M-CHiPS is well suited especially for correspondence analysis and for correlating the samples to all given biological factors. In addition, there is a variety of commercial tools available for data analysis.

4. *Spotting volume*: The volume used in the spotting plates is dependent on the spotting robot and the number of replicates and slides. For one spot, usually between 0.5 and 10 nL are used depending on pin size (contact printing) or number of drops (non-contact printing). The volume in each sample uptake is 0.25–1.25 µL for pin spotters. Usually, the minimum volume, which can be handled by a microarraying robot is around 5 µL. A volume of 15 µL should be highly sufficient for most spotting projects.
5. *Fluorescently labelled protein in test runs*: Fluorescently labelled proteins should not be used for optimisation of the spotting process. Their increased hydrophobicity that is due to the dye hides problems that could occur with antibodies. In addition, artefacts are introduced, which would never occur with unlabelled proteins. Therefore, it is recommended to use unlabelled BSA for optimising the spotting parameters in a test run and to visualise immobilised proteins by additional staining with Sypro Ruby (Subheading 3.1.4).
6. *Pin handling*: Always use powder-free nitrile gloves for handling pins and pin tool. Clean pin tool and pins exactly as recommended by the manufacturer. In a last step, dip them in 100% ethanol and dry them completely using a stream of air. Do not use pressurised air canisters, which might contain organic propellants.
7. *Spots are missing*: If spots are missing at random, in most cases spotting pins got stuck in the pin tool. Clean and dry pins and pin tool thoroughly. If the problem is persistent, decrease the humidity in the spotter to a value of 35–45%. If spotting stops after a certain number of replicates or samples, perform an intensive pin cleaning protocol using special reagents according to the protocol of the pin manufacturer (e.g., http://arrayit.com/Products/MicroarrayI/PPCK80/ppck80.html). If problems persist, increase the pin washing time between sample uptakes, reload pins more often or increase the concentration of detergent in the spotting buffer. If spots can be observed optically after array production, e.g., making them visible by "breathing" at the surface, but no or few proteins can be detected after staining, make sure that the epoxy groups of the array surface are still active. Avoid TRIS or betaine additions to the spotting solution.
8. *Carryover*: If carryover in negative controls is observed, increase washing time between sample uptakes and make sure that pins are completely dried after washing. Exchange washing buffer more often, add detergents (e.g., 0.1% Tween-20) or use additional sonication.

9. *Inhomogeneous spot morphology*: To avoid doughnut-shaped spots, increase humidity during spotting process or add compounds to the spotting solution which are reducing the evaporation speed or are increasing surface tension. To avoid blurry spots, increase the detergent concentration of the spotting buffer.
10. *Inconsistent spot morphology*: Make sure that buffer formulation and concentration of the antibodies is consistent. If there is a systematic pattern, test the surface coating prior to spotting by breathing carefully at the surface. The absorption of vapour on the surface should be completely homogenous. Additionally, surface inhomogeneities (which sometimes pass quality control by the manufacturers) can be detected by scanning a slide at full laser power and PMT.
11. *Alternative to tissue homogenisation*: Alternative to the mortar and pestle method, minced tissue samples can be homogenised using Potter Elvehjem homogeniser. In this mode, minced tissues are first transferred into the homogeniser, topped with 10 μL lysis buffer for each 1 mg of tissue, and are then homogenised with 20–30 strokes.
12. *Preventing protein degradation*: Cell lysis is usually accompanied by the release of proteases and phosphatases, which may compromise protein structure and integrity. In addition to the favourable effect of including protease and phosphatase inhibitors to the extraction buffer, lowering the temperature contributes further to keeping the remaining activities to the minimum.
13. *Traces of medium*: Traces of medium may contain serum proteins that may introduce false-positive results on the array.
14. *Volume of extraction buffer*: We generally add 175, 350, and 1,000 μL of extraction buffer for 25T, 75T, and 175T culture flasks, respectively, followed by tilting the flask from side to side to spread the solution on the full surface. Depending on the type of cells, the protein concentration obtained this way is in the range of 1.75–3.75 mg/mL, which is sufficient for a successful labelling.
15. *Concentrating protein samples*: If the concentration of samples is too low but samples are available in sufficient quantity, they can be concentrated either by vacuum concentration or by filtration using Microcons (Millipore, Billerica, USA).
16. *Down-scaling of labelling reaction*: It is possible to perform label reactions with less starting material of your sample, although sensitivity may suffer. Reduce the overall reaction volume but keep the concentrations of protein and label reagent. Eventually, use smaller columns for the removal of unreacted dye.

17. *Dye handling*: Dissolve the label reagent in H_2O and use immediately in order to prevent hydrolysis of the NHS-esters. If the label reagent is delivered in larger quantities, it can be dissolved in DMSO or DMF and stored in aliquots under dry conditions at –20°C.
18. *Blocking and incubation buffer*: If increased background is observed, use "the blocking solution" (Candor Biosciences GmbH, Weißensberg, Germany) for blocking and incubation steps.
19. *Slidebooster*: If no slidebooster instrument is available, incubations can be performed in Quadriperm chambers (Greiner-Bio One, Germany) (18). However, larger incubation buffer volumes of 3–5 mL may be needed to cover the whole surface of a slide.
20. *Combining information derived from two scanner settings*: If it is not possible to obtain a representative majority of spots from one scanner setting, it is possible to perform multiple scans at different intensities and combine them using Masliner prior to data analysis (19).
21. *Array normalisation*: See the limma guide for additional normalisation methods or for the possibility to weight spots in the normalisation according to their quality flags introduced during spot recognition. For arrays with a small number of features and a high degree of differentially expressed features a normalisation based on selected non-differentially proteins can be beneficial (20).

References

1. Borrebaeck, C. A. K., and Wingren, C. (2007) High-throughput proteomics using antibody microarrays: an update. *Expert Rev Mol Diagn* 7, 673–686.
2. Ekins, R. P. (1998) Ligand assays: from electrophoresis to miniaturized microarrays. *Clin Chem* **44**, 2015–2030.
3. Alhamdani, M. S., Schröder, C., and Hoheisel, J. D. (2009) Oncoproteomic profiling with antibody microarrays. *Genome Med* **1**, 68.
4. Wingren, C., Ingvarsson, J., Dexlin, L., Szul, D., and Borrebaeck, C.A.K. (2007) Design of recombinant antibody microarrays for complex proteome analysis: choice of sample labeling-tag and solid support. *Proteomics* 7, 3055–3065.
5. Kusnezow, W., Banzon, V., Schröder, C., Schaal, R., Hoheisel, J.D., Rüffer, S., Luft, P., Duschl, A., and Syagailo, Y.V. (2007) Antibody microarray-based profiling of complex specimens: systematic evaluation of labeling strategies. *Proteomics* 7, 1786–1799.
6. Kusnezow, W., Syagailo, Y. V., Rüffer, S., Baudenstiel, N., Gauer, C., Hoheisel, J. D., Wild, D., and Goychuk, I. (2006) Optimal design of microarray immunoassays to compensate for kinetic limitations: theory and experiment. *Mol Cell Proteomics* **5**, 1681–1696.
7. Schröder, C., Jacob, A., Tonack, S., Radon, T. P., Sill, M., Zucknick, M., Rüffer, S., Costello, E., Neoptolemos, J. P., Crnogorac-Jurcevic, T., Bauer, A., Fellenberg, K., and Hoheisel, J. D. (2010) Dual-color proteomic profiling of complex samples with a microarray of 810 cancer-related antibodies. *Mol Cell Proteomics* **9**, 1271–80.
8. Churchill, G. A. (2002) Fundamentals of experimental design for cDNA microarrays. *Nat. Genet* **32** *Suppl*, 490–495.
9. Angenendt, P., and Glökler, J. (2004) Evaluation of antibodies and microarray coatings as a prerequisite for the generation of optimized antibody microarrays. *Methods Mol Biol* **264**, 123–134.

10. Alhamdani, M. S. S., Schröder, C., Werner, J., Giese, N., Bauer, A., and Hoheisel, J. D. (2010) Single-step procedure for the isolation of proteins at near-native conditions from mammalian tissue for proteomic analysis on antibody microarrays. *J Proteome Res* **9**, 963–971.

11. Saeed, A. I., Sharov, V., White, J., Li, J., Liang, W., Bhagabati, N., Braisted, J., Klapa, M., Currier, T., Thiagarajan, M., Sturn, A., Snuffin, M., Rezantsev, A., Popov, D., Ryltsov, A., Kostukovich, E., Borisovsky, I., Liu, Z., Vinsavich, A., Trush, V., and Quackenbush, J. (2003) TM4: a free, open-source system for microarray data management and analysis. *BioTechniques* **34**, 374–378.

12. Smyth, G. K. (2004) Linear models and empirical Bayes methods for assessing differential expression in microarray experiments. *Stat Appl Genet Mol Biol* **3**, 3.

13. Gentleman, R. C., Carey, V. J., Bates, D. M., Bolstad, B., Dettling, M., Dudoit, S., Ellis, B., Gautier, L., Ge, Y., Gentry, J., Hornik, K., Hothorn, T., Huber, W., Iacus, S., Irizarry, R., Leisch, F., Li, C., Maechler, M., Rossini, A. J., Sawitzki, G., Smith, C., Smyth, G., Tierney, L., Yang, J. Y. H., and Zhang, J. (2004) Bioconductor: open software development for computational biology and bioinformatics. *Genome Biol* **5**, R80.

14. Kapushesky, M., Kemmeren, P., Culhane, A. C., Durinck, S., Ihmels, J., Körner, C., Kull, M., Torrente, A., Sarkans, U., Vilo, J., and Brazma, A. (2004) Expression Profiler: next generation – an online platform for analysis of microarray data. *Nucleic Acids Res* **32**, W465–W470.

15. Fellenberg, K., Hauser, N. C., Brors, B., Neutzner, A., Hoheisel, J. D., and Vingron, M. (2001) Correspondence analysis applied to microarray data. *Proc Natl Acad Sci USA* **98**, 10781–10786.

16. Fellenberg, K., Hauser, N. C., Brors, B., Hoheisel, J. D., and Vingron, M. (2002) Microarray data warehouse allowing for inclusion of experiment annotations in statistical analysis. *Bioinformatics* **18**, 423–433.

17. Fellenberg, K., Busold, C. H., Witt, O., Bauer, A., Beckmann, B., Hauser, N. C., Frohme, M., Winter, S., Dippon, J., and Hoheisel, J. D. (2006) Systematic interpretation of microarray data using experiment annotations. *BMC Genomics* 7, 319.

18. Alhamdani, M. S., Schröder, C., and Hoheisel, J. D. (2010) Analysis conditions for proteomic profiling of mammalian tissue and cell extracts with antibody microarrays. *Proteomics* **10**, 3203–3207.

19. Dudley, A. M., Aach, J., Steffen, M. A., and Church, G. M. (2002) Measuring absolute expression with microarrays with a calibrated reference sample and an extended signal intensity range. *Proc Natl Acad Sci USA* 99, **11**, 7554–7559.

20. Sill, M., Schröder, C., Hoheisel, J. D., Benner, A., and Zucknick, M. (2010) Assessment and optimisation of normalisation methods for dual-colour antibody microarrays. *BMC Bioinformatics.* **11**, 556.

Chapter 15

High-Throughput Studies of Protein Glycoforms Using Antibody–Lectin Sandwich Arrays

Brian B. Haab and Tingting Yue

Abstract

The antibody–lectin sandwich arrays (ALSA) is a powerful new tool for glycoproteomics research. ALSA enables precise measurements of the glycosylation states of multiple proteins captured directly from biological samples. The platform can be used in a high-throughput mode with low sample consumption, making it well suited to biomarker research exploring glycan alterations on specific proteins. This article provides detailed descriptions of the use of ALSA, with a particular focus on biomarker research. The preparation and selection of antibodies and lectins, the preparation and use of the arrays and samples, and special considerations for using the platform for biomarker research are covered.

Key words: Antibody arrays, Antibody–lectin sandwich arrays, Glycan profiling, Lectin detection, Serum biomarkers

1. Introduction

This chapter covers the use of antibody–lectin sandwich arrays (ALSA) to probe glycosylation levels on multiple proteins captured directly from biological samples, with a particular focus of the use of ALSA for biomarker studies (1, 2). The use of altered glycoforms of specific proteins as biomarkers has great potential (3), due to the fact that altered glycosylation can be more specifically associated with disease as compared to changes in protein abundance. The key to harnessing that information for biomarker studies is the ability to sensitively and reproducibly detect changes in glycosylation on specific proteins in biological samples. Furthermore, biomarker studies benefit from high-throughput sample processing and low consumption of clinical samples. Conventional glycobiology methods

Ulrike Korf (ed.), *Protein Microarrays: Methods and Protocols*, Methods in Molecular Biology, vol. 785,
DOI 10.1007/978-1-61779-286-1_15,

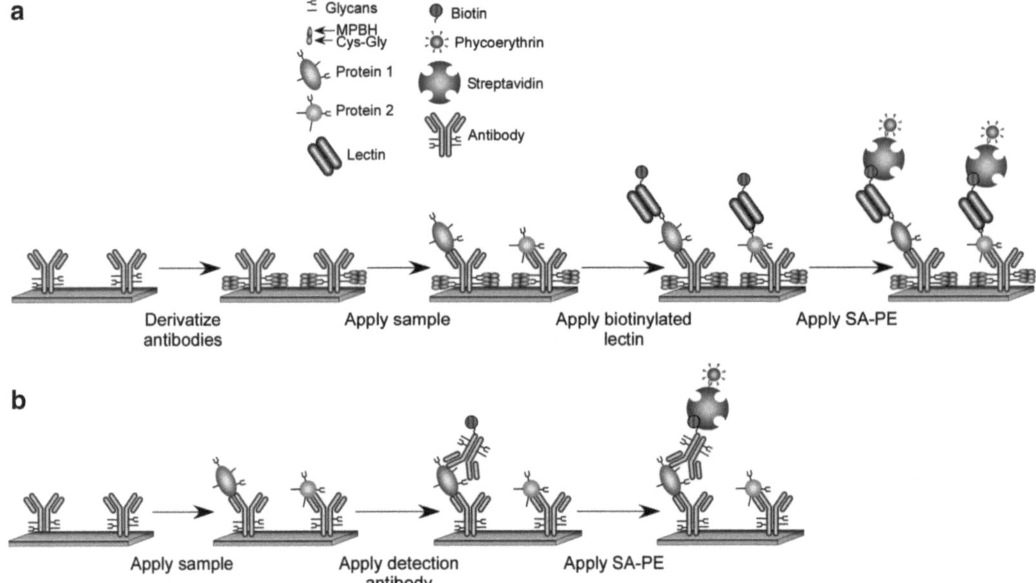

Fig. 1. Glycan and protein detection on antibody arrays. (**a**) Glycan detection. The drawing depicts antibodies immobilized on a planar surface. The glycans on the antibodies are derivatized to prevent lectin binding; a sample is incubated on the antibody array; proteins are captured by the antibodies; biotinylated lectins bind to the glycans on the captured proteins; and the level of bound lectin is determined by scanning for fluorescence from streptavidin–B-phycoerythrin. (**b**) Protein detection. This approach provides measurements of the levels of the core proteins detected in (**a**). Antibody derivatization is not required, and individual proteins are detected using specific antibodies.

based on separations or mass spectrometry, although providing invaluable information on structure, do not score well on these points, since throughput can be low, sample requirements high, with no ability to precisely measure changes between samples.

A graphical overview of the method is given in Fig. 1. A biological sample, such as serum, is incubated on the surface of a microarray of immobilized antibodies, and proteins bind to the antibodies according to their specificities. The levels of specific glycan structures on the captured proteins are probed using lectins (proteins with glycan-binding activity) or antibodies targeting glycan epitopes. Different types of lectins and glycan-binding antibodies can be used to probe various glycan structures. An important first step in this procedure is a method to chemically derivatize the glycans on the immobilized antibodies. This step alters the glycans so that they are no longer recognized by the lectins or glycan-binding antibodies, ensuring that only the glycans on the captured proteins are probed.

Some of the advantages of ALSA for biomarker studies stem from the use of affinity reagents – molecules that can be used to detect particular targets through specific binding interactions. Affinity reagents enable reproducible and sensitive detection in the presence of highly complex biological backgrounds, such as from blood serum.

The ability to directly detect analytes in biological samples reduces the time and variability of assays, due to the reduced number of experimental steps. The use of lectins – carbohydrate-binding proteins – as reagents to detect glycan levels has been explored in many different settings (4).

Other advantages of ALSA stem from the use of the microarray platform (5). The usefulness of the microarray platform is in its multiplexing capability, enabling the acquisition of many data points in parallel, and its miniaturization, resulting in very small consumption of reagents and samples. These qualities are valuable for biomarker research because multiple candidate biomarkers can be evaluated in parallel with low consumption of precious clinical samples.

This chapter covers the procedures and important considerations for using this technology for biomarker studies. We do not cover the fabrication of antibody arrays. Several robotic microarrayers are available for producing arrays, each with particular performance features that might influence parameters, such as the composition of the print solution or the substrate onto which the antibodies are printed. Previous chapters give some practical instructions and considerations for printing antibody microarrays and the handling and preparation of antibodies (6–9). Here, we cover the selection and preparation of antibodies and lectins; the derivatization of antibody arrays to prevent lectin binding; sample incubation and detection; and high-throughput processing methods.

2. Materials

2.1. Reagents

1. $NaIO_4$ (Pierce Biotechnology, Rockford, IL).
2. 4-(4-*N*-Maleimidophenyl) butyric acid hydrazide hydrochloride (MPBH) (Pierce Biotechnology, Rockford, IL).
3. Cysteine–Glycine (CysGly) dipeptide (Sigma-Aldrich, St. Louis, MO).
4. Streptavidin–B-phycoerythrin (Invitrogen, Carlsbad, CA).
5. Protease Inhibitors (Complete Tablet, Roche Applied Science, Indianapolis, IN).
6. Biotinylated lectins (Vector Labs, Burlingame, CA, and other suppliers).
7. Mouse, goat, sheep, and rabbit IgG antibodies, and chicken IgY antibodies (Jackson ImmunoResearch Labs, West Grove, PA).
8. Tween-20 (Sigma-Aldrich, St. Louis, MO).
9. Brij-35 (Sigma-Aldrich, St. Louis, MO).
10. NHS-biotin (Pierce Biotechnology, Rockford, IL).

2.2. Solutions

1. 1× coupling buffer (0.02 M sodium acetate, pH 5.5).
2. Coupling buffer + 0.1% Tween-20.
3. Phosphate-buffered saline (PBS), pH 7.4 (137 mM NaCl, 2.7 mM KCl, 4.3 mM Na_2HPO_4, 1.4 mM KH_2PO_4).
4. PBST 0.1: PBS + 0.1% Tween-20.
5. PBST 0.5: PBS + 0.5% Tween-20.
6. PBST 0.1 + 1 mM CysGly (prepare immediately before use).
7. Coupling buffer + 200 mM $NaIO_4$ (prepare immediately before use).
8. Coupling buffer + 1 mM MPBH + 1 mM CysGly (prepare immediately before use).
9. PBST 0.5 + 1% bovine serum albumin (BSA).
10. PBST 0.1 + 0.1% BSA + 1 μg/ml streptavidin–B-phycoerythrin.

2.3. Hardware and Instruments

1. Dialysis chambers (Slide-A-Lyzer Mini Dialysis Units, Pierce Biotechnology, 69550-69574).
2. Microscope slide staining chambers with slide racks (Shandon Lipshaw, cat. No. 121).
3. Microscope slide boxes (several versions available).
4. Wafer handling tweezers (Techni-Tool, Worcester, PA, cat. No. 758TW178, style 4WF).
5. Slide Imprinter, for printing wax partitions on slides (The Gel Company, San Francisco, CA).
6. Clinical centrifuge with flat swinging buckets for holding slide racks (Beckman Coulter, Fullerton, CA, among others).
7. Microarray scanner (several models available).
8. Ultracentrifuge (several models available).

3. Methods

The methods are divided into the following sections: (1) Selecting and preparing antibodies and lectins; (2) Derivatizing antibody arrays; (3) Sample incubation and detection; and (4) High-throughput sample processing.

3.1. Selecting the Capture Antibodies and the Detection Lectins

3.1.1. Selecting and Preparing the Capture Antibodies

Antibody performance can vary widely with respect to affinity, nonspecific binding, and information content in the relevant application. The microarray platform is ideal for making side-by-side comparisons of antibody performance. The affinities and specificities of multiple antibodies can be compared by running antibody array experiments (using the methods described below) on solutions containing various concentrations of the targeted proteins in a complex background, such as fetal bovine serum. The signal at each antibody with respect to protein concentration provides information on performance. A linear response combined with low signal when the protein is absent indicates acceptable performance while high background and lack of consistent responses indicate poor performance. Affinity and specificity alone do not predict the ultimate value of particular antibodies in the actual applications; some antibodies may provide more valuable information because they target distinct isoforms of proteins that have varying associations with particular disease conditions. Therefore, it is useful to include multiple functional antibodies against each candidate biomarker.

The spotted capture antibodies should be as pure as possible (see Note 1). Considerations on the purification of antibodies and additives to their solutions were provided earlier (6, 8). In addition to standard purification steps, we recommend dialysis and ultracentrifugation to obtain optimal purity. Dialysis removes impurities that are smaller than the molecular weight cutoff of the dialysis membrane. This removal can be important for fluorescence detection, since some small contaminants are weakly fluorescent and contribute to background signal. Dialyze the antibody solutions for 2 h against PBS at 4°C. Ultracentrifugation removes protein aggregates. Aggregates, which can be caused by protein denaturation due to age or repeated freezing and thawing, can be a source of nonspecific binding within an antibody spot. If an ultracentrifuge is available, spin the antibody solutions at 85,000 × g for 1 h at 4°C (see Note 2). After preparation and purification, we recommend aliquotting the antibody solutions so that a fresh aliquot can be used with every round of microarray production.

3.1.2. Selecting the Detection Lectins

Lectins are proteins that have carbohydrate-binding activities. The selection of the lectins to use as detection reagents in the ALSA assay depends on the glycan targets, the analytical performance of the lectins, and the availability of the lectins. The determination of the glycans to target is based on experimental or literature information about which glycans are overexpressed or involved in the disease of interest. Once the target glycans are determined, one may search for lectins or glycan-binding antibodies that bind those glycans. Basic information on the specificities of many lectins can be found from lectin suppliers and literature reports (10).

A tool that has facilitated detailed explorations of lectin specificities is the glycan microarray (11, 12). Glycan microarrays are composed of numerous biologically relevant glycans. By incubating a lectin on a glycan microarray and detecting the level of lectin binding at each glycan, one may determine the preferred ligands for that lectin. The availability of glycan microarray data has been promoted by the Consortium for Functional Glycomics (CFG) (13). Participating researchers submit lectins and glycan-binding antibodies to the CFG for glycan array analysis, and the results of the experiments are made available on the CFG Web site (http://www.functionalglycomics.org).

This extensive experimental information is extremely valuable in providing detailed information about lectin specificities. With this information in hand, researchers may be better able to search for lectins with defined specificities to use as analytical reagents. The key for using glycan microarray data in this way is the ability to extract specificity information and to make it available in a searchable form. We recently developed an automated method for extracting specificity information from glycan microarray data (14) and have assembled the results from these analyses in a searchable database (http://vai-apps.tgen.org/haab/). Researchers may search for lectins that bind a user-defined glycan motif or obtain information on the binding specificities of given lectins. This tool provides access to detailed information about lectin specificity and may be useful for identifying lectins that are valuable as analytical reagents to detect specific glycans.

3.2. Sample Incubation and Detection

Prior to running the experiments, three experimental factors should be determined: the level of cross-reactivity between the detection lectins and the capture antibodies; the amount by which the experimental samples should be diluted; and the optimal concentration of the detection reagents.

3.2.1. Blocking Cross-reactivity to the Capture Antibodies

Since antibodies contain glycans, one must confirm that the detection lectins do not cross-react with the glycans on the spotted capture antibodies. The amount of cross-reactivity between the detection lectins and the capture antibodies can be determined in preliminary experiments. Use the protocol given in Subheading 3.2.3, except use PBS buffer as the sample instead of an experimental sample, and use all the detection lectins and antibodies planned for the actual experiments. Record the amount of signal at each capture antibody for each detection lectin. If low signal is observed, the targeted glycan of that detection lectin is not present, and it is not necessary to block cross-reactivity (see Note 3). If strong signal is observed, for example greater than twice that seen when no lectin is present, it will be necessary to treat the glycans on the capture antibodies to prevent lectin binding. If a mixture of cross-reactive and non-cross-reactive antibodies is present on a single array, the entire array could be blocked for the sake of the cross-reactive antibodies, or the antibodies could be split into two arrays, one to be blocked and the other to be left untreated.

A convenient method to prevent lectin binding to the glycans on the capture antibodies is mild oxidation followed by reaction with hydrazide reagents (1). Hydrazide (N_3) reacts with the aldehyde groups that are produced after oxidation using sodium periodate.

We use a hydrazide reagent that is conjugated to another molecule to sterically hinder lectin binding to nearby nonoxidized glycans. We prefer the cysteine–glycine dipeptide as the conjugate since it is large enough to block access to the glycan but not so large as to interfere with antibody activity. We have shown that the oxidation step can slightly reduce antibody affinity, but that the drop in affinity is minor relative to the drop in nonspecific binding of detecting lectins (1). The details of this blocking protocol were given earlier (15).

3.2.2. Determining the Sample Dilution Factor and the Detection Reagent Concentration

Prior to running the experimental samples, it is critical to determine the proper dilution level of the sample and the optimal concentration of the detection reagent (the lectin or antibody). It is important that the concentrations of the targeted analytes are in the linear response range of the assay. The linear range of the assay depends upon the antibodies and lectins used and, therefore, must be determined independently for each new assay. In order to match the analyte concentrations in the experimental samples to the linear ranges of the assays, the samples must be diluted by an appropriate amount. High-concentration analytes require high dilution, and low-concentration analytes require little or no dilution.

Pools of experimental samples are useful for this determination, since they contain each analyte at its average concentration over the sample set. The sample pools should be serially diluted and analyzed on antibody arrays using the detection reagents of interest. We recommend testing each detection reagent at several concentrations, since the detection reagent concentration can affect the linear range of the assay (see Note 4). The signal from each assay (each lectin–antibody combination) should be plotted with respect to dilution factor for each detection reagent concentration (Fig. 2). Record the dilution factors that produce signal in the middle of the linear range and the detection reagent concentrations at which linearity is achieved. These conditions should be selected for future experiments. A challenge for array experiments is to find conditions that are optimal for multiple assays. If analytes that have greatly divergent concentrations are targeted, it may be necessary to use separate arrays to detect the high- and low-abundance analytes, respectively.

3.2.3. Sample Preparation

Once the optimal dilution factors are determined, the samples can be prepared for use. Biological samples should be stored in frozen aliquots so that a fresh aliquot can be used for each experiment (see Notes 5 and 6). After thawing fresh aliquots of sample, dilute the samples using the buffers and additives given in the table below to the final concentrations of each. We recommend preparing a stock solution of each at the indicated concentrations.

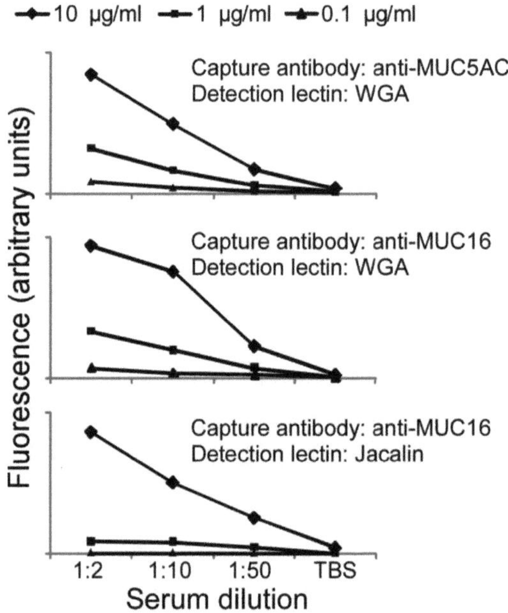

Fig. 2. Dilution curves to determine optimal experimental parameters. A pool of serum samples was diluted 2-fold, 10-fold, and 50-fold, incubated on antibody arrays, and detected with either wheat germ agglutinin (WGA) or the Jacalin lectin at three different lectin concentrations each. Each set of curves shows the signal at a given capture antibody with respect to serum dilution. Detection of MUC16 using WGA shows saturation at the 1:2 serum dilution using the highest lectin concentration, but the other assays are still in the linear range at the 1:2 dilution and highest lectin concentration.

	Final concentration	**Stock concentration**
Sample buffer	1×: 0.1% Tween-20 and 0.1% Brij-35 in 1× PBS	10×: 1% Tween-20 and 1% Brij-35 in 10× PBS
IgG/IgY cocktail	1×: 100 µg/ml each of mouse, goat, and sheep IgG and chicken IgY, 200 µg/ml of rabbit IgG	4×: 400 µg/ml each of mouse, goat, and sheep IgG and chicken IgY, 800 µg/ml of rabbit IgG
Protease inhibitor cocktail	1×	10×: 1 tablet into 1 ml PBS

The detergents are important for reducing adsorption of serum proteins to the array surface, and the IgG/IgY cocktail function to prevent nonspecific binding of various components of the serum samples to the capture antibodies. The protease inhibitor cocktail is necessary because certain serum proteases can be activated upon dilution of the sample.

3.2.4. Sample Incubation and Detection

The next step is to incubate the samples on the arrays and to detect the levels of proteins or glycans captured at each antibody.

1. If the spotted antibodies are chemically derivatized to block the glycans, perform the procedure provided earlier (1, 15) immediately before the steps described below. After derivatization, wash the microarray slides in three baths of PBST 0.5 for 3 min each with gentle rocking. If not derivatizing the spotted antibodies, begin with the next step.
2. Incubate the slides in a blocking solution of PBST 0.5 + 1% BSA for 1 h with gentle rocking.
3. Wash the slides in three baths of PBST 0.5 for 3 min each.
4. Spin the slides at 1,000 × *g* for 1 min to dry the slides. Use a clinical centrifuge with swinging buckets and a slide rack to hold the slides vertically. The coated surface of the slide should be facing out, allowing the liquid to easily spin off the slide.
5. Place the slides horizontally, coated side up, in a slide box containing a moist paper towel. The paper towel will humidify the chamber and prevent evaporation during the sample incubation.
6. Apply the serum solutions to their designated arrays, cover the slide box, and incubate at room temperature for 1–2 h with gentle shaking. The volume applied to each array depends on the size of the array. Using the 48-array/slide design shown in Fig. 3c, about 7 μl thoroughly covers each array.
7. Prepare the glycan-binding detection reagents. Biotinylated lectins should be prepared in PBST 0.1 with 0.1% BSA at the previously determined optimal concentration. Biotinylated antibodies (see Note 7) should be prepared at 1 μg/ml in the same buffer as the lectins.
8. Wash the slides in three baths of PBST 0.1 for 3 min each with gentle rocking.
9. Spin the slides at 1,000 × *g* for 1 min to dry the slides.
10. Apply the biotinylated lectin or antibody solutions to each array, and incubate the slides in a humidified slide box at room temperature for 1 h with gentle shaking.
11. Prepare streptavidin–B-phycoerythrin in PBST 0.1 + 0.1% BSA at a concentration of 1 μg/ml.
12. Wash the slides in three baths of PBST 0.1 for 3 min each.
13. Spin the slides at 1,000 × *g* for 1 min to dry the slides.
14. Apply streptavidin–B-phycoerythrin to each array, and incubate the slides in a humidified box at room temperature for 1 h with gentle shaking.
15. Wash the slides in three baths of PBST 0.1 for 3 min each.
16. Spin the slides at 1,000 × *g* for 1 min to dry the slides.

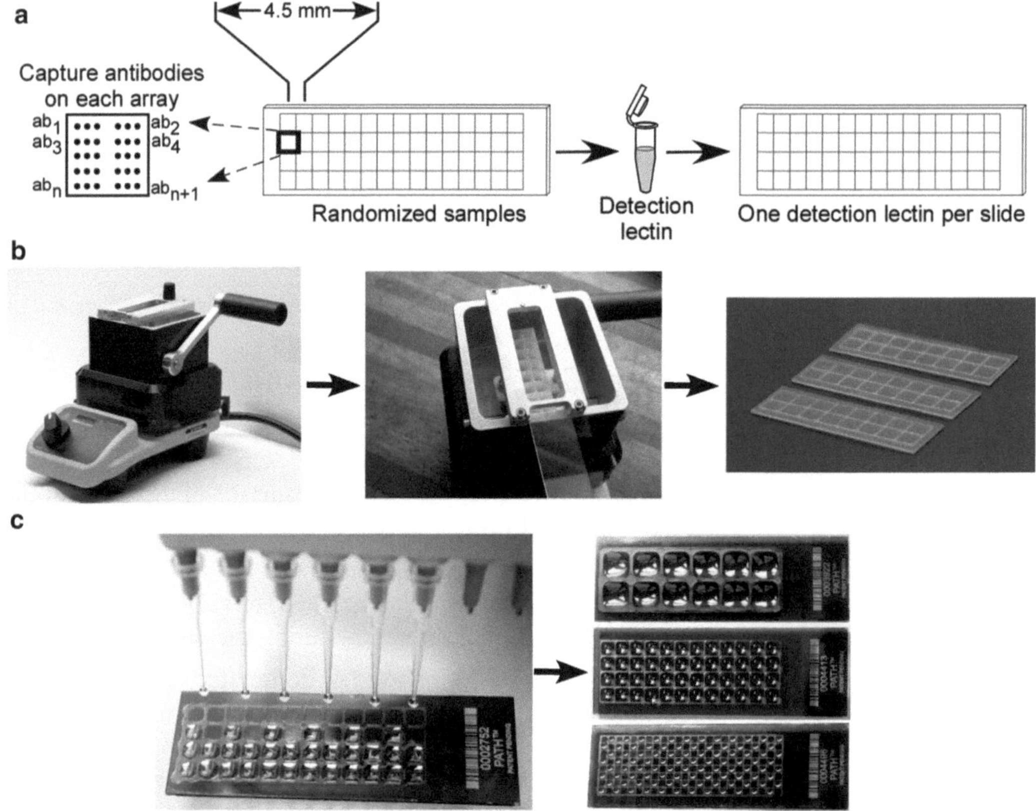

Fig. 3. High-throughput processing of antibody arrays. (**a**) Multiple arrays can be printed on single slides using 4.5 mm spacing between arrays. A convenient approach is to incubate randomized samples on one slide followed by detection using a single lectin. (**b**) Partitioning the arrays. A SlideImprinter device (The GelCompany, San Francisco, CA) is used to imprint wax patterns onto microscope slides. A bath of wax is melted on a hotplate, and a slide is inserted upside-down into a holder above the bath. Pulling the level forward elevates a stamp out of the bath to contact the slide, leaving a pattern of wax on the slide. (**c**) Multichannel liquid handling can reduce loading times and variability between the arrays. Various designs of stamps can be used, for example that partition 12, 48, or 192 arrays (*right*), with smaller array sizes holding proportionately smaller liquid volumes. The slide printed with 192 arrays contains samples in alternating arrays.

17. Scan the slides using a microarray scanner with appropriate resolution (10 μm or better) and emission and excitation settings. If the slides are not scanned immediately, store them vacuum-sealed with desiccant in the refrigerator. Most fluorescent reagents can be stored under these conditions for months without appreciable loss of signal.

3.3. Using Antibody Arrays in Biomarker Experiments

3.3.1. High-Throughput Processing of Antibody Arrays

The ALSA platform can have greatly increased value when coupled with the ability to process the arrays in high throughput. Certain research projects call for the analysis of many samples or conditions. For example, in biomarker research, measurements from dozens to hundreds of subjects can be required to properly assess the performance of a candidate marker. Samples may need to be run multiple times to characterize reproducibility, or multiple conditions may need to be tested to identify their effects on performance. Furthermore, experimental optimization can require the systematic and repeated testing of many different conditions. In these cases, the ability to efficiently and rapidly process many samples is valuable.

The approach presented here is practical and can be implemented with limited investment in new equipment. Begin by printing many replicate microarrays on single microscope slides (Fig. 3a). The number of replicate arrays depends on the size of each array. Antibody microarray experiments are often designed to test specific candidate molecules, which may be just a few or up to a few dozen, and sometimes the size of the array is limited by the availability or cost of the antibodies. A useful array size is a 12×12 grid of 144 spots, which could contain 48 different elements each spotted in triplicate. At a typical spacing of 250 μm between spots, each array has a width of 2.75 mm. At this width, a convenient spacing between the arrays is 4.5 mm, the spacing of a 384-well microtiter plate. This spacing is compatibility with multichannel pipettes for parallel loading of the arrays or automation using liquid-handling robots. A 1×3 in. microscope slide can fit a grid of 4×15 arrays, or 60 arrays, at a spacing of 4.5 mm between the arrays. We have found this to be the optimal arrangement for balancing the ability to print many spots on each array and the ability to print many arrays on each slide.

The next step is to segregate the arrays so that samples do not spill over between the arrays. Segregation can be achieved either by placing an appropriate gasket on top of the slide or by imprinting hydrophobic borders on the slide in between the arrays. Hydrophobic borders can be imprinted on microscope slides using a stamping device that imprints wax patterns (SlideImprinter, The Gel Company, San Francisco, CA) (Fig. 3b). The device lifts a stamp out of a melted wax bath to contact the surface of a microscope slide, thus depositing the pattern of wax. Any design of border can be imprinted, depending on the stamp (Fig. 3c). The thin borders remain on the slide throughout the experiment and the scanning. Gaskets also work well but are not as suitable for small arrays and in our experience are not as convenient, as they require extra steps to assemble and remove pressure-based or adhesive seals to ensure the lack of leaking between arrays. An advantage of small arrays segregated by hydrophobic borders is the small sample volume required for each array; the design using arrays spaced at 4.5 mm requires only 6 μl of solution, which is a significant advantage for biomarker work using clinical samples.

3.3.2. Experimental Considerations for Biomarker Research

For biomarker research, a primary consideration is to minimize experimental variability between samples, especially between the case and control samples, and between experiment sets run on different days. A basic rule is to treat all samples identically, and especially to eliminate experimental differences between the case and control samples. The best way to eliminate differences between the case and control samples is to blind the researchers to the sample identities and to randomize the samples. If the samples are not blinded, they should be randomized and coded with identifiers that do not indicate their disease status.

It is useful to use a single detection lectin or antibody for all the arrays on a microscope slide, with case and control samples randomized on that slide (Fig. 3a). If more samples than will fit on one microscope slide, the samples can be randomized over multiple slides are analyzed, ideally run in the same experimental set to minimize variability. Automated liquid handling and pipetting (Fig. 3c), if available, can facilitate high-throughput studies and further reduce variability between samples.

For a high-throughput profiling experiment design, it is useful to include a dilution curve within an experimental unit. For a 48-pad microarray slide, a typical design is to use four arrays for pooled samples diluted at three dilutions plus a TBS buffer sample. The dilutions should span the dilution used for the samples. The curve serves as a quality control measure to ensure proper functioning of the assay and the detection of the analytes in the linear response range.

4. Notes

1. A useful method to evaluate the purity, concentration, and structural integrity of antibodies is one-dimensional SDS-PAGE in both denaturing and nondenaturing modes. Gels can be stained with Coomassie Blue to reveal protein content. A pure, intact antibody should show a single band at 155 kDa in the nondenaturing gel and single bands at 50 kDa (the heavy chain) and 25 kDa (the light chain) in denaturing gels. Extra bands could indicate contamination or degradation of the antibody.
2. An ultracentrifuge that accepts low-volume tubes (25–100 μl) is useful because high volumes of antibody might not be available. When removing the antibody solution from the tube with a pipette tip, avoid touching the sides or bottom so that precipitated material is not also removed.
3. To properly assess lection performance, it is desirable to have proper positive and negative controls spotted on the arrays. Such controls might be proteins that are known to either contain or not contain the glycan targeted by each lectin. Commercially available and inexpensive control glycoproteins can be tested in preliminary experiments for their reactivity to the lectins of interest.
4. Typical optimal concentrations of detection reagents range from 0.1 to 2 μg/ml. Several factors affect performance, including affinity, avidity, nonspecific binding, and self-interference effects, so the optimal range must be determined experimentally for each reagent.

5. Repeated freezing and thawing can be damaging to proteins and therefore should be minimized. The damage is variable between proteins, but proteomics studies have shown in general that up to three thaws is acceptable for most proteins (16). The proper aliquotting of biological samples enables the use of a fresh aliquot for each experiment with no more than three thaws for each aliquot. The volume of each aliquot should be similar to that used in each experiment to reduce waste of sample.
6. The thawing process should be carried out slowly to minimize damage. Remove samples from −80°C storage and place on ice, allowing the samples to gradually thaw over the course of an hour.
7. Lectins can be purchased biotinylated or can be biotinylated using the appropriate labeling reagents (see Subheading 2). We recommend modifications to the manufacturer-supplied protocols for labeling with *N*-hydroxysuccinimide reagents, in which we lower the pH, temperature, and time of the reaction to limit the number of amine groups that are labeled. See ref. 6 for details.

References

1. Chen, S., LaRoche, T., Hamelinck, D., Bergsma, D., Brenner, D., Simeone, D., Brand, R. E., and Haab, B. B. (2007) Multiplexed analysis of glycan variation on native proteins captured by antibody microarrays. *Nature methods* 4, 437–444.
2. Yue, T., Goldstein, I. J., Hollingsworth, M. A., Kaul, K., Brand, R. E., and Haab, B. B. (2009) The prevalence and nature of glycan alterations on specific proteins in pancreatic cancer patients revealed using antibody-lectin sandwich arrays. *Mol Cell Proteomics* 8, 1697–1707.
3. Dube, D. H., and Bertozzi, C. R. (2005) Glycans in cancer and inflammation--potential for therapeutics and diagnostics. *Nat Rev Drug Discov* 4, 477–488.
4. Hirabayashi, J. (2004) Lectin-based structural glycomics: glycoproteomics and glycan profiling. *Glycoconjugate journal* 21, 35–40.
5. Yue, T., and Haab, B. B. (2009) Microarrays in glycoproteomics research. *Clin Lab Med* 29, 15–29.
6. Haab, B. B., and Zhou, H. (2004) Multiplexed protein analysis using spotted antibody microarrays. *Methods in molecular biology Clifton, N.J* 264, 33–45.
7. Chen, S., and Haab, B. B. (2007) Antibody Microarrays for Protein and Glycan Detection. In: Van Eyk, J., and Dunn, M., eds. *Clinical Proteomics*, Wiley, VCH, Weinheim, Germany.
8. Haab, B. B. (2005) Multiplexed protein analysis using antibody microarrays and label-based detection. *Methods Mol Med* 114, 183–194.
9. Haab, B. B., and Lizardi, P. M. (2006) RCA-enhanced protein detection arrays. *Methods in molecular biology Clifton, N.J* 328, 15–29.
10. Hirabayashi, J. (2008) Concept, strategy and realization of lectin-based glycan profiling. *Journal of biochemistry* 144, 139–147.
11. Culf, A. S., Cuperlovic-Culf, M., and Ouellette, R. J. (2006) Carbohydrate microarrays: survey of fabrication techniques. *Omics* 10, 289–310.
12. Wang, D. (2003) Carbohydrate microarrays. *Proteomics* 3, 2167–2175.
13. Blixt, O., Head, S., Mondala, T., Scanlan, C., Huflejt, M. E., Alvarez, R., Bryan, M. C., Fazio, F., Calarese, D., Stevens, J., Razi, N., Stevens, D. J., Skehel, J. J., van Die, I., Burton, D. R., Wilson, I. A., Cummings, R., Bovin, N., Wong, C. H., and Paulson, J. C. (2004) Printed covalent glycan array for ligand profiling of diverse glycan binding proteins. *Proceedings of the National Academy of Sciences of the United States of America* 101, 17033–17038.

14. Porter, A., Yue, T., Heeringa, L., Day, S., Suh, E., and Haab, B. B. (2009) A motif-based analysis of glycan array data to determine the specificities of glycan-binding proteins. *Glycobiology.*
15. Chen, S., and Haab, B. B. (2009) Analysis of glycans on serum proteins using antibody microarrays. *Methods in molecular biology Clifton, N.J* 520, 39–58.
16. Rai, A. J., Gelfand, C. A., Haywood, B. C., Warunek, D. J., Yi, J., Schuchard, M. D., Mehigh, R. J., Cockrill, S. L., Scott, G. B., Tammen, H., Schulz-Knappe, P., Speicher, D. W., Vitzthum, F., Haab, B. B., Siest, G., and Chan, D. W. (2005) HUPO Plasma Proteome Project specimen collection and handling: towards the standardization of parameters for plasma proteome samples. *Proteomics* 5, 3262–3277.

Chapter 16

Microspot Immunoassay-Based Analysis of Plasma Protein Profiles for Biomarker Discovery Strategies

Johanna Sonntag, Heiko Mannsperger, Anika Jöcker, and Ulrike Korf

Abstract

To expedite the development of personalized medicine, new and reliable biomarkers are required to facilitate early diagnosis, to determine prognosis, predict response or resistance to different therapies, and to monitor disease progression or recurrence. Human body fluids, such as blood, present a promising resource for biomarker discovery, in every sense. Microspot immunoassays allow the simultaneous quantification of multiple analytes from a minute amount of samples in a single measurement. The experimental design of microspot immunoassays is based on antibody pairs recognizing different epitopes of the analyte. The first antibody is used to capture the analyte from the complex sample, and the second antibody is used for detection. As with traditional enzyme-linked immunosorbent assays, highly reliable and reproducible results are obtained.

Key words: Microspot immunoassay, Antibody microarray, Multiplex ELISA, Biomarker, Plasma, Personalized medicine, QuantProReloaded

1. Introduction

A biomarker has been defined as a characteristic that is objectively measured and evaluated as an indicator of normal biological processes, pathogenic processes, or pharmacologic responses to a therapeutic indication (1). Biomarkers can be used in various applications, for example in diagnosis, prognosis, or prediction. As diagnostic tool, biomarkers can identify patients with disease or abnormal conditions. Diagnosis of diabetes mellitus, for example, can be facilitated by detection of an elevated blood glucose level (2). Prognostic biomarkers can give valuable hints on the severity of a certain disease and can aid therapy decisions (3). Especially in cancer treatment, this is a fundamental problem, since many patients are overtreated, whereas others are treated ineffectively (4, 5).

Ulrike Korf (ed.), *Protein Microarrays: Methods and Protocols*, Methods in Molecular Biology, vol. 785,
DOI 10.1007/978-1-61779-286-1_16,

Thus, there is a strong interrelation between prognostic and predictive biomarkers. A predictive biomarker has the potential to give hints as to which subgroup of patients will benefit from a special treatment, and which subgroup will not. As the field of targeted therapies for personalized medicine is constantly growing, the identification of suitable biomarkers becomes more and more important (6). Clinically useful predictive biomarkers in breast cancer, for example, are the ones indicating a positive estrogen receptor status, or a positive HER2 status, to select patients for endocrine therapy, or for therapy with trastuzumab, respectively (7). In addition, a biomarker can be used to monitor treatment response or recurrence of disease over time. For example, the prostate-specific antigen (PSA) is well-accepted as biomarker that monitors the recurrence of prostate cancer (8), whereas in the field of breast cancer, carcinoembryonic antigens (CEA) and CA15.3 can serve as biomarkers to monitor disease recurrence (9).

Blood is a preferred source for biomarker discovery studies because it should reflect the various physiological or pathological states due to its heterogeneous composition, including metabolites, peptides, proteins, cell-free DNA and RNA, circulating tumor cells, and auto-antibodies. In addition, blood can be sampled in a comparatively less-invasive way, as well as in outpatient settings (10). Traditionally, mass spectrometry and immunoassays are commonly used for biomarker discovery and validation. Enzyme-linked immunosorbent assays (ELISA) are still the gold standard for single-analyte detection and quantification in the clinical routine, but are not well-suited for the discovery process, where multiple candidate proteins are investigated in parallel (11). However, microspot immunoassays overcome the technical limitations of ELISA techniques by saving sample volume, time, and reagent consumption. Figure 1 illustrates the experimental design of microspot immunoassays. Capture antibodies are spotted as technical replicates, and in a predefined number of identical subarrays on nitrocellulose-coated glass slides. Afterward, the slides are mounted in an incubation chamber to create distinct wells for each subarray. These wells are used to either incubate with a serially diluted standard protein mix, or with plasma samples. The detection of the analyte is done in a three-step procedure by incubation with sample, a biotinylated detection antibody mix, and finally, with a near-infrared-dye labeled streptavidin. In between the different working steps, a thorough wash procedure is employed to remove even minute amounts of materials that could potentially increase experimental noise. Next, all slides are scanned on an infrared imaging system to determine the signal intensities of the different spots. The resulting data is then used to calculate a calibration curve for each of the different analytes present in the multiplexed standard mixture, and to assess the different analyte concentration in all samples. This data processing step can be facilitated with a tailored software program, such as QuantProReloaded (12). A crucial step for the development

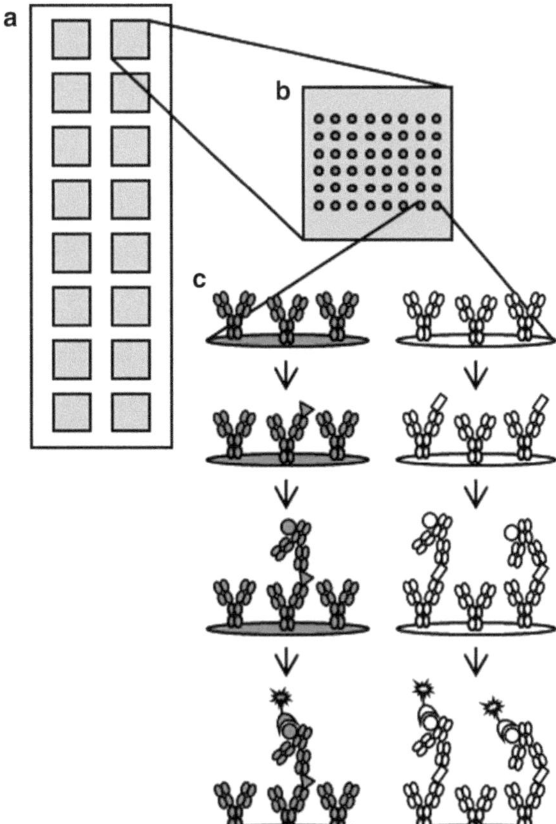

Fig. 1. Experimental design of a microspot immunoassay. Nitrocellulose-coated glass slides with 16 identical capture antibody subarrays as the basis for a miniaturized 16-well incubation plate. The first six subarrays are dedicated for incubation with calibrator proteins, with remaining available for incubation with sample (**a**). Capture antibodies are printed in technical replicates on the subarray (**b**). The detection of analytes in the sample is performed in a three-step procedure by incubation with samples, biotinylated-detection antibodies, and finally with near-infrared-dye labeled streptavidin. Between these three steps, a thorough washing procedure is employed (**c**).

of a new microspot immunoassay is the identification of suitable antibody pairs. Besides common requirements of an immunoassay, such as accuracy, linearity, recovery, specificity, and sensitivity (13), suitable antibody pairs must also work when used in a multiplexed set-up. This can best be verified by tests proposed by Gonzalez and coworkers (14).

2. Materials

2.1. Slide Printing

1. 384-well microtiter plate with lid (see Note 1).
2. 10× phosphate-buffered saline (PBS) stock solution: 1.37 M NaCl, 27 mM KCl, 18 mM KH_2PO_4, 100 mM Na_2HPO_4,

pH 7.4. Autoclave and store at room temperature. Dilute with nine parts of deionized water to prepare a 1× PBS solution.

3. PBS-T: 0.1% Tween-20 (Sigma-Aldrich Laborchemikalien GmbH, Seelze, Germany) in PBS.
4. Capture antibodies recognizing the targets of interest and validated to work in multiplexed microspot immunoassay.
5. Sealing film for microtiter plate (Steinbrenner Laborsysteme GmbH, Heidelberg, Germany).
6. Nitrocellulose-coated glass slides (ONCYTE® Avid, nitrocellulose film-slides, Grace Bio-Labs, Bend, OR, USA) (see Note 2).
7. Aushon 2470 arrayer equipped with 185 μm pins (Aushon BioSystems, Billerica, MA, USA).

2.2. Microspot Immunoassay

1. Blocking buffer for fluorescent Western blotting (Rockland Immunochemicals, Inc., Gilbertsville, PA, USA).
2. Bovine serum albumin (BSA).
3. Wash box (LI-COR Bioscience, Lincoln, Nebraska, USA) (see Note 3).
4. 8-channel pipette.
5. 8-channel aspiration device.
6. Incubation chamber 3/16 (Metecon GmbH, Mannheim, Germany) (see Note 4).
7. PBS-T wash buffer: 0.1% Tween-20 (Sigma-Aldrich Laborchemikalien GmbH, Seelze, Germany) in phosphate-buffered saline.
8. Plasma samples (see Note 5).
9. Antigens of known concentration, matching the specificity of capture antibodies.
10. Fetal bovine serum (FBS) diluted 1:5 with deionized water.
11. 96-well microtiter plate (Greiner Bio-One GmbH, Frickenhausen, Germany).
12. Biotinylated detection antibody for each analyte, which was validated to work in multiplexed microspot immunoassay.
13. Tris-buffered saline with 0.1% Tween-20 (Sigma-Aldrich Laborchemikalien GmbH, Seelze, Germany) and 0.1% BSA (Sigma-Aldrich Laborchemikalien GmbH, Seelze, Germany).
14. Streptavidin Alexa Flour® 680 conjugate 2 mg/ml (Invitrogen GmbH, Karlsruhe, Germany).

2.3. Slide Scanning and Data Analysis

1. Infrared imaging system (Odyssey®, LI-COR Bioscience, Lincoln, Nebraska, USA).
2. Microarry image analysis software (GenePixPro, Molecular Devices, Sunnyvale, CA, USA).
3. QuantProReloaded (http://code.google.com/p/quantpro-reloaded/).

3. Methods

3.1. Slide Printing

Slides are printed employing the Aushon 2470 contact arrayer, equipped with eight 185 μm solid pins. On each slide, 16 identical subarrays consisting of sixfold technical replicates of capture antibodies are printed, using a printhead configuration of 2 × 4 with 9 × 9 μm pin spacing.

1. Fill the wash water container with deionized water, and empty the waste water container.
2. Load the desired number of nitrocellulose-coated glass slides onto slide platens and place them into the arrayer (see Note 6).
3. Dilute the capture antibody stock solutions to an appropriate capture antibody concentration (see Note 7), and adjust the Tween-20 concentration to a final concentration of 0.05% (see Note 8).
4. Transfer 3–5 μl capture antibody solution per well to a 384-well microtiter plate, cover with sealing film and centrifuge for 1 min at 2,000 × *g*. The 8-pins printhead configuration requires that each capture antibody has to be present in eight wells. For example: A1, A3, A5, A7, C1, C3, C5, and C7 are used for the first capture antibody, and A2, A4, A6, A8, C2, C4, C6, and C8 are used for the second capture antibody, and so on.
5. Place the covered 384-well microtiter plate on the plate holder and position the plate holder into the arrayer.
6. Start the software and design the array layout with the following settings. Top offset: 6.2 mm; left offset: 3.15 mm; x-spacing and y-spacing: at least 400 μm; replication type: horizontal linear; super-array replicates: 2; number of replicates: 5; deposition order: by feature; super-array order: column–row.
7. Set the wash procedure parameters.
8. Set humidity to 80%.
9. Start printing run.
10. Printed slides can be used immediately or stored at 4°C for up to 1 month.

3.2. Microspot Immunoassay

The following protocol describes the processing of the three slides in parallel in a single incubation chamber (see Note 9). This set-up allows the determination of analyte concentrations for ten plasma samples in triplicate measurements.

1. Transfer three slides to the wash box, and block the free-binding sites with 7.5 ml blocking buffer overnight at 4°C, on a rocking platform.

2. Prepare the standard mix from recombinant proteins used as reference. Make an 11-step twofold dilution series with FBS/deionized water mixture (1:5) (see Note 10). The first and the last step of the dilution series requires a minimum of 360 μl standard mix, all other dilution steps at least 160 μl.
3. Dilute plasma samples 1:6 in PBST and centrifuge for 10 min at 16,000 × *g* (see Note 11).
4. Mount slides in incubation chamber and add 300 μl wash buffer to each well to prevent the nitrocellulose surface from drying. The incubation chamber creates 16 distinct wells on each of the three slides.
5. Prepare the standard mix/sample template plate. Transfer 160 μl of the standard mix or plasma sample to a 96-microtiter plate in the same order as they should be incubated on the slides. The first six wells of each slide are dedicated to incubation with standard mix, and the remaining ten wells to incubation with plasma samples. The first and the last step of the dilution series, as well as the antigen-free blank, should be incubated on all three slides (see Note 12).
6. Remove the wash buffer by aspiration using an 8-channel aspiration device, and transfer 150 μl per well standard mix, or sample with an 8-channel pipette from the template plate, to the slides in the incubation chamber.
7. Cover the incubation chamber with a lid and incubate for 2 h at room temperature on a rocking platform.
8. Aspirate the standard mix/samples and wash each well 2× 5 min with 300 μl wash buffer on a rocking platform.
9. Remove the slides from the incubation chamber, and place them upside up into the wash box. Subsequently, wash slides twice for 5 min with 15 ml wash buffer on a rocking platform.
10. Dilute the biotinylated detection antibodies in Tris-buffered saline containing 0.1% Tween-20 and 0.1% BSA, and cover slides with 7 ml detection antibody mix. Incubate for 1 h at room temperature on a rocking platform.
11. Aspirate off the detection antibody mix, and wash slides 4× 5 min with 15 ml wash buffer on a rocking platform.
12. Add 7.5 ml Streptavidin Alexa Flour® 680 conjugate diluted 1:5,000 in wash buffer and incubate for 30 min on a rocking platform.
13. Aspirate off the Streptavidin Alexa Flour® 680 conjugate and wash the slides 4× 5 min, with 15 ml wash buffer on a rocking platform. Afterward, wash the slides 2× 5 min with deionized water.
14. Air-dry slides in the dark.

3.3. Scanning and Data Analysis

1. Scan the slides on the Odyssey® infrared imaging system. Use the 700-channel (solid-state diode laser at 680 nm) and a resolution of 21 μm. Start with a scan intensity of five, and adjust the scan intensity, if necessary.
2. Quantify the spot intensities using the resulting TIFF-file with a microarray image analysis software, e.g., GenePix Pro.
3. The software QuantProReloaded, or a comparable software program, can be used for a standard curve fitting, and the estimation of analyte concentrations.

4. Notes

1. The microtiter plate must meet the specifications of the Aushon 2470 microtiter plate holder. The Aushon instrument requires a microtiter plate with lid to avoid evaporation during spotting runs. Lids are removed just before printing.
2. Nitrocellulose-coated glass slides are available with single-pad and multi-pad coatings. To our experience, single-pad slides yield better results than multi-pad slides, due to a more even coating. Thus, the size of capture antibody spots is uniform, and merely controlled by the drop size, and not by the thickness of the nitrocellulose coating. In addition, detachment of single pads was occasionally observed when working with multi-pad slides.
3. Use a wash box which protects the slides from light.
4. The incubation chamber was developed at the German Cancer Research Center (Heidelberg, Germany) in collaboration with Metecon GmbH (Mannheim, Germany; http://www.metecon.info/). The chamber is designed to hold three slides and creates 16 distinct wells on each slide. The incubation chamber wells match the measurements of a standard 96-well microtiter plate, and therefore allow liquid handling with an 8-channel pipette/8-channel aspiration device; these are compatible with microplate shakers. In addition, the wells of the Metecon-16-pad frame can take up to 300 μl of wash buffer, whereas other commercially available incubation chambers often have a much smaller wash-buffer capacity. We have noticed that a high wash-buffer volume in this type of miniaturized assay largely improves the efficacy of the washing procedure.
5. Analyte properties should be considered to decide whether rather serum or plasma samples are the preferred source to answer the clinical question of interest. For example, VEGF should not be assayed from human serum samples (15). Similarly, the decision on the anti-coagulant used for sample

preparation is equally important (16). In order to yield robust results, the protocol for sample collection and sample preparation should be evaluated carefully, before starting to collect large numbers of patient samples.

6. A single-slide platen can hold up to ten nitrocellulose-coated glass slides. We obtained improved spotting morphology by using the last slide of a platen to remove excess wash water from the pins, and by blotting the pins after each wash cycle.
7. The optimal capture antibody concentration depends on a particular type of antibody and has to be determined as one of the initial steps during assay development. As a guideline, we recommend starting with a capture antibody concentration of 250 μg/ml.
8. Best results with respect to spot morphology were obtained by adding Tween-20 in a final concentration of 0.05% (w/v) to the capture antibody solution.
9. Processing several incubation chambers in parallel is possible so that 30–40 different samples can easily be measured per day by a single person.
10. FBS can be used as diluent to mimic the complex matrix of the human plasma samples.
11. The appropriate dilution factor of the plasma samples needs to be determined experimentally during assay set-up, and depends on the sensitivity of the particular assay, the linear range of detection procedure, the expected concentration range of a particular analyte, as well as the concentration range of other analytes assayed in parallel.
12. The lowest and the highest concentration of a serial dilution of the standard proteins are included on all three slides. The other nine concentration points are distributed on the three slides and are assayed only once. This way, slide-to-slide irregularities are detected and can be corrected as part of a slide normalization procedure.

Acknowledgments

This work was supported by the German Federal Ministry for Education and Science in the framework of the Program for Medical Genome Research (grants 01GS0890 and 01GS0864), the Program for Medical Systems Biology (grant 0315396B), as well as the Helmholtz Systems Biology Initiative (SBCancer). Thanks go also to Liz Segal for her help with proofreading.

References

1. Biomarkers Definitions Working Group (2001) Biomarkers and surrogate endpoints: preferred definitions and conceptual framework, *Clin Pharmacol Ther 69*, 89–95.
2. American Diabetes Association (2010) Diagnosis and classification of diabetes mellitus, *Diabetes Care 33 Suppl 1*, S62-69.
3. Donegan, W. L. (1992) Prognostic factors. Stage and receptor status in breast cancer, *Cancer 70*, 1755–1764.
4. Donnelly, J. G. (2004) Pharmacogenetics in cancer chemotherapy: balancing toxicity and response, *Ther Drug Monit 26*, 231–235.
5. Dunn, L., and Demichele, A. (2009) Genomic predictors of outcome and treatment response in breast cancer, *Mol Diagn Ther 13*, 73–90.
6. Duffy, M. J., and Crown, J. (2008) A personalized approach to cancer treatment: how biomarkers can help, *Clin Chem 54*, 1770–1779.
7. Goldhirsch, A., Ingle, J. N., Gelber, R. D., Coates, A. S., Thurlimann, B., and Senn, H. J. (2009) Thresholds for therapies: highlights of the St Gallen International Expert Consensus on the primary therapy of early breast cancer 2009, *Ann Oncol 20*, 1319–1329.
8. Lilja, H., Ulmert, D., and Vickers, A. J. (2008) Prostate-specific antigen and prostate cancer: prediction, detection and monitoring, *Nat Rev Cancer 8*, 268–278.
9. Molina, R., Barak, V., van Dalen, A., Duffy, M. J., Einarsson, R., Gion, M., Goike, H., Lamerz, R., Nap, M., Soletormos, G., and Stieber, P. (2005) Tumor markers in breast cancer- European Group on Tumor Markers recommendations, *Tumour Biol 26*, 281–293.
10. Hanash, S. M., Pitteri, S. J., and Faca, V. M. (2008) Mining the plasma proteome for cancer biomarkers, *Nature 452*, 571–579.
11. Rai, A. J. (2007) Biomarkers in translational research: focus on discovery, development and translation of protein biomarkers to clinical immunoassays, *Expert Rev Mol Diagn 7*, 545–553.
12. Jöcker, A., Sonntag, J., Henjes, F., Gotschel, F., Tresch, A., Beissbarth, T., Wiemann, S., and Korf, U. (2010) QuantProReloaded: quantitative analysis of microspot immunoassays, *Bioinformatics 26*, 2480–2481.
13. Sweep, F. C. G. J., Thomas, C. M. G., and Schmitt, M. (2006) Analytical aspects of biomarker immunoassays in cancer research, in *Biomarkers in breast cancer: Molecular diagnostics for predicting and monitoring therapeutic effect* (Gasparini, G., and Hayes, D. F., Eds.), pp 17–30, Humana Press.
14. Gonzalez, R. M., Seurynck-Servoss, S. L., Crowley, S. A., Brown, M., Omenn, G. S., Hayes, D. F., and Zangar, R. C. (2008) Development and validation of sandwich ELISA microarrays with minimal assay interference, *J Proteome Res 7*, 2406–2414.
15. Jelkmann, W. (2001) Pitfalls in the measurement of circulating vascular endothelial growth factor, *Clin Chem 47*, 617–623.
16. Rai, A. J., Gelfand, C. A., Haywood, B. C., Warunek, D. J., Yi, J., Schuchard, M. D., Mehigh, R. J., Cockrill, S. L., Scott, G. B., Tammen, H., Schulz-Knappe, P., Speicher, D. W., Vitzthum, F., Haab, B. B., Siest, G., and Chan, D. W. (2005) HUPO Plasma Proteome Project specimen collection and handling: towards the standardization of parameters for plasma proteome samples, *Proteomics 5*, 3262–3277.

Chapter 17

Recombinant Antibodies for the Generation of Antibody Arrays

Carl A.K. Borrebaeck and Christer Wingren

Abstract

Affinity proteomics, mainly represented by antibody microarrays, has in recent years been established as a powerful tool for high-throughput (disease) proteomics. The technology can be used to generate detailed protein expression profiles, or protein maps, of focused set of proteins in crude proteomes and potentially even high-resolution portraits of entire proteomes. The technology provides unique opportunities, for example biomarker discovery, disease diagnostics, patient stratification and monitoring of disease, and taking the next steps toward personalized medicine. However, the process of designing high-performing, high-density antibody micro- and nanoarrays has proven to be challenging, requiring truly cross-disciplinary efforts to be adopted. In this mini-review, we address one of these key technological issues, namely, the choice of probe format, and focus on the use of recombinant antibodies vs. polyclonal and monoclonal antibodies for the generation of antibody arrays.

Key words: Recombinant antibodies, Antibody arrays, Microarrays, Nanoarrays, Antibody design, scFv, Protein expression profiling, Disease proteomics

1. Introduction

Affinity proteomics, mainly represented by antibody arrays, is a recently established proteomic technology providing unique opportunities for protein expression profiling of crude, nonfractionated proteomes that further enhance our fundamental knowledge of biological processes in both health and disease (1–6). The current concept of generating miniaturized (mm^2 to cm^2 range) antibody arrays is based on either printing (pL scale or less) (1, 6–8), self-addressing (9–11), or self-assembling (12–17) small amounts (fmol range) of individual antibodies (ranging from a few to several hundreds) with the desired specificity in discrete positions (nm to

Ulrike Korf (ed.), *Protein Microarrays: Methods and Protocols*, Methods in Molecular Biology, vol. 785, DOI 10.1007/978-1-61779-286-1_17, © Springer Science+Business Media, LLC 2011

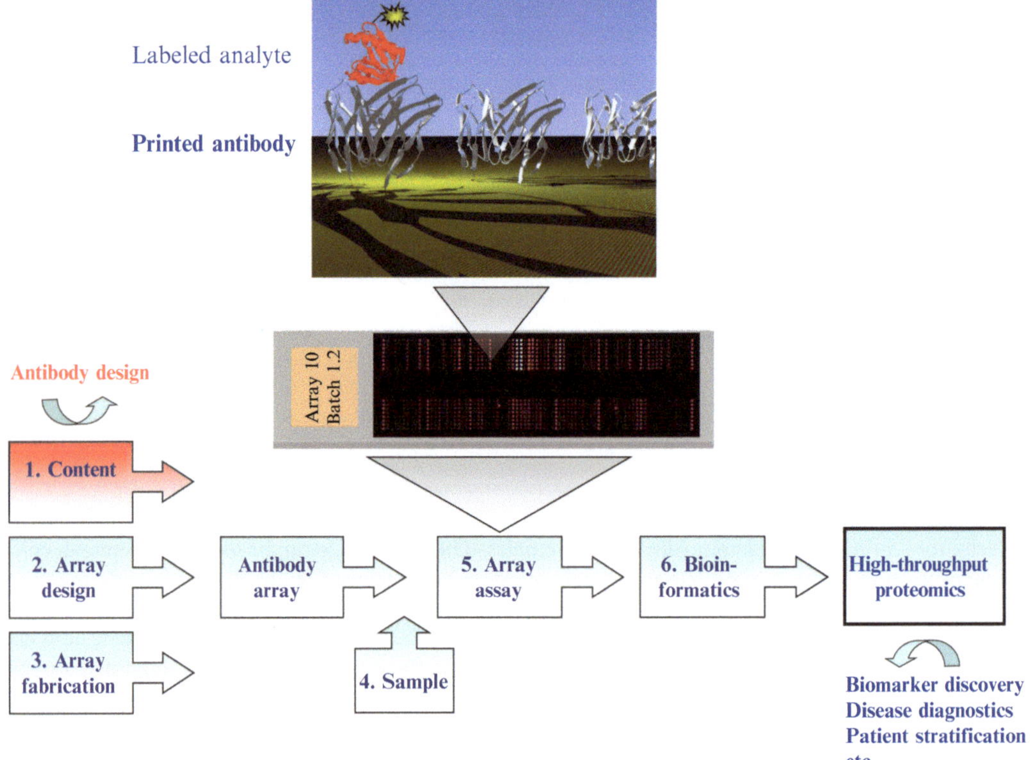

Fig. 1. Schematic illustration of the recombinant antibody microarray setup and key technological issues to consider when designing arrays.

μm sized spot features) in an ordered pattern, an array, onto a solid support (Fig. 1). As an alternative to planar affinity arrays, the so-called bead arrays, based on antibody-functionalized beads in solution, have also been developed (4, 18–20). The immobilized antibodies will then act as highly specific capture molecules, or probes, for the targeted analytes. The arrays are then exposed to minute amounts (μL scale) of (labeled) crude biological sample, e.g., serum, before any specifically bound analytes are detected and quantified, mainly using fluorescence as mode of detection. These multiplexed and rapid (<3 h assay time) analysis regularly display an assay sensitivity in the pM to fM range, enabling also low-abundant (pg/ml) analytes to be profiled in crude proteomes (21–23). The generated array images are then transformed into protein expression profiles, or protein maps, deciphering the composition of the sample at the molecular level.

To date, a range of antibody array setups have been developed and established as a key tool for affinity proteomic-based efforts (1, 5–7, 21–24), as illustrated by several successful applications within disease proteomics (2, 3, 25–29). From a technical point of view, this development work has clearly shown that six key issues should be addressed in parallel, multidisciplinary initiatives to

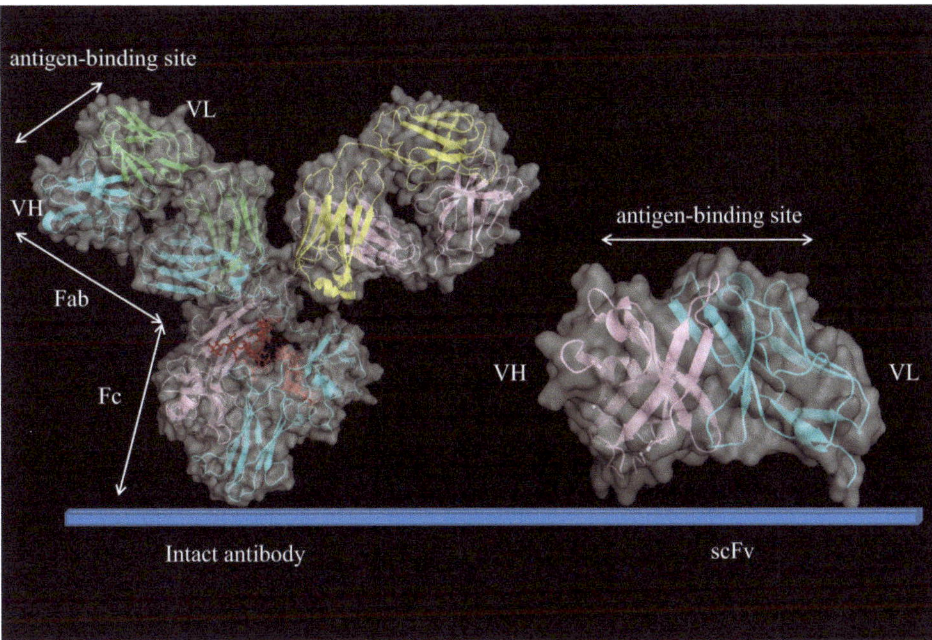

Fig. 2. Structural representation of arrayed intact antibody versus recombinant scFv antibody fragment.

successfully generate high-performing array technology platforms suitable for high-throughput proteomics (1, 30), including the (1) content, (2) array design, (3) array fabrication, (4) sample, (5) array assay, and (6) bioinformatics (Fig. 1). Based on these efforts, the repertoire of currently available state-of-the-art array platforms allows the end-user to perform rapid, selective, and sensitive assays commonly targeting 10–400 analytes per assay, for review see (1, 3, 5–7, 30, 31). But despite this success, there is still room for significant technological advances to be achieved within each of these six key areas (1, 30), in particular since efforts have been launched to extend the methodology beyond the current state-of-the-art to enable global proteomics, setting a novel standard for affinity proteomics. In this mini-review, we address on one of these points, namely, the choice of probe format, and focus on the use of recombinant antibody fragments vs. other antibody formats for the generation of antibody arrays (Fig. 2).

2. Probe Format

2.1. Key Issues

So far, antibodies are by far the most well-characterize and commonly used affinity reagent within biomedical and biotechnical research. Consequently, they have been applied in a wide variety of applications, such as ELISA, Western blot, immunohistochemistry (IHC), flow cytometry, and more recently also affinity proteomics.

Considering their intrinsic specificity and affinity for their target antigen, combined with their wide diversity, i.e., range of specificities, generated in vivo, antibodies are and will continue to be a highly preferred choice as affinity reagent (32). In this review, we discuss only the use of different antibody formats as probes for arrays. The use of antibodies vs. affinity reagents based on other scaffolds, such as affibodies (33, 34) and aptamers (35–38), is outside the scope of this article, and has been reviewed elsewhere (1, 6, 30, 31).

To date, antibody arrays have been produced using different antibody formats as probes, including (1) intact monoclonal antibodies (mAb) (of different isotypes), (2) intact polyclonal antibodies (pAb), (3) fragments of intact antibodies, e.g., $F(ab')_2$, and (4) recombinant scFv and Fab antibody fragments (Fig. 2) (1, 5, 26, 30, 39–41). While functional arrays have been obtained based on all these probe formats, the choice is essential, setting the stage for the downstream applications. In more detail, the choice of probe format directly affects and/or reflects the (1) on-chip probe performances, (2) range of specificities, (3) scaling-up, (4) renewability, (5) choice of scaffold, and (6) redesign. Based on these six essential points, it is thus essential to select the probe format that best matches the requirements of the specific application at hand. So far, a majority of the arrays have been produced using intact mAb and pAb as probes. However, as is outlined and discussed below, this bias has probably reflected the general availability of the various antibody formats to the broader life science community, rather than reflecting strict, rational decisions. As a matter of fact, recombinant antibodies have been found to constitute an excellent probe source for array-based applications (1, 21, 23, 30, 42–44).

2.2. On-Chip Performances

The process of producing antibody arrays by printing the probes in pL scale drops, that will evaporate within a few seconds after hitting the surface adopting the most commonly used printing protocols, will place high demands on the protein structure (42). It is well known that proteins in general tend to denature and lose their function when immobilized onto a solid support in a dried out state. Based on early work, concerns were also raised regarding the functionality of antibodies for array-based applications (8, 45, 46). In fact, the work showed that the on-chip performances of readily available off-the-shelf intact mAb and pAb varied significantly, with up to 95% displaying poor performances in some studies (8, 47). In addition, it is critical that the arrayed probes behave as similar as possible with respect to, e.g., their on-chip functionality and stability, to make sure that any observed differences are due to true differences in analyte levels and not simply reflecting differences in probe performances. Since pAb and mAb are produced in vivo, the choice of scaffold, the variable and constant domains building up the molecule, cannot be selected and/or predetermined, and could

Table 1
Clans and subclasses of human IgG

IGHV	I	IGHV1, IGHV5, IGHV7
	II	IGHV2, IGHV4, IGHV6
	III	IGHV3
IGKV	I	IGKV1
	II	IGKV2, IGKV3, IGKV4, IGKV6
	III	IGKV5, IGKV7
IGLV	I	IGLV1, IGLV2, IGLV6, IGLV10
	II	IGLV3
	III	IGLV7, IGLV8
	IV	IGLV5, IGLV11
	V	IGLV4, IGLV9
Subclasses	IgG1, IgG2, IgG3, IgG4	

thus vary significantly from one antibody clone to another. In the case of human IgG antibodies, 3 × 8 × 4 scaffolds can be utilized in vivo (48, 49). These 96 scaffolds, all generating functional proteins, are built up of three clans (a set of subgroups which appear related on phylogenetic trees) of the variable domains of the heavy chain (VH), eight clans of the variable domain of the light chain (VL) (3 κ and 5 λ), and four subclasses (Table 1). Consequently, it is likely that these molecules, based on different scaffolds (Fig. 3a), also will display different on-chip properties. Hence, the data clearly demonstrated that the functionality of intact mAb and pAb in array-based applications have to be tested and validated prior to use a scenario common to many other antibody-based assays, e.g., ELISA and Western blot. Albeit manageable, this poses a key logistical bottleneck, clearly highlighting the key advantage of using probes that have been predesigned to perform well in array applications.

In this context, it should be noted that a number of large (>10^{10} members) recombinant single-chain Fv (scFv) and Fab antibody phage display libraries have been constructed (50–55) potentially constituting promising probe sources. The access of these libraries to the life science community in general could, however, be restricted based on several factors reflecting, e.g., technical, commercial, and IPRs issues. Still, recombinant scFv (and to some extent Fab) antibodies have frequently been evaluated and used in various array applications (21, 23, 30, 42, 43, 56–61). While scFv antibodies selected from different libraries have been found to display varied on-chip performances, such as functional stability, some libraries have been validated to provide probes displaying excellent

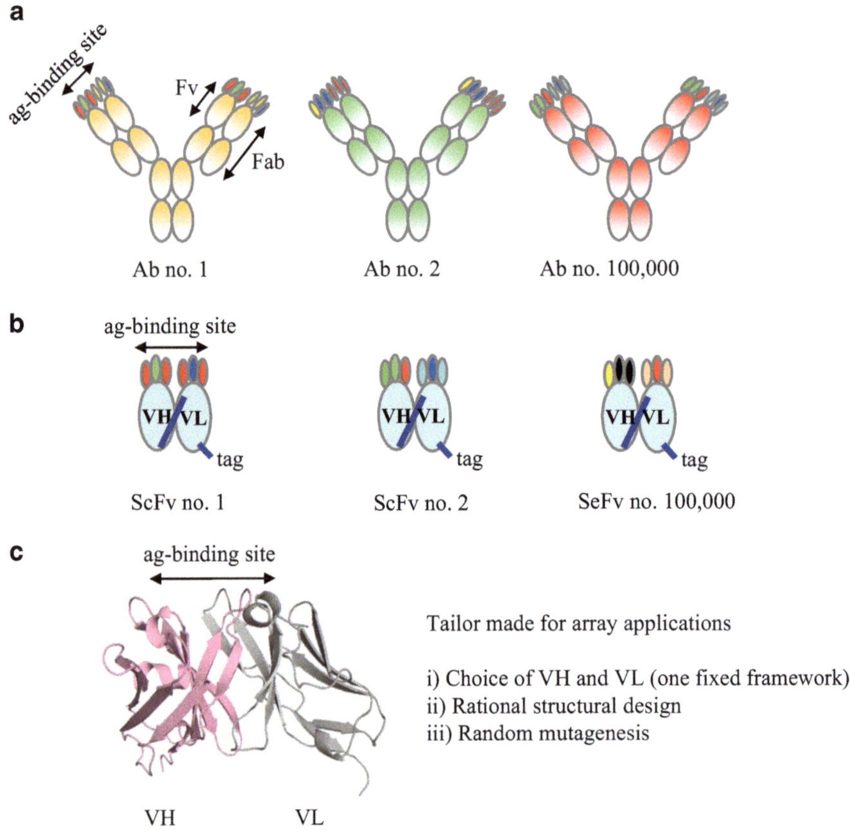

Fig. 3. Correlation between antibody format and choice of scaffold. (**a**) The choice of scaffold cannot be selected and/or predetermined when generating intact pAb and mAb, but will vary significantly from clone to clone. (**b**) Large recombinant antibody libraries can be generated based on a single, fixed scaffold (VH–VL). (**c**) Schematic outline of redesign strategies for recombinant scFv antibodies in order to enhance their on-chip performances even further.

on-chip performances, for review see (1, 2, 42, 60). As for example, scFv antibodies selected from the n-CoDeR library (55), all based on a single, fixed scaffold (VH-3-23/VL-1-47), i.e., constant set of VH and VL domains (Fig. 3b), have been thoroughly validated as probes for array-based applications (1, 21, 23, 30, 42, 62). In more detail, the probes have been demonstrated to display high on-chip functionality, meaning e.g., that the arrays could be produced on day 0, and then be stored in a dehydrated state at room temperature for up to 150 days while the arrayed antibodies still maintained almost full activity (63, 64). More recently, these data have been extended, outlining a shelf-life of arrayed dehydrated scFv arrays up to 475 days (Wingren et al., unpublished observations). On-chip properties like these are essential for enabling practical handling of the arrays, long-term storage, etc. Further, the probes have been found to be compatible with array designs in the end providing high selectivity and specificity, targeting low-abundant protein analytes (pM to fM range) in crude, directly labeled proteomes containing thousands of irrelevant proteins (21, 44, 65). Hence, recombinant probe sources can be generated providing numerous recombinant antibodies displaying superior on-chip performances, setting the stage for antibody array-based applications.

2.3. Availability/Range of Specificities/ Scaling-Up

The availability of antibodies and range of antibody specificities at hand have frequently been highlighted as a limiting factor for large-scale antibody endeavors, including antibody arrays (4, 17, 66, 67). This is the case even though numerous mAb and pAb are available from several commercial and academic sources. Several factors can be pin-pointed explaining this apparent contradiction. Firstly, a majority of these antibodies have not been tested in array applications, and considering the high frequency (up to 95%) of antibodies frequently failing to meet the required on-chip performances (6, 8, 42, 47, 68), many of these antibodies could be disqualified. This could, in turn, mean that several antibodies would have to be remade in order to have binders displaying the desired specificities. Secondly, a wide range of specificities are still not at hand. However, generating the large number of mAb and pAb that would be required for large-scale affinity proteomics, i.e., to cover for all missing antibody specificities and for those probes that would have to be remade, by conventional animal immunization technologies (69), will be logistically overwhelming, labor intensive, and a costly approach to undertake (43, 60, 70). Thirdly, the limited accessibility of purified proteins for all desired target antigens is another key bottleneck. In this context, it could be of interest to highlight the effort launched by the Swedish Human Proteome Resource program, aiming to generate affinity purified polyclonal rabbit antibodies against the nonredundant human proteome (about 20,400 target proteins) (http://www.ensembl.org) (71–73). These antibodies have been generated against protein epitope signature tags (PrEST) antigens, consisting of 100–150 amino acid residues. So far, about 8,800 antibodies have been validated, predominantly by determining their reactivity patterns against a variety of tissues in IHC, and released in the protein atlas (version 5) (http://www.proteinatals.org). Fourthly, the specificity and functionality of many antibodies have been tested and evaluated in other applications than array-based setups, such as Western blot and IHC. More precisely, adopting antibodies originally optimized for setups targeting denatured or semi-denatured protein could cause serious issues in most array setups, since their reactivity against the native counterpart protein could pose a potential issue. In practice, this means that finding existing, array-validated mAb and pAb with the desired specificities could be challenging. These key limitations also pose a major bottleneck when scaling up, aiming for high-density arrays for use in large discovery projects.

By adopting large recombinant antibody libraries, composed of up to 10^{10} of members displaying an extended range of specificities (55), the supply of antibodies and range of specificities per se is not the limiting factor. In fact, recombinant antibody libraries have been designed to provide numerous probes containing rearrangements beyond what is normally found in the natural repertoire (74). In contrast to the situation for mAb and pAb, these circumstances make recombinant antibody libraries a probe source compatible with generating high-density arrays. Still, the costs for and/or logistics associated with performing large-scale selections are substantial (75). The lack of purified proteins as target antigen for selections could of course cause the same issue as for the large-scale production efforts of mAb and pAb.

The need for high-density arrays, i.e., numerous of antibodies, when aiming for global proteomics could, however, be significantly reduced by adopting the Global Proteome Survey (GPS) approach (76). We have recently developed a novel affinity proteomic discovery platform providing us with unique means to perform (in-depth) global proteome analysis using a minimal number of binders (76) (Olsson et al. unpublished observations). The technology platform is based on interfacing affinity proteomics with a mass spectrometry-based read-out. Briefly, we have defined the first generation of a set of four or six amino acid long peptide motifs, each motif being shared among 5–100 different proteins. Next, we have selected recombinant scFv antibodies specific for these motifs. In this way, 200 antibodies, each targeting a motif shared among 50 different proteins, would potentially target almost half the nonredundant human protein. Furthermore, this novel class of motif-specific antibodies could be used to enrich motif containing peptides from digested proteomes in a specie-independent manner, providing yet another competitive edge. To run the setup, the target proteome would first be digested, exposed to immobilized antibodies, whereby motif containing peptides would be enriched. Next, eluted peptides would then be detected and identified using tandem mass spectrometry allowing us to back-track the original protein even in a quantitative manner. In other words, this approach would significantly reduce the number of antibodies required, by using antibodies specific for a set of proteins, instead of one unique probe per protein as in classical affinity proteomic.

2.4. Renewability

Although pAb constitutes a useful and valuable resource for many applications, its applicability is hampered by the fact that the source can only be renewed by a novel, additional animal immobilization, introducing batch-to-batch variations (32, 77). In contrast, mAb and recombinant antibodies are readily renewable probe sources. Recombinant antibody fragments can readily be produced in, e.g., *Escherichia coli* and *Pichia pastoris*, and simple, established production protocols, commonly resulting in high protein yields, are already at hand.

In this context, it should also be considered how the arrays are produced. First, adopting a standard printing protocol, in which the antibodies are deposited one by one by mainly using ink-jet printers, requires access to predominantly prepurified antibodies compatible with all antibody probe formats. Second, fabricating the arrays by self-addressing, where the antibodies are first specifically functionalized with short anti-zipcode tags, short stretches of DNA, i, and sorted on a matching complementary DNA array, is in principle also compatible with all antibody formats (9–11). But to be able to generate high-density arrays in this manner, the sorting, i.e., hybridization, must be performed at elevated temperatures in order to achieve sufficiently selective DNA binding on high-density DNA arrays. This has frequently been pointed out as a key bottleneck, as the DNA-tagged antibodies must survive at elevated temperatures, which is not always the case (1, 30). Furthermore, it would be essential that nonpurified antibody could be specifically tagged with DNA, e.g., via an affinity tag, to eliminate the need of having to prepurify the probes prior to tagging and sorting them. While recombinant libraries can be designed to carry almost any affinity-tag, pAb and mAb are not compatible with such an approach. Again, this highlights the advantageous use of recombinant antibodies, microarray adapted by molecular design, engineered with increased thermal stability and/or favorable affinity-tags as probe source for array applications (1, 30, 60). Third, producing the arrays by self-assembling is based on cell-free expression of the individual antibody probes directly on the array in their unique spot (6, 12–17). To date, proof-of-principle has been published for at least three technology platforms, including protein in situ array (PISA) (14), nucleic acid programmable protein array (NAPPA) methodology (15–17, 78), and DNA to protein array (DAPA) (12, 13). In all cases, access to the DNA encoding each antibody is essential. Hence, this approach of fabricating arrays is mainly compatible with mAb and recombinant antibodies only. However, it could be favorable to produce smaller probe formats (28 vs. 150 kDa) when adopting a self-assembling approach, further promoting the choice of recombinant antibody fragments as probe source.

2.5. Choice of Scaffold

The choice of scaffold is critical, since it is a major factor determining the on-chip performances of the arrayed probes (1, 30, 42). As outlined above, the choice of scaffold cannot be selected and/or predetermined for pAbs and mAbs produced in vivo while the option is open for recombinant antibodies. Hence, this section focuses entirely on recombinant antibodies.

Different approaches have been adopted when designing recombinant scFv antibody libraries. The libraries can be designed around a single, constant scaffold, i.e., one combination of VH and VL (Fig. 3b) (55), or a mixture of a few scaffolds, i.e., several combinations of VH and VL (52). However, from an array point of view, the choice of library design is critical (1, 30). Early work has shown that recombinant scFv antibodies based on different scaffolds displayed large differences in on-chip properties, such as functional stability (31, 63). As for example, the functional on-chip half-life of arrayed, dehydrated scFv based on various scaffolds was found to vary significantly (63). In more detail, the on-chip half-life of the recombinant antibodies was found to increase in the order of 7 days (VH5-51/VL2-23) < 39 days (VH3-30/VKIIIa) < 42 days (VH5-51/VKIIIb) < 180 days (VH-3-23/VL-1-47). The data implied that the interface between VH and VL was central for the observed stability, as reflected by a correlation between the functional on-chip stability and the size of the inter-domain interface and the number of inter-domain Van der Waals interactions. Since then, scFv antibodies based on VH-3-23/VL-1-47 have repeatedly been shown to display outstanding functional on-chip properties, clearly outlining their potential as probes for array-based applications (21, 23, 30). In this context, it should be noted that the biophysical properties of the human antibody variable domains recently were mapped in detail by Pluckthun and coworkers (79, 80). The domain constructs were first evaluated individually and then in a systematic series of VH–VL combinations in the scFv format. The results showed that the overall stability of scFv fragments depended on the intrinsic stability of VH and VL as well as on the extrinsic stability provided by their interaction. In fact, the data showed that the VH-3-23/VL-1-47 combination were among the top candidates to be used from a structural point of view, when aiming for stable, functional molecules that would express well. Notable, although the stability of the isolated VL domains was low, with VH3, these domains built up the most stable scFv fragments tested (79). Further, the data showed that the VH3 consensus domain tested was found to be the thermodynamically most stable VH domain, both alone and in combination with a VL domain. The superior quality of VH3 was implied to be based on a number of small, but critical contributions, seen in all parts of the structure, e.g., hydrophobic core and residues involved in ionic interactions.

The advantage of using a recombinant scFv antibody library based on a single, fixed scaffold means that all probes are as similar as possible, differing only by their complementarity determining regions (CDRs) governing their specificity. Consequently, the probes display more similar on-chip characteristics, allowing the probes to be viewed as similar chemicals rather than heterogeneous biological material. This is a central feature for minimizing any probe-dependent assay variations.

2.6. Redesign

The solution to rescue and/or reoptimize probes that displayed the desired specificity, but poor on-chip performances could be to redesign their scaffold. However, this approach would only be compatible with renewable probe formats for which its corresponding DNA is at hand, i.e., for mAb and recombinant antibodies. From a practical point of view, antibodies based on different scaffolds, such as most mAb, would have to be optimized individually, making mAb less compatible with the approach. In contrast, recombinant antibody libraries are well suited for even iterative redesign efforts. In particular, this is the case for libraries based around a single, constant scaffold, which enables modifications, performed in the framework region, to be applied in the whole antibody repertoire in one step (Fig. 3c). In the end, this would enable the probes to be truly array adapted by protein design.

A redesign could be initiated to improve specific on-chip prestanda, such as functional stability. As for example, we have launched both directed and random mutagenesis approaches to enhance the functional on-chip stability of our recombinant scFv antibodies, all based on the VH-3-23/VL-1-47 scaffold (Fig. 3c) (1, 6, 30). We have shown that a few single mutations could be delineated that improved the stability, providing cooperative effects if introduced simultaneously (Wingren et al., unpublished observations). Other engineering efforts could be directed toward, e.g., the affinity tag to further improve its binding properties, for example immobilization or functionalization via the tag. As for example, we have shown that replacing the single his_6-tag by a double his_6-tag significantly enhanced the binding to Ni^{+2}-NTA (64). In a similar manner, we have taken the first steps toward introducing nonnatural amino acid derivatives to functionalize the probes with light-sensitive residues for enhanced immobilization and/or functionalization (Wingren et al., unpublished observations).

3. Applications Based on Recombinant Array

Applications based on recombinant antibody arrays are outside the scope of this mini-review, and is therefore briefly described. To date, a set of recombinant scFv micro- (21, 23) and nanoarray (65, 81, 82) technology platforms have been designed and used in a variety of applications, ranging from dedicated arrays targeting a few analytes to more semi-global protein expression efforts, for review see (1, 30, 60, 65). Briefly, sensitive profiling of water-soluble proteins and membrane proteins have been demonstrated targeting a wide variety of complex, nonfractionated sample formats, including intact cells (83), cell supernatants (84), cell lysates (Dexlin-Mellby et al., unpublished observations), tissue extracts (25) (Dexlin-Mellby et al., unpublished observations), serum and plasma (21, 23, 27, 85), and urine (Wingren et al., unpublished observations). The potential of the technology for disease proteomics and for deciphering disease-associated candidate biomarker signatures have been outlined targeting cancers, such as metastatic breast cancer (85), pancreatic cancer (27), glioblastoma multiforme (Carlsson et al., (86)), gastric adenocarcinoma (25), and mantle cell lymphoma (Dexlin-Mellby et al., unpublished observations), as well as various inflammatory conditions, including *Helicobacter pylori* infection (25), systemic lupus erythematosus (Carlsson et al., unpublished observations), pancreatitis (Sandström et al., unpublished observations), and pre-eclampsia (Wingren et al., unpublished observations).

4. Conclusions

In the coming years, the role of recombinant antibody libraries as the premium probe source for arrays is further manifested and extended, in particular for large-scale discovery projects targeting numerous analytes. The use of recombinant libraries, microarray adapted by molecular design, become even more important as the (specific) demands on the array performances continue to increase. These discovery projects are then followed-up by prevalidation and validation studies confirming and refining the initial data, resulting in focused biomarker signatures, ready to be transferred and implemented within clinical settings (refs). These condensed signatures, based on a limited number of probes (<25), are then converted into clinical assays, which could still be in the array format, although it could be even more favorable to adopt technology platforms already established in the clinical laboratories, such as ELISA. In such setups, monoclonal (and polyclonal) antibodies continue to play an important role as affinity reagent.

Acknowledgments

This study was supported by grants from the Swedish National Science Council (VR-NT), SSF, Strategic Center for Translational Cancer Research (CREATE Health) (http://www.createhealth.lth.se), and the Crafoord Research Foundation.

References

1. Borrebaeck, C. A., and Wingren, C. (2007) High-throughput proteomics using antibody microarrays: an update, *Expert review of molecular diagnostics 7*, 673–686.
2. Borrebaeck, C. A., and Wingren, C. (2009) Transferring proteomic discoveries into clinical practice, *Expert review of proteomics 6*, 11–13.
3. Haab, B. B. (2006) Applications of antibody array platforms, *Current opinion in biotechnology*.
4. Hartmann, M., Roeraade, J., Stoll, D., Templin, M. F., and Joos, T. O. (2009) Protein microarrays for diagnostic assays, *Analytical and bioanalytical chemistry 393*, 1407–1416.
5. Kingsmore, S. F. (2006) Multiplexed protein measurement: technologies and applications of protein and antibody arrays, *Nature reviews 5*, 310–320.
6. Wingren, C., and Borrebaeck, C. (2006) Antibody Microarrays - Current Status and Key Technological Advances, *OMICS 10*, 411–427.
7. Angenendt, P. (2005) Progress in protein and antibody microarray technology, *Drug discovery today 10*, 503–511.

8. MacBeath, G. (2002) Protein microarrays and proteomics, *Nat Genet 32 Suppl*, 526–532.
9. Svedhem, S., Pfeiffer, I., Larsson, C., Wingren, C., Borrebaeck, C., and Hook, F. (2003) Patterns of DNA-labeled and scFv-antibody-carrying lipid vesicles directed by material-specific immobilization of DNA and supported lipid bilayer formation on an Au/SiO2 template, *Chembiochem 4*, 339–343.
10. Wacker, R., and Niemeyer, C. M. (2004) DDI-microFIA--A readily configurable microarray-fluorescence immunoassay based on DNA-directed immobilization of proteins, *Chembiochem 5*, 453–459.
11. Wacker, R., Schroder, H., and Niemeyer, C. M. (2004) Performance of antibody microarrays fabricated by either DNA-directed immobilization, direct spotting, or streptavidin-biotin attachment: a comparative study, *Anal Biochem 330*, 281–287.
12. He, M., Stoevesandt, O., Palmer, E. A., Khan, F., Ericsson, O., and Taussig, M. J. (2008) Printing protein arrays from DNA arrays, *Nature methods 5*, 175–177.
13. He, M., Stoevesandt, O., and Taussig, M. J. (2008) In situ synthesis of protein arrays, *Current opinion in biotechnology 19*, 4–9.
14. He, M., and Taussig, M. J. (2001) Single step generation of protein arrays from DNA by cell-free expression and in situ immobilisation (PISA method), *Nucleic Acids Res 29*, E73–73.
15. Ramachandran, N., Hainsworth, E., Bhullar, B., Eisenstein, S., Rosen, B., Lau, A. Y., Walter, J. C., and LaBaer, J. (2004) Self-assembling protein microarrays, *Science 305*, 86–90.
16. Ramachandran, N., Hainsworth, E., Demirkan, G., and LaBaer, J. (2006) On-chip protein synthesis for making microarrays, *Methods in molecular biology Clifton, N.J 328*, 1–14.
17. Ramachandran, N., Raphael, J. V., Hainsworth, E., Demirkan, G., Fuentes, M. G., Rolfs, A., Hu, Y., and LaBaer, J. (2008) Next-generation high-density self-assembling functional protein arrays, *Nature methods 5*, 535–538.
18. Joos, T. O., Stoll, D., and Templin, M. F. (2002) Miniaturised multiplexed immunoassays, *Current opinion in chemical biology 6*, 76–80.
19. Schwenk, J. M., Gry, M., Rimini, R., Uhlen, M., and Nilsson, P. (2008) Antibody suspension bead arrays within serum proteomics, *Journal of proteome research 7*, 3168–3179.
20. Templin, M. F., Stoll, D., Bachmann, J., and Joos, T. O. (2004) Protein microarrays and multiplexed sandwich immunoassays: what beats the beads?, *Combinatorial chemistry & high throughput screening 7*, 223–229.
21. Ingvarsson, J., Larsson, A., Sjoholm, A. G., Truedsson, L., Jansson, B., Borrebaeck, C. A., and Wingren, C. (2007) Design of recombinant antibody microarrays for serum protein profiling: targeting of complement proteins, *Journal of proteome research 6*, 3527–3536.
22. Kusnezow, W., Banzon, V., Schroder, C., Schaal, R., Hoheisel, J. D., Ruffer, S., Luft, P., Duschl, A., and Syagailo, Y. V. (2007) Antibody microarray-based profiling of complex specimens: systematic evaluation of labeling strategies, *Proteomics 7*, 1786–1799.
23. Wingren, C., Ingvarsson, J., Dexlin, L., Szul, D., and Borrebaeck, C. A. (2007) Design of recombinant antibody microarrays for complex proteome analysis: choice of sample labeling-tag and solid support, *Proteomics 7*, 3055–3065.
24. Haab, B. B. (2005) Multiplexed protein analysis using antibody microarrays and label-based detection, *Methods Mol Med 114*, 183–194.
25. Ellmark, P., Ingvarsson, J., Carlsson, A., Lundin, B. S., Wingren, C., and Borrebaeck, C. A. (2006) Identification of protein expression signatures associated with Helicobacter pylori infection and gastric adenocarcinoma using recombinant antibody microarrays, *Mol Cell Proteomics 5*, 1638–1646.
26. Haab, B. B. (2005) Antibody arrays in cancer research, *Mol Cell Proteomics 4*, 377–383.
27. Ingvarsson, J., Wingren, C., Carlsson, A., Ellmark, P., Wahren, B., Engstrom, G., Harmenberg, U., Krogh, M., Peterson, C., and Borrebaeck, C. A. (2008) Detection of pancreatic cancer using antibody microarray-based serum protein profiling, *Proteomics 8*, 2211–2219.
28. Orchekowski, R., Hamelinck, D., Li, L., Gliwa, E., vanBrocklin, M., Marrero, J. A., Vande Woude, G. F., Feng, Z., Brand, R., and Haab, B. B. (2005) Antibody microarray profiling reveals individual and combined serum proteins associated with pancreatic cancer, *Cancer research 65*, 11193–11202.
29. Sanchez-Carbayo, M., Socci, N. D., Lozano, J. J., Haab, B. B., and Cordon-Cardo, C. (2006) Profiling bladder cancer using targeted antibody arrays, *Am J Pathol 168*, 93–103.
30. Borrebaeck, C. A., and Wingren, C. (2009) Design of high-density antibody microarrays for disease proteomics: Key technological issues, *Journal of proteomics.*
31. Wingren, C., and Borrebaeck, C. A. (2004) High-throughput proteomics using antibody microarrays, *Expert review of proteomics 1*, 355–364.
32. Saerens, D., Ghassabeh, G. H., and Muyldermans, S. (2008) Antibody technology in proteomics, *Brief Funct Genomic Proteomic 7*, 275–282.
33. Renberg, B., Nordin, J., Merca, A., Uhlen, M., Feldwisch, J., Nygren, P. A., and Karlstrom, A. E.

(2007) Affibody molecules in protein capture microarrays: evaluation of multidomain ligands and different detection formats, *Journal of proteome research 6*, 171–179.

34. Renberg, B., Shiroyama, I., Engfeldt, T., Nygren, P. K., and Karlstrom, A. E. (2005) Affibody protein capture microarrays: synthesis and evaluation of random and directed immobilization of affibody molecules, *Anal Biochem 341*, 334–343.
35. Lao, Y. H., Peck, K., and Chen, L. C. (2009) Enhancement of Aptamer Microarray Sensitivity through Spacer Optimization and Avidity Effect, *Anal Chem.*
36. Walter, J. G., Kokpinar, O., Friehs, K., Stahl, F., and Scheper, T. (2008) Systematic investigation of optimal aptamer immobilization for protein-microarray applications, *Anal Chem 80*, 7372–7378.
37. Cho, E. J., Collett, J. R., Szafranska, A. E., and Ellington, A. D. (2006) Optimization of aptamer microarray technology for multiple protein targets, *Analytica chimica acta 564*, 82–90.
38. Collett, J. R., Cho, E. J., Lee, J. F., Levy, M., Hood, A. J., Wan, C., and Ellington, A. D. (2005) Functional RNA microarrays for high-throughput screening of antiprotein aptamers, *Anal Biochem 338*, 113–123.
39. Campbell, C. J., O'Looney, N., Chong Kwan, M., Robb, J. S., Ross, A. J., Beattie, J. S., Petrik, J., and Ghazal, P. (2006) Cell interaction microarray for blood phenotyping, *Anal Chem 78*, 1930–1938.
40. Haab, B. B. (2003) Methods and applications of antibody microarrays in cancer research, *Proteomics 3*, 2116–2122.
41. Song, S., Li, B., Wang, L., Wu, H., Hu, J., Li, M., and Fan, C. (2007) A cancer protein microarray platform using antibody fragments and its clinical applications, *Mol Biosyst 3*, 151–158.
42. Wingren, C., Ingvarsson, J., Lindstedt, M., and Borrebaeck, C. A. (2003) Recombinant antibody microarrays–a viable option?, *Nat Biotechnol 21*, 223.
43. Pavlickova, P., Schneider, E. M., and Hug, H. (2004) Advances in recombinant antibody microarrays, *Clin Chim Acta 343*, 17–35.
44. Wingren, C., and Borrebaeck, C. A. (2008) Antibody microarray analysis of directly labelled complex proteomes, *Current opinion in biotechnology 19*, 55–61.
45. Kingsmore, S. F., and Patel, D. D. (2003) Multiplexed protein profiling on antibody-based microarrays by rolling circle amplification, *Current opinion in biotechnology 14*, 74–82.
46. Mitchell, P. (2002) A perspective on protein microarrays, *Nature biotechnology 20*, 225–229.
47. Haab, B. B., Dunham, M. J., and Brown, P. O. (2001) Protein microarrays for highly parallel detection and quantitation of specific proteins and antibodies in complex solutions, *Genome Biol 2*, RESEARCH0004.
48. Lefranc, M. P. (2004) IMGT-ONTOLOGY and IMGT databases, tools and Web resources for immunogenetics and immunoinformatics, *Molecular immunology 40*, 647–660.
49. Lefranc, M. P. (2008) IMGT, the International ImMunoGeneTics Information System for Immunoinformatics : methods for querying IMGT databases, tools, and web resources in the context of immunoinformatics, *Mol Biotechnol 40*, 101–111.
50. Barbas, C. F., 3rd, Kang, A. S., Lerner, R. A., and Benkovic, S. J. (1991) Assembly of combinatorial antibody libraries on phage surfaces: the gene III site, *Proceedings of the National Academy of Sciences of the United States of America 88*, 7978–7982.
51. Hanes, J., Schaffitzel, C., Knappik, A., and Pluckthun, A. (2000) Picomolar affinity antibodies from a fully synthetic naive library selected and evolved by ribosome display, *Nat Biotechnol 18*, 1287–1292.
52. Knappik, A., Ge, L., Honegger, A., Pack, P., Fischer, M., Wellnhofer, G., Hoess, A., Wolle, J., Pluckthun, A., and Virnekas, B. (2000) Fully synthetic human combinatorial antibody libraries (HuCAL) based on modular consensus frameworks and CDRs randomized with trinucleotides, *J Mol Biol 296*, 57–86.
53. Lee, C. V., Liang, W. C., Dennis, M. S., Eigenbrot, C., Sidhu, S. S., and Fuh, G. (2004) High-affinity human antibodies from phage-displayed synthetic Fab libraries with a single framework scaffold, *J Mol Biol 340*, 1073–1093.
54. Marks, J. D., Hoogenboom, H. R., Bonnert, T. P., McCafferty, J., Griffiths, A. D., and Winter, G. (1991) By-passing immunization. Human antibodies from V-gene libraries displayed on phage, *J Mol Biol 222*, 581–597.
55. Soderlind, E., Strandberg, L., Jirholt, P., Kobayashi, N., Alexeiva, V., Aberg, A. M., Nilsson, A., Jansson, B., Ohlin, M., Wingren, C., Danielsson, L., Carlsson, R., and Borrebaeck, C. A. (2000) Recombining germline-derived CDR sequences for creating diverse single-framework antibody libraries, *Nat Biotechnol 18*, 852–856.
56. Angenendt, P., Wilde, J., Kijanka, G., Baars, S., Cahill, D. J., Kreutzberger, J., Lehrach, H., Konthur, Z., and Glokler, J. (2004) Seeing

better through a MIST: evaluation of monoclonal recombinant antibody fragments on microarrays, *Anal Chem 76*, 2916–2921.

57. Cahill, D. J. (2001) Protein and antibody arrays and their medical applications, *J Immunol Methods 250*, 81–91.
58. Cahill, D. J., and Nordhoff, E. (2003) Protein arrays and their role in proteomics, *Adv Biochem Eng Biotechnol 83*, 177–187.
59. Wingren, C., and Borrebaeck, C. (2006) Recombinant Antibody Microarrays, *Screening - Trends in Drug Discovery 2*, 13–15.
60. Wingren, C., and Borrebaeck, C. A. (2009) Antibody-based microarrays, *Methods in molecular biology Clifton, N.J 509*, 57–84.
61. Seurynck-Servoss, S. L., Baird, C. L., Miller, K. D., Pefaur, N. B., Gonzalez, R. M., Apiyo, D. O., Engelmann, H. E., Srivastava, S., Kagan, J., Rodland, K. D., and Zangar, R. C. (2008) Immobilization strategies for single-chain antibody microarrays, *Proteomics 8*, 2199–2210.
62. Wingren, C., Steinhauer, C., Ingvarsson, J., Persson, E., Larsson, K., and Borrebaeck, C. A. (2005) Microarrays based on affinity-tagged single-chain Fv antibodies: sensitive detection of analyte in complex proteomes, *Proteomics 5*, 1281–1291.
63. Steinhauer, C., Wingren, C., Hager, A. C., and Borrebaeck, C. A. (2002) Single framework recombinant antibody fragments designed for protein chip applications, *BioTechniques Suppl*, 38–45.
64. Steinhauer, C., Wingren, C., Khan, F., He, M., Taussig, M. J., and Borrebaeck, C. A. (2006) Improved affinity coupling for antibody microarrays: Engineering of double-(His)(6)-tagged single framework recombinant antibody fragments, *Proteomics 6*, 4227–4234.
65. Wingren, C., and Borrebaeck, C. A. (2007) Progress in miniaturization of protein arrays--a step closer to high-density nanoarrays, *Drug discovery today 12*, 813–819.
66. Gulmann, C., Sheehan, K. M., Kay, E. W., Liotta, L. A., and Petricoin, E. F., 3rd. (2006) Array-based proteomics: mapping of protein circuitries for diagnostics, prognostics, and therapy guidance in cancer, *The Journal of pathology 208*, 595–606.
67. Utz, P. J. (2005) Protein arrays for studying blood cells and their secreted products, *Immunological reviews 204*, 264–282.
68. MacBeath, G., and Schreiber, S. L. (2000) Printing proteins as microarrays for high-throughput function determination, *Science 289*, 1760–1763.
69. Bailey, G. S. (1994) The raising of a polyclonal antiserum to a protein, *Methods in molecular biology Clifton, N.J 32*, 381–388.
70. Phelan, M. L., and Nock, S. (2003) Generation of bioreagents for protein chips, *Proteomics 3*, 2123–2134.
71. Agaton, C., Galli, J., Hoiden Guthenberg, I., Janzon, L., Hansson, M., Asplund, A., Brundell, E., Lindberg, S., Ruthberg, I., Wester, K., Wurtz, D., Hoog, C., Lundeberg, J., Stahl, S., Ponten, F., and Uhlen, M. (2003) Affinity proteomics for systematic protein profiling of chromosome 21 gene products in human tissues, *Mol Cell Proteomics 2*, 405–414.
72. Nilsson, P., Paavilainen, L., Larsson, K., Odling, J., Sundberg, M. r., Andersson, A.-C., Kampf, C., Persson, A., Szigyarto, C. A.-K., Ottosson, J., Bjorling, E., Hober, S., WernÃ©rus, H., Wester, K., PontÃ©n, F., and Uhlen, M. (2005) Towards a human proteome atlas: High-throughput generation of mono-specific antibodies for tissue profiling, *Proteomics 5*, 4327–4337.
73. Uhlen, M., Bjorling, E., Agaton, C., Szigyarto Cristina, A.-K., Amini, B., Andersen, E., Andersson, A.-C., Angelidou, P., Asplund, A., Asplund, C., Berglund, L., Bergstrom, K., Brumer, H., Cerjan, D., Ekstrom, M., Elobeid, A., Eriksson, C., Fagerberg, L., Falk, R., Fall, J., Forsberg, M., Bjorklund Marcus, G., Gumbel, K., Halimi, A., Hallin, I., Hamsten, C., Hansson, M., Hedhammar, M., Hercules, G., Kampf, C., Larsson, K., Lindskog, M., Lodewyckx, W., Lund, J., Lundeberg, J., Magnusson, K., Malm, E., Nilsson, P., Odling, J., Oksvold, P., Olsson, I., Oster, E., Ottosson, J., Paavilainen, L., Persson, A., Rimini, R., Rockberg, J., Runeson, M., Sivertsson, A., Skollermo, A., Steen, J., Stenvall, M., Sterky, F., Stromberg, S., Sundberg, M. r., Tegel, H., Tourle, S., Wahlund, E., WaldÃ©n, A., Wan, J., WernÃ©rus, H., Westberg, J., Wester, K., Wrethagen, U., Xu Lan, L., Hober, S., and PontÃ©n, F. (2005) A Human Protein Atlas for Normal and Cancer Tissues Based on Antibody Proteomics, *Molecular & Cellular Proteomics 4*, 1920–1932.
74. Borrebaeck, C. A., and Ohlin, M. (2002) Antibody evolution beyond Nature, *Nature biotechnology 20*, 1189–1190.
75. Hallborn, J., and Carlsson, R. (2002) Automated screening procedure for high-throughput generation of antibody fragments, *BioTechniques Suppl*, 30–37.
76. Wingren, C., James, P., and Borrebaeck, C. A. (2009) Strategy for surveying the proteome using affinity proteomics and mass spectrometry, *Proteomics 9*, 1511–1517.
77. Stoevesandt, O., and Taussig, M. J. (2007) Affinity reagent resources for human proteome detection: initiatives and perspectives, *Proteomics 7*, 2738–2750.

78. Hartmann, M., Schrenk, M., Dottinger, A., Nagel, S., Roeraade, J., Joos, T. O., and Templin, M. F. (2008) Expanding assay dynamics: a combined competitive and direct assay system for the quantification of proteins in multiplexed immunoassays, *Clinical chemistry 54*, 956–963.
79. Ewert, S., Huber, T., Honegger, A., and Pluckthun, A. (2003) Biophysical properties of human antibody variable domains, *J Mol Biol 325*, 531–553.
80. Worn, A., and Pluckthun, A. (2001) Stability engineering of antibody single-chain Fv fragments, *J Mol Biol 305*, 989–1010.
81. Ellmark, P., Ghatnekar-Nilsson, S., Meister, A., Heinzelmann, H., Montelius, L., Wingren, C., and Borrebaeck, C. A. (2009) Attovial-based antibody nanoarrays, *Proteomics 9*, 5406–5413.
82. Ghatnekar-Nilsson, S., Dexlin, L., Wingren, C., Montelius, L., and Borrebaeck, C. A. (2007) Design of atto-vial based recombinant antibody arrays combined with a planar wave-guide detection system, *Proteomics 7*, 540–547.
83. Dexlin, L., Ingvarsson, J., Frendeus, B., Borrebaeck, C.A.K., and Wingren, C. (2008) Design of recombinant antibody microarrays for cell surface membrane proteomics. *J. Proteome Research 7*, 319–327.
84. Lundberg, K., Lindstedt, M., Larsson, K., Dexlin, L., Wingren, C., Ohlin, M., Greiff, L., and Borrebaeck, C. A. (2008) Augmented Phl p 5-specific Th2 response after exposure of dendritic cells to allergen in complex with specific IgE compared to IgGl and IgG4, *Clinical immunology (Orlando, Fla 128*, 358–365.
85. Carlsson, A., Wingren, C., Ingvarsson, J., Ellmark, P., Baldertorp, B., Ferno, M., Olsson, H., and Borrebaeck, C. A. (2008) Serum proteome profiling of metastatic breast cancer using recombinant antibody microarrays, *Eur J Cancer 44*, 472–480.
86. Carlsson, A., Persson, O., Widegren, B., Salford, L.G., Borrebaeck, C.A.K., and Wingren, C. (2010) Plasma protein profiling reveals biomarker patterns associated with prognosis and therapy selection in glioblastoma multiforme patients. *Proteomics Clinical Applications 4*, 591–602.

Part III

Protein Microarrays

Chapter 18

Producing Protein Microarrays from DNA Microarrays

Oda Stoevesandt, Michael J. Taussig, and Mingyue He

Abstract

The development of protein microarrays makes possible interaction-based protein assays in miniaturised, multiplexed formats. A major requirement determining their uptake and use is the availability and stability of purified, functional proteins for immobilisation. With conventional methods, involving individual expression and purification of recombinant proteins, the cost of commercial high-content protein arrays is often found to be prohibitively high. Moreover, due to the need for specialised microarray production equipment, custom-made protein arrays containing more focussed sets of proteins of interest are also in little use. In the DNA array to protein array technology described herein, repeated economical printing of protein microarrays from a reusable template DNA microarray is performed on demand by cell-free protein synthesis. Once the template DNA microarray has been obtained, protein microarrays are made using purely macro-handling procedures, making protein arraying accessible without sophisticated microarraying apparatus.

Key words: Protein array, Protein microarray, DNA microarray, Cell-free protein synthesis, Protein immobilisation

1. Introduction

Protein microarrays are miniaturised multiplexed formats of interaction-based solid-phase protein assays, lending themselves to applications in proteomics and systems biology, where extensive multi-parameter datasets are required. Using minimal amounts of samples and reagents, protein–protein, protein–antibody, or protein–non-protein (nucleic acids, lipids, small molecules, and other compounds) interactions can be probed (1). Despite this potential for diverse cutting-edge applications, protein arrays are not as widely used as other methods, due to the effort and cost involved in preparing large numbers of purified proteins for

Ulrike Korf (ed.), *Protein Microarrays: Methods and Protocols*, Methods in Molecular Biology, vol. 785,
DOI 10.1007/978-1-61779-286-1_18,

immobilisation. Another major problem is the limited half-life of functional proteins, especially when supply logistics necessitate storage of arrays over prolonged time periods before their application. Compared to the proteins themselves, protein-expression constructs containing coding sequences (CDS) of tagged proteins are inexpensive to prepare and can be amplified by means of PCR. In addition, DNA arrays can be stored dry without deterioration. In recognition of this fact, a variety of methods has been developed to transform microarrays of protein-coding DNA into the corresponding protein microarrays by cell-free protein synthesis (2–8). Cell-free protein expression systems are based on homogeneous cell lysates, which perform protein synthesis directly from externally added DNA templates (9).

The existing systems for protein arraying by cell-free expression use different strategies for confining the location of template DNA and protein expressed from it and for protein immobilisation. In this chapter, we describe the details of the DNA array to protein array (DAPA) procedure (7) (Fig. 1).

In DAPA, a slide with a DNA microarray encoding a set of proteins, including a double hexahistidine-tag [$2(His)_6$] is assembled face-to-face with an Ni-NTA-functionalised slide. A membrane

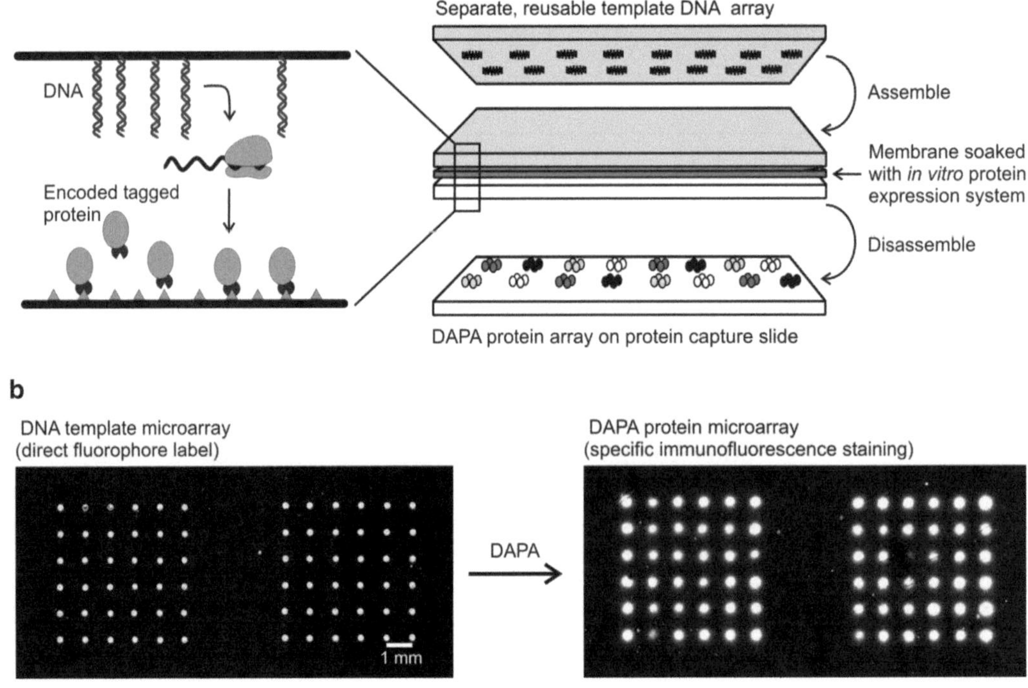

Fig. 1. Principle and example of the DAPA procedure. (**a**) Schematic molecular (*left*) and macroscopic (*right*) view of protein array printing by DAPA. (**b**) Sample results of DAPA.

soaked with an *Escherichia coli*-based in vitro transcription and translation system is positioned between the two slide surfaces. $2(His)_6$-tagged proteins synthesised from the immobilised DNA diffuse towards the Ni-NTA slide surface, where they become immobilised, creating the protein array corresponding to the template DNA array. The morphology of the protein spots is defined by the diffusional process, resulting in reproducible intensity profiles which are well described by Gaussian curves. The DNA template slide can be reused up to 20 times in further cycles of printing DAPA protein arrays (7).

This following protocol comprises:

1. Sequences of the required elements of DNA templates for microarraying and *E. coli* based cell-free protein expression. A strategy for the PCR-assembly of suitable constructs is described (see Note 1).
2. Microarraying and immobilisation of the linear dsDNA expression constructs on epoxy-activated slides, yielding template DNA microarrays.
3. The DAPA procedure itself for printing a protein microarray from the template DNA array, using an *E. coli* lysate-based system for cell-free protein synthesis.

2. Materials

2.1. DNA-Encoding Proteins of Interest

Any source of DNA amenable to PCR-amplification of the CDS of the proteins of interest is suitable.

2.2. Templates Containing Generic Elements for Cell-Free Protein Expression

2.2.1. Template for Upstream Elements

Upstream elements required for cell-free expression include the T7 promoter, the prokaryotic ribosome-binding site and the start codon (uppercase, with bold translation). The following 101 nucleotide sequence can be found in the control plasmid included in the RTS100 *E. coli* HY kit (5 Prime) or in the pIVEX2.3d vector (5 Prime) for in vitro protein expression:

```
 1  gatctcgatcccgcgaaattaatacgactcactatagggagaccacaacggtttccctct
 1                                                         M
61  agaaataattttgtttaactttaagaaggagatataccATG
```

2.2.2. Template for Downstream Elements

Elements to be inserted downstream of the CDS of the protein of interest include (1) the CDS for a flexible linker and $2(His)_6$ tag (nucleotide positions 1–153 below, upper case, with translation) and (2) two consecutive stop codons followed by the transcription termination region (positions 154–332 below, lower case, which can be obtained from the pIVEX2.3d vector). The full sequence is constructed using standard molecular biology methods and for convenience can be cloned into a plasmid to serve as a stable template for later PCR amplification. The detailed nucleotide and encoded amino acid sequences are:

1	R S R G G G S G G G S G G G T G G G S G
1	CGCTCTAGAGGCGGTGGCTCTGGTGGCGGTTCTGGCGGTGGCACCGGTGGCGGTTCTGGC
21	G G K R A D A A H H H H H H S R A W R H
61	GGTGGCAAACGGGCTGATGCTGCACATCACCATCACCATCACTCTAGAGCTTGGCGTCAC
41	P Q F G G H H H H H H * *
121	CCGCAGTTCGGTGGTCACCACCACCACCACCACtaataaaagggcgaattccagcacact
181	ggcggccgttactagtggatccggctgctaacaaagcccgaaaggaagctgagttggctg
241	ctgccaccgctgagcaataactagcataaccccttggggcctctaaacgggtcttgaggg
301	gtttttgctgaaaggaggaactatatccgga

2.3. Primers for PCR-Assembly of DNA Templates for DAPA

For an overview of primer binding sites with respect to the generated expression constructs, see Fig. 2. All primer sequences are shown in 5′→3′ orientation.

1. Primer pair for generic upstream fragment, including T7 promoter:

   T7-for: GATCTCGATCCCGCG

   T7-rev: CATGGTATATCTCCTTCTTAAAG

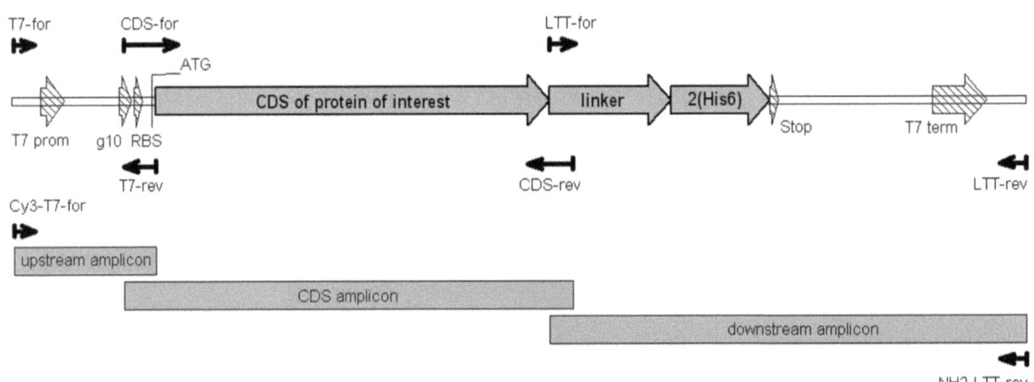

Fig. 2. Overview of template structure for cell-free expression and PCR assembly strategy. *Black arrows* indicate primer-binding sites and primer extension directions.

2. Primer pair for generic downstream fragment, including linker, tag and T7 terminator:

   LTT-for: CGCTCTAGAGGCGGTGGC

   LTT-rev: TCCGGATATAGTTCCTCC

3. Individual primer pairs for fragments encoding the different CDS of interest, introducing 5′ and 3′ overhangs for assembly with generic upstream and downstream fragments:

   CDS-for: CTTTAAGAAGGAGATATACCATG$(N)_{15-25}$

   CDS-rev: CACCGCCTCTAGAGCG$(N)_{15-25}$

   $(N)_{15-25}$ are the first amino-acid coding 15–25 nucleotides from either end of the CDS of interest. At the 3′ end, the stop codon of the CDS must be omitted.

4. Primers for final amplification of assembled expression constructs, introducing modifications for immobilisation and detection on array slides:

   Cy3-T7-for: Cy3-GATCTCGATCCCGCG (see Note 2)

   NH_2-LTT-rev: NH_2-TCCGGATATAGTTCCTCC

2.4. General Reagents and Kits for Molecular Biology

1. Desoxy Nucleotides (Sigma).
2. Taq DNA polymerase and buffers (Qiagen).
3. GenElute™ Gel Extraction kit (Sigma).
4. GenElute™ PCR Clean-Up kit (Sigma).

2.5. Materials for Template DNA Microarray Spotting

1. Nexterion slide E (epoxysilane coated) (Schott Nexterion).
2. 6× spotting buffer: 300 mM sodium phosphate, pH 8.5.
3. 0.1% (v/v) Tween-20.
4. 1 mM HCl.
5. 100 mM KCl.
6. Quenching buffer: 0.1 M Tris–HCL, pH 9.0. Ethanolamine (Sigma) is added to a final concentration of 50 mM immediately before use.
7. Saturated NaCl solution for humidified incubation chamber.

2.6. Material for DAPA Cell-Free Protein Arraying

1. RTS100 *E. coli* HY (5 Prime, Hamburg, Germany) cell-free protein expression system for production of up to 20 μg protein/50 μl reaction.
2. Durapore 0.22-μm membrane filters (Millipore).
3. Ni-NTA-coated microscope slide (Xenopore, USA).

3. Methods

3.1. Generation of Linear Templates for Cell-Free Expression in DAPA

3.1.1. Overview of Template Structure and Construction Strategy

The overall structure of the linear expression templates for cell-free protein synthesis is shown in Fig. 2. Expression templates are assembled from three PCR-generated fragments (amplicons): (1) A generic upstream amplicon, introducing T7 promoter, ribosome-binding site and start codon. (2) A construct-specific amplicon containing the CDS of the protein of interest, with the stop codon deleted. The 5′- and 3′-ends of this amplicon overlap with the ends of the generic upstream and downstream fragments. (3) A generic downstream amplicon, introducing the CDS for a C-terminal linker and double $2(His)_6$-tag, double stop codon, and T7 terminator (see Note 3).

These three PCR-fragments are joined by assembly PCR, and then amplified with the primers located at the overall 5′ and 3′ ends. For the amplification of full-length constructs for DNA microarray spotting, a fluorophore-labelled forward primer (for detection) and an NH_2-modified reverse primer (for immobilisation) are used. The $2(His)_6$ tag used here has improved affinity for Ni-NTA-modified surfaces compared to a conventional single $(His)_6$ tag (10, 11). The linker is a sequence rich in Ser and Gly which has been optimised for conveying stability to single-chain proteins (12).

3.1.2. PCR-Based Strategy for Template Assembly and Amplification

3.1.2.1. Generation of PCR Fragments for Assembly

1. Set up standard PCR reactions (see Note 4):

	25 μl reaction for each CDS amplicon	100 μl reaction for upstream and downstream amplicon
10× PCR buffer	2.5 μl	10 μl
5× Q solution	5 μl	20 μl
dNTP (2.5 mM each)	2 μl	8 μl
Forward primer (16 μM)	0.75 μl	3 μl
Reverse primer (16 μM)	0.75 μl	3 μl
Template	0.5–5 ng	2–20 ng
Taq	0.63 U	2.5 U
Nuclease-free H_2O	to 25 μl	to 100 μl

with the following combinations of primers and templates:

Forward primer	Reverse primer	Template	PCR product (Fig. 2)
T7-for	T7-rev	see Subheading 2.2.1	Upstream amplicon (101 bp)
CDS-for	CDS-rev	respective CDS of interest	CDS amplicon (encoding protein of interest with deleted stop codon and short flanking 5′ and 3′ sequences)
LTT-for	LTT-rev	see Subheading 2.2.2	Downstream amplicon (332 bp)

PCR programme:

30 cycles	94°C, 30 s→54°C, 30 s→72°C, 80 s
1 cycle	72°C, 480 s
End	Hold 10°C

2. Check the PCR products by electrophoresis on a 1% agarose gel and isolate the bands of expected lengths by extraction from the gel (GenElute™ Gel Extraction kit, Sigma). Use nuclease-free H_2O for the final elution step, rather than the elution buffer provided in the kit.
3. Determine the concentration and purity of the cleaned PCR product by absorption at 260 and 280 nm.

3.1.2.2. Assembly of PCR Fragments to Complete Construct for Cell-Free Expression

1. For each CDS amplicon, set up a 25 μl assembly PCR reaction:

10× PCR buffer	2.5 μl
5× Q solution	5 μl
dNTP (2.5 mM each)	1 μl
Equimolar mix of upstream amplicon, CDS amplicon, downstream amplicon	50–100 ng total
Taq	0.63 U
H_2O	to 25 μl

PCR programme for fragment assembly:

8 cycles	94°C, 30 s→54°C, 60 s→72°C, 60 s
End	Hold 10°C

2. Amplify the assembled products by a second 50 μl PCR:

10× PCR buffer	5 μl
5× Q solution	10 μl
dNTP (2.5 mM each)	4 μl
T7-for (16 μM)	1.5 μl
LTT-rev (16 μM)	1.5 μl
Template: Product of assembly PCR above	2 μl
Taq	1.25 U
Nuclease-free H_2O	to 50 μl

PCR programme for amplification of assembled products:

30 cycles	94°C, 30 s→54°C, 60 s→72°C, 80 s
1 cycle	72°C, 480 s
End	Hold 10°C

3. Analyse the PCR products by electrophoresis on a 1% agarose gel and in case of multiple bands purify by gel excision (see Note 5). The resulting PCR constructs may be stored at –20°C for at least 6 months and will serve as templates for the following re-amplification step with labelled primers.

3.1.2.3. Reamplification of Construct with Labelled Primers for Immobilisation on Arrays

1. Set up standard 50 μl PCR reactions as in step 2 of the assembly PCR above, but with modified primers: Cy3-T7-for (instead of T7-for) and NH_2-LTT-rev (instead of LTT-rev).
2. Check the PCR product for correct size by agarose gel electrophoresis, purify through a spin column to remove excess primers (GenElute™ PCR Clean-Up kit), and elute in H_2O (see Note 6).
3. Determine the concentration and purity of the cleaned PCR product by absorption at 260 and 280 nm. A DNA concentration of 50–100 ng/μl is recommended for DAPA template array spotting (see Subheading 3.2) (see Note 7).

3.2. Generation of DNA Arrays as Templates for DAPA

1. Add 1 volume of 6× spotting buffer to 5 volumes of the labelled PCR product (Subheading 3.1.2).
2. Mark epoxysilane slides with a diamond pen to simplify their identification and orientation in later steps (see Note 8).
3. Spot DNA samples on epoxysilane slides with centre-to-centre spot distances of 1 mm and volumes per spot of 2–5 nl.
4. Incubate spotted slides in a humidified box at RT for 16 h.
5. Incubate slides at 60°C for 30 min (see Note 9).

6. Wash the slides at RT on a rocking table:
   (a) 1× with 0.1% Tween-20 for 5 min
   (b) 2× with 1 mM HCl for 2 min
   (c) 1× with 100 mM KCl for 10 min
   (d) 1× with ddH_2O for 1 min
7. Quench remaining epoxy groups by incubating slides in 0.1 M Tris–HCl pH 9.0, 50 mM ethanolamine at 50°C for 15 min.
8. Wash slides with dd H_2O for 1 min and dry by stripping off the liquid film and droplets by pressurised air (Note 10).
9. Scan slides in microarray scanner to confirm immobilisation of Cy3-labelled DNA.

   The template DNA arrays are now ready for use in DAPA and can be stored in the dark at 4°C.

3.3. DAPA: Printing Protein Arrays from the Template DNA Array

Use a slide holder similar to the prototype shown schematically in cross section in Fig. 3. Its functions are (1) to align DNA template slide and protein capture slide, (2) to enable sealing of the slide sandwich against evaporation, and (3) to provide even pressure over the slide area. Fig. 3 also illustrates the order of assembling the DAPA slide sandwich.

1. Cut a Durapore membrane filter large enough to cover the area of the DNA template array.
2. Mark Ni-NTA slides with a diamond pen to simplify their identification and orientation in later steps (see Note 8).

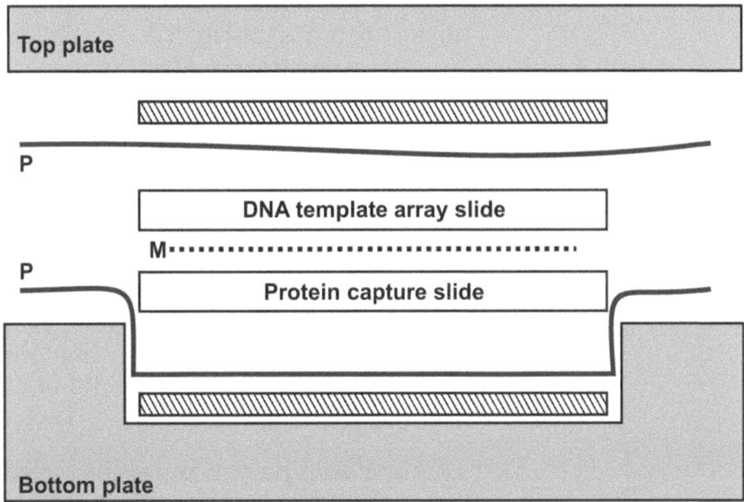

Fig. 3. Schematic cross section of a simple DAPA apparatus with exploded view of DAPA slide sandwich assembly. *M* membrane filter soaked with cell-free protein expression system, *P* parafilm, *hatched* rubber spacer.

3. For every 1 cm^2 of the membrane, prepare 10 μl *E. coli* cell-free protein expression system (RTS100 *E. coli* HY, mixture of reconstituted components according to the manufacturer's instructions). Prepare an extra 10 μl to set up controls for expression.
4. Assemble the DAPA sandwich in the slide holder in the following order
   (a) Bottom plate of slide holder
   (b) Rubber spacer
   (c) Layer of parafilm
   (d) Ni-NTA-coated slide
   (e) Cell-free protein expression system, distributed on the surface of the Ni-NTA slide in the area equivalent to the DNA template array
   (f) Membrane filter, allowing it to soak up the lysate (see Note 11)
   (g) DNA array template slide, with DNA surface facing down
   (h) Layer of parafilm (see Note 12)
   (i) Rubber spacer
   (j) Top plate

   Ensure even pressure on the slide sandwich.
5. Split the extra 10 μl of cell-free protein expression system into two aliquots, add 0.5 μl of GFP-encoding control vector (also provided in the kit) to one and 0.5 μl nuclease-free H_2O to the other. These are the positive and negative controls for functionality of the cell-free expression system.
6. Incubate the assembled slide holder and the tubes with the control reactions at 30°C for 4 h.
7. Check the control reaction tubes on a UV table. The tube with added GFP-encoding vector should now fluoresce green, confirming expression of GFP and functionality of the cell-free system.
8. Disassemble the slide sandwich and immerse the Ni-NTA slide with the DAPA protein array on it into PBS-Tween for washing. Do not dry the DAPA array before application in order to avoid denaturation of the immobilised proteins. Wash the DNA template slide with ddH_2O, dry by pressurised air, and store at 4°C for use in further DAPA cycles.

The DAPA protein array is now ready for immediate use in an assay of choice, with handling and detection protocols dependent on the individual application. To control for expression of proteins and their immobilisation on the array, immunofluorescence staining with appropriate reagents against the arrayed proteins can be performed (7).

4. Notes

1. With gene synthesis becoming ever more affordable, it may be well worth considering it as an alternative to the PCR-construction process described herein.
2. The main consideration for fluorophore choice is detectability on an available microarray scanner. Cy3-range dyes may be preferable over Cy5-range dyes for reasons of higher stability, especially when considering that the DNA template array will be used for repeated rounds of protein array printing.
3. The presence and location of a tag may affect the expression, solubility, or function of a fusion protein. Also, location of the tag may affect its accessibility. The C-terminal immobilisation tag described here is strategically preferable, as its presence guarantees that only full length proteins can be immobilised by tag-capturing surfaces. However, it might be necessary for some proteins to place the immobilisation tag at the N terminus.
4. The generic upstream and downstream amplicons which are required for every construct can be produced in larger quantities and stored at –20°C for future construction rounds.
5. In general, unpurified PCR products can be directly used as templates for protein synthesis in cell-free systems. In this case, since a further re-amplification step for product labelling is following, it is advisable to remove side product bands at this stage.
6. Elution in the elution buffer supplied with the PCR clean-up kit might interfere with the immobilisation of the DNA in the subsequent arraying step.
7. If the eluted PCR product is below this range, it can be concentrated in a vacuum centrifuge. The fluorophore-label of the purified PCR product is usually not detectable by absorption, as there is only one fluorophore per dsDNA, and thus molar concentrations are very low.
8. Be sure to remove any glass splinters or other dust on the slide surfaces by pressurised air.
9. This step, although not mentioned in the instructions of the slide manufacturer, has been found to improve immobilisation efficiency greatly.

10. Avoid drying slides by evaporation, as this would leave more droplet marks on the slides, which increase background when scanning.
11. The soaking process takes a few seconds. Swift handling is crucial at this stage to avoid compromising the activity of the cell-free system by partial drying within the membrane filter.
12. The parafilm must form an airtight seal around the slide sandwich (Fig. 3) in order to prevent evaporation of cell-free lysate soaked in the membrane filter between the two slides.

Acknowledgments

Research at the Babraham Institute is supported by Biotechnology and Biological Sciences Research Council (BBSRC), UK. The Protein Technology Group at Babraham Bioscience Technologies is a partner in the EC FP6 026008 ProteomeBinders and EC FP7 222635 AffinityProteome.

References

1. Hall D A, Ptacek J, Snyder M. (2007) Protein Microarray Technology. Mech. *Ageing Dev.* **128**, 161–167.
2. He M, Taussig M J. (2001) Single step generation of protein arrays from DNA by cell-free expression and *in situ* immobilization (PISA method). *Nucleic Acid Res.* **29**, e73.
3. He M, Taussig M J. (2003) DiscernArray™ technology: a cell-free method for the generation of protein arrays from PCR DNA. *J. Immunol. Methods* **274**, 265–270.
4. Angenendt P, Kreutzberger J, Glokler J, Hoheisel J D. (2006) Generation of high density protein microarrays by cell-free in situ expression of unpurified PCR products. *Mol. Cell. Proteomics* **5**, 1658–1666.
5. Ramachandran N, Hainsworth E, Bhullar B, Eisenstein S, Rosen B, Lau AY, Walter JC, LaBaer J. (2004) Self-assembling protein mircoarrays. *Science* **305**, 86–90.
6. Ramachandran N, Raphael J V, Hainsworth E, Demirkan G, Fuentes M G, Rolfs A, Hu Y, LaBaer J. (2008) Next-generation high-density self-assembling functional protein arrays. *Nat. Methods* **5**, 535–538.
7. He, M., Stoevesandt, O., Palmer, E.A., Khan, F., Ericsson, O. and Taussig, M.J (2008) Printing protein arrays from DNA arrays. *Nat. Methods* **5**, 175–177.
8. He M, Stoevesandt O, Taussig M J. (2008) In situ synthesis of protein arrays. *Curr. Opin. Biotechnol.* **19**, 4–9.
9. He M. (2008) Cell-free protein synthesis: applications in proteomics and biotechnology. *N. Biotechnol.* **25**, 126–32.
10. Khan F, He M, Taussig M J. (2006) A double-His tag with high affinity binding for protein immobilisation, purification, and detection on Ni-NTA surfaces. *Anal. Chem* **78**, 3072–3079.
11. Steinhauer C, Wingren C, Khan F, He M, Taussig MJ, Borrebaeck CA. (2006) Improved affinity coupling for antibody microarrays: engineering of double-(His)6-tagged single framework recombinant antibody fragments. *Proteomics* **6**, 4227–34.
12. Robinson CR, Sauer RT. (1998) Optimizing the stability of single-chain proteins by linker length and composition mutagenesis. *Proc. Natl. Acad. Sci. USA.* **95**, 5929–34.

Chapter 19

Cell Arrays and High-Content Screening

Holger Erfle, Anastasia Eskova, Jürgen Reymann, and Vytaute Starkuviene

Abstract

Endocytosis is one of the most essential cellular processes, which enables cells to internalise diverse material. It is crucial for regulation of receptor activity and signalling, cell polarisation, attachment and motility, and a great number of other cellular functions. A number of diverse endocytosis pathways are described by now; however, their specificity for different cellular cargoes is poorly resolved. Only few of endocytosis regulators are well-characterised and even less are attributed to the specific cargo. That is very true for the integrin endocytosis pathway, which is a key process in cell migration, adhesion, and signalling. The recent advent of quantitative fluorescent microscopy and cell arrays opened an exciting possibility to systematically characterise molecules playing a role in this crucially important process. Here, we describe a fluorescent screening microscopy-based assay to identify regulators of integrin $\alpha 2$ internalisation. The experimental procedure is the best suited for a highly parallel screening format, such as cell arrays, albeit can be used in single experiments. We provide protocols for sample preparation, fabrication of cell arrays and quantification of integrin $\alpha 2$ internalisation. The approach can be modified to quantify endocytosis of other cargo, and can be used under the conditions of knock-down and knock-in as well as for chemical screening.

Key words: Reverse transfection, Cell arrays, Fluorescent screening microscopy, RNA interference, Endocytosis, Integrins

1. Introduction

Integrins are heterodimeric receptors, localised at the plasma membrane (PM), that play significant roles in cell adhesion, signalling, and physical attachment of cells. Their function is regulated through endocytosis, recycling back to the cell surface, or degradation (1). The regulators of integrin endocytosis and recycling are known to some extent; however, the comprehensive picture is far from being understood. Moreover, different integrins pursue

Ulrike Korf (ed.), *Protein Microarrays: Methods and Protocols*, Methods in Molecular Biology, vol. 785, DOI 10.1007/978-1-61779-286-1_19, © Springer Science+Business Media, LLC 2011

different routes, and the same integrins under different conditions can also undertake different pathways (2). The situation is further complicated as it is not clear how specific or redundant these pathways are.

There are several studies addressing endocytosis of integrins (2, 3). However, no comprehensive network can be deduced from these data due to low throughput of methods and/or not sufficient specificity. Both shortages can be overcome by utilising high-content and high-throughput microscopy-based screens, particularly, when cellular phenotypes are resolved at the level of individual cells (4), and information about behaviour of cell populations is collected (5). One of the most essential components of high-resolution large-scale data collection is nevertheless attributed to the appropriate experimental format (6). Here, cell arrays give a competitive edge in contrast to multi-well plates, and the advantages cover both technical and biological aspects (7–9). Due to the homogeneous surface properties of the glass slide resulting in an improved autofocussing, cell arrays enable a significantly increased data acquisition speed by microscopy. Especially for time-lapse experiments this is of utmost interest (10). Standard cell arrays furthermore do not feature physically separated experiments/spots. As a consequence, parallel cell seeding and the simultaneous add-on of biochemical stimuli enable advanced experiment setups to be addressed. Importantly, cell arrays are of significant advantage over multi-well plates when synchronisation of assay is needed, particularly within a short period of time.

2. Materials

2.1. Cell Arrays

1. Probe substrates for the cell array: LabTek 1-well chamber slides (further LabTek) (Nalge Nunc, Rochester, NY, USA) consisting of borosilicate cover glass (0.16–0.19-mm thick) with 8.6-cm^2 culture area per well.
2. Chemicals for spotting solution of transfection array: siRNA oligonucleotides (for example, Qiagen or Thermofisher-Ambion); Lipofectamine 2000 (Invitrogen, California, USA); sucrose (USB, Cleveland, USA); OptiMEM I+GlutaMAX I (Invitrogen, California, USA); drying pearls, orange – heavy metal free (Sigma–Aldrich, Missouri, USA); gelatin (Sigma–Aldrich, Missouri, USA); human fibronectin (Sigma–Aldrich, Missouri, USA).
3. Automated liquid handling robot for automated sample preparation, "MICROLAB STAR" (Hamilton, Reno, NV, USA); equipped with 96-channel head and cooling carrier blocks for

multi-well plates, 384-well low-volume plates (Nalge Nunc, Rochester, NY, USA), 96-well standard plates (Kisker, Germany), reservoir (Nalge Nunc, Rochester, NY, USA), 300 µl standard volume tips without filter and 10 µl low-volume tips without filter (Hamilton, Reno, NV, USA).

4. Spotting of the transfection array is performed via a contact printer (ChipWriter "Pro," Bio-Rad Laboratories, California, USA) featuring solid pins for spots with diameters of 400 µm.

2.2. Integrin Internalisation Assay

1. Assay is performed in HeLa cells ATCC (CCL-2).
2. Cells are grown in 8-well µ-slide (Ibidi, Germany) or 8-well LabTek (Nalge Nunc, Rochester, NY, USA).
3. Growth medium for quantitative integrin internalisation assay: MEM (Sigma–Aldrich, Missouri, USA) pH 7.2–7.4 buffered with 30 mM HEPES and containing 10% foetal bovine serum, 2 mM glutamine, 100 U/ml penicillin, and 100 µg/ml streptomycin.
4. Solution for binding antibody to integrins on PM (MEM-BSA): MEM (Sigma–Aldrich, Missouri, USA) pH 7.2–7.4 buffered with 30 mM HEPES, with 0.01% (w/v) bovine serum albumin (BSA) or fraction V (Roth, Germany) and 2 mM glutamine. 1% BSA stock solution in water should be prepared freshly, filtered and added to MEM together with 2 mM glutamine. Prepared MEM-BSA can be stored at 4°C for 3 weeks.
5. Monoclonal antibody mAB1950 (Clone P1E6, Millipore, Germany) was used for binding to the endogenous integrin $\alpha 2$ on the PM.
6. The buffer to remove bound antibodies on PM (stripping buffer): 0.5 M NaCl, 0.5% (v/v) acetic acid in water, pH 2.6. The solution is stable at 4°C for 1 month.
7. Reagents for immunostaining: Paraformaldehyde (Polysciences, Pennsylvania, USA); glycin; saponin (Sigma–Aldrich, Missouri, USA); secondary antibody labelled with Alexa Fluor® dyes (Invitrogen, California, USA); Hoechst 33342.
8. Permeabilisation buffer for immunostaining: 0.2% (w/v) saponin, 10% (v/v) foetal calf serum in PBS. The solution is aliquoted and stored at −20°C.
9. siRNAs targeting clathrin heavy chain and dynamin2 were used as positive controls and "All Stars" siRNA as negative control in the internalisation assay (all purchased from Qiagen, Germany).

2.3. Imaging on an Automated Wide-Field Screening Microscope

Image acquisition is done with a fully automated scanning microscope running Scan^R acquisition software (Olympus IX81 Scan^R; Olympus Biosystems, Germany), but it can be done on any other inverted fluorescence microscope. The major parts of Scan^R system are as described previously (11), with the exception of a stabilised 150 W Hg/Xe light source for the excitation of the fluorophores. Imaging of cell arrays is performed using 10×/0.4 NA air objective lens (UPlanSApo; Olympus Biosystems, Germany).

2.4. Automated Data Evaluation

For fluorescence intensity-based integrin endocytosis assay the Scan^R Analysis software (http://www.olympus.com) was used. However, the automated data analysis is not restricted to this program, and other software capable of automated batch processing of images can be used.

3. Methods

We describe here a fluorescent microscopy-based assay to study the internalisation of human integrin α2. We have chosen this particular molecule because of several reasons. Firstly, it is ubiquitously expressed in different tissues. Secondly, integrin α2 builds the unique complex with integrin β1, and therefore we are able to investigate a specific endocytosis pathway. Thirdly, integrin α2β1 localises to PM domains enriched in GPI-anchored proteins, and can be internalised through caveolae upon clustering – the pathway, which is still little investigated (2).

It is worthy to mention, that only with small modifications, the assay can be used to analyse endocytic trafficking of other cellular cargoes, when antibodies to bind extracellular epitopes are available.

There are several microscopy-based endocytosis assays described in literature (3, 12). Similar to our assay, primary antibodies are bound to the extracellular epitope of the protein of interest, and cells are incubated further to allow internalisation of protein-antibody complex. Internalisation rate is calculated from the difference between secondary antibody staining in permeabilised and non-permeabilised cells (Fig. 1a, b). Unfortunately, for the internalisation of integrin α2 such method does not give a sufficient dynamic range that would be a prerequisite for a high-throughput screening. That is so because only in ~30% of HeLa cells endocytosis of integrin α2 is detectable by immunostaining on a wide-field microscope. In addition, the above described method requires treating two sets of samples in parallel: permeabilised and non-permeabilised. In order to overcome these problems, we have modified an experimental procedure so that only the intracellular pool of integrins is visualised after internalisation is completed

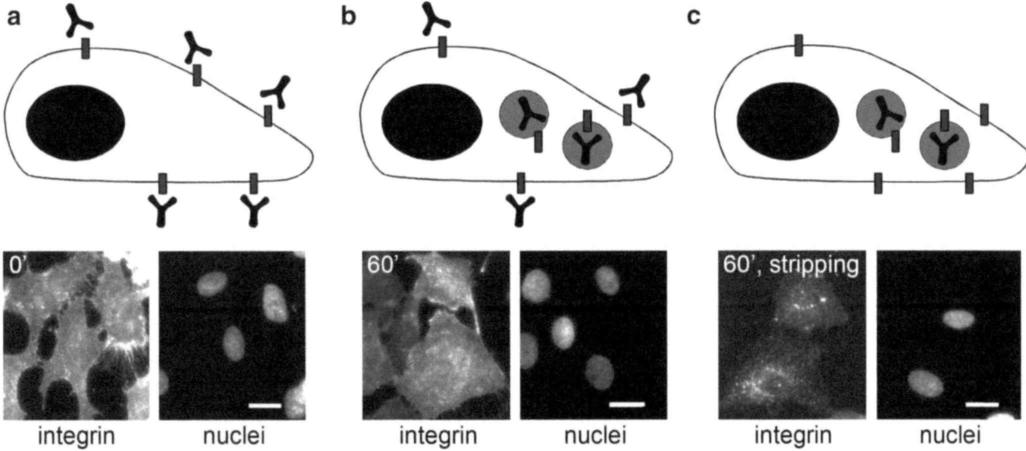

Fig. 1. Integrin internalisation assay. The *upper panel* shows the scheme of the assay. The *lower panel* illustrates every step of the assay. The internalisation of integrin α2 and corresponding nuclei staining is shown. *Bar* 20 μm. (**a**) Antibodies bind surface integrin α2 for 50 min on ice. (**b**) Cells internalise antibody-bound integrin for 1 h at 37°C. (**c**) After short stripping with acidic buffer antibodies remaining on the cell surface are removed, and internalised integrin is visualised.

(Fig. 1c). To achieve that, cells are shortly treated with an acidic buffer to remove non-internalised antibodies from the cell surface. Then, only the intracellular pool of the antibody-integrin complex is detected in permeabilised cells. The amount of internalised protein can be compared across different cells and serves as a read-out of the assay. Quantifying only the intracellular pool of integrin improves the dynamic range of the assay, reduces the amount of work and reagents and, importantly, allows better discrimination of intracellular structures. That is particularly useful when performing an in-depth analysis of how morphology of endocytic structures has been altered.

3.1. Preparation and Spotting of Cell Arrays

For RNAi screening, spotting solution for transfection arrays is designed and produced as described in ref. 6. Spotting of the array at LabTeks is performed as follows:

3.1.1. Preparation of the Spotting Solution

1. Transfer 6.5 μl of transfection solution to each well of a 384-well low-volume plate containing 3 μl OptiMEM, 0.4 M sucrose, and 3.5 μl Lipofectamine 2000 and mix thoroughly.
2. Prepare a 96-well gelatin/fibronectin stock solution plate manually, with 48 μl gelatin/fibronectin stock solution and place it into the pipetting robot. Keep it at room temperature (RT). Prepare the gelatin 0.2% solution freshly. For this, dissolve the gelatin powder at 56°C for 20 min in milliQ water. Cool it down to RT and filter the solution with a sterile filter (0.45 μm).

3. Prepare 30 μM siRNA stock solution by dissolving lyophilised siRNAs in milliQ water on the cooled carrier blocks (14°C). Use standard 300 μl volume tips without filters.
4. Transfer 6.5 μl of the transfection stock solution (see step 1) into each well of a 384-well low-volume plate using 10 μl low-volume tips without filters at RT.
5. Transfer 5 μl of the siRNA stock solution (see step 3) into the 384-well low-volume plate. The liquid handler is thereby adjusted to a mix volume of 7 μl and eight cycles of mixing (using 96-channel head) at RT. Incubate for 20 min at RT.
6. Add 7.25 μl of the gelatin/fibronectin stock solution (see step 2) into each well of the 384-well low-volume plate. The liquid handler is thereby adjusted to a mix volume of 10 μl and eight cycles of mixing (using the 96-channel head).

3.1.2. Spotting of the Transfection Solution on LabTeks and Storage of the Ready to Transfect Substrates

1. Adjust the number of pins in the contact printer. We routinely use eight solid pins PTS 600. The spot diameter of the solid pins PTS 600 pins is about 400 μm.
2. Adjust the temperature of the 384-well plate to 12°C to avoid evaporation of the sample.
3. Set the spot-to-spot distance in the contact printer menu with respect to the application in mind. We routinely use 1,125 μm, what allows to spot 384 samples per LabTek, however, spot-to-spot distance could be set to be 750 μm or 900 μm depending on the application.
4. Set the dwell time of the pins in the 384-well low-volume plate (time pins stay in the spotting solution) in the contact printer menu to 0.5 s.
5. Set the LabTek dwell time (time pins stay on LabTek) in the contact printer menu to 0.3 s.
6. Carry-out solid pin washing between individual samples. The procedure is set-up in the following way: pins remain in the washing container with milliQ water for 10 s at RT. Pins remain in the sonication container with milliQ water for 10 s at RT. Move the pins above the holes of the vacuum drying array of the contact printer and vacuum dry the pins for 10 s at RT.
7. After printing of the array keep the LabTeks in a gel drying box containing 50 g of drying gels for at least 24 h. After drying, the LabTeks can be stored more than 18 months at RT without losing transfection efficiency.

3.2. Integrin Internalisation Assay

1. Plate 2.5×10^4–3×10^4 HeLa cells/well in the 8-well μ-slide or 8-well LabTek containing 300 μl of the growth medium/well and incubate for 24 h at 37°C (see Notes 1 and 2).

2. Depending on the experiment, transfect cells with siRNAs or cDNAs using Lipofectamine 2000 according to the protocol provided by the manufacture. As a rule, 20 pmol of siRNA or 0.2 μg of plasmid is sufficient for the efficient transfection in 100 μl of the growth medium/well. Incubate transfected cells for 24–72 h before the assay depending on the plasmid/siRNA used.
3. 14 h before the start of the assay wash cells with MEM-BSA and incubate in this solution remaining time at 37°C (see Note 3).
4. Dilute antibody to bind integrin α2 (see Subheading 2.2) in MEM-BSA to final concentration 10 μg/ml. Remove MEM-BSA from cells and overlay with 80–100 μl/well of the freshly diluted antibody solution. Keep on a pre-cooled metal plate on ice for 50 min (Fig. 1a); wash twice with ice-cold MEM-BSA, add 300 μl/well pre-warmed MEM-BSA and incubate cells at 37°C for 1 h (Fig. 1b; see Note 4).
5. Wash cells with PBS, remove antibodies bound to the PM by incubation of cells with 100–300 μl/well stripping buffer for 30–40 s and wash once with PBS (Fig. 1c). All steps should be done at RT (see Note 5).
6. Fix cells with 2% paraformaldehyde in PBS for 20 min at RT and quench paraformaldehyde with 30 mM glycin in PBS for 5 min at RT.
7. Permeabilise cells with 0.2% saponin in 10% foetal calf serum in PBS for 10–15 min at RT.
8. Stain cells with the appropriate secondary antibodies and nuclei with Hoechst 33342 stain (final concentration is 0.1 μg/ml). For more information, see ref. 11.

3.3. Integrin Internalisation Assay on Cell Arrays

1. Plate HeLa cells at densities of 5×10^5–6×10^5 in the 1-well LabTek with the spotted siRNAs and incubate for 48 h. 4.5 ml growth medium is sufficient.
2. All the other steps are as described in Subheading 3.2, except different volumes of solutions used when working in 8-well substrates. Namely, 800 μl of diluted primary and secondary antibodies and Hoechst stain is needed to cover cells in 1-well LabTek.

3.4. Imaging Integrin Internalisation Assay on Cell Arrays

1. To image integrin internalisation, set-up two separate channels for recording nuclei and intracellular integrin. To image nuclei stained with Hoechst 33342, use excitation wavelength 360–370 nm, emission wavelength 460 nm. When imaging of internalised anti-integrin α2 antibody, the choice of the channel depends on dye coupled to the secondary antibody. For instance, having secondary antibody labelled with Alexa-488

use excitation wavelength 480–500 nm, emission wavelength 520–550 nm. It is important that excitation/emission spectra of two fluorophores are sufficiently resolved.

2. Adjust image acquisition settings like exposure time and intensity of excitation light for each fluorophore separately.
3. Set-up the layout of positions to be imaged. Imaging with 10×/0.4 NA air objective lens allows capturing the area of 867×661 μm which covers one spot/image. Take each spot as a separate well in Scan^R, set 1 imaged position per well.
4. Set software gradient-based autofocus using 1×1 binning. A coarse autofocus (25 layers with a step width of 6.4 μm) and fine autofocus (20 layers with a step width of 1.0 μm) are performed for each spot.
5. Label the first and the last spot on the LabTek before seeding cells to be able to later identify the start position for imaging. Labelling of the last spot serves as a control for the proper positioning throughout the whole experiment (see Notes 6 and 7).

3.5. Automated Data Evaluation

The following procedure describes image analysis with Scan^R Analysis software (see Note 8).

1. Use the so-called rolling ball algorithm to correct background for all images taken. For integrin internalisation assay imaged with 10×/0.4 NA air objective lens, the size of filter was set to 100.
2. Identify the cell nuclei by manually setting-up the threshold value in the images of stained nuclei. For the images taken with 10×/0.4 NA air objective, thresholded objects smaller than 200 pixels and bigger than 10,000 pixels are excluded from further analysis. Cell nuclei touching the image border are not considered for further analysis (see Note 9).
3. Create and dilate the binary mask to encompass as much cell as possible. Multiple the mask with the non-threshold fluorescence channel, used to image integrins. Calculate the average intensity of integrin-specific fluorescence within the mask.
4. Use the module "Cut Image" applied to both imaged channels to take into consideration only cells in the central circular area of the image (400 μm in diameter), where a spot lies (see Note 10).
5. Average integrin-specific fluorescence in one cell across all cell recorded.

3.6. Quantitative Data Analysis

The first important step of the data analysis is the quality control of images, which can be done either manually or automatically, i.e. discarding the images with unusually low cell counts, out of focus images, bad staining quality, etc. It is meaningful to remove 1–2%

of the brightest cells from further calculations as they could potentially represent experimental artefacts or abnormal cells.

In the described integrin endocytosis assay, the rate of the large part of cell population (up to 70%) is not high enough to be detected by screening fluorescence microscopy. Thus, scoring only of the sub-population of cells, namely, the brightest 30% is used. To achieve higher dynamic range of the assay and quantification being more robust, another approach has been tried. It considers both amount of endocytosed material, represented by the brightness of integrin-specific fluorescence, as well as number of cells having this fluorescence. Then, the average mean intensity of bright cells is multiplied to the percentage of bright cells in a spot. The resulting parameter is used to score endocytosis rate of integrin. The threshold to separate bright and dim cells can be derived by visual inspection in every experiment individually. However, automation of the process is highly needed for medium- and high-throughput assays.

One of the most important parameters in getting reliable data when working with cell arrays is positioning and number of positive and negative controls. It is recommended to have randomly positioned siRNA spots with at least three negative and three positive control siRNAs. In the assay described here, siRNAs targeting clathrin heavy chain and siRNAs-targeting dynamin2 were used as positive controls. Reliable separation of positive and negative controls is yet another important quality control parameter.

Image analysis by Scan^R Analysis software is a fast and robust one. However, not many features can be analysed with this software. Mean intensity of the internalised integrin in each cell is the sufficient parameter to consider for the primary screen. Yet, more advanced image analysis potent to resolve intracellular structures and their numbers, shape, brightness, and spatial distribution would be highly beneficial. Particularly, much information can be extracted while measuring other associated features like cell shape and cytoskeleton organisation.

4. Notes

1. Both Nunc and Ibidi 8-well chambered cover glasses can be used in assay, but the number of plated cells should be optimised for each type of cover glass separately. Take care about the thickness of the cover glass: for image acquisition at higher resolutions (NA > 1.0), the objective lenses require thickness of typically 170 μm.
2. Cell density needs to be optimised before the assay for better performance and reproducibility. Take into consideration that

relatively many cells might be detached and washed away during stripping, and therefore high cell numbers should be plated.

3. Replacement of serum with BSA serves for better control of medium components. Otherwise, integrin endocytosis could potentially be influenced by growth factors and hormones present in serum.
4. Binding of antibodies on ice prevents the immediate internalisation of bound integrin and serves for pulse-chase procedure.
5. Stripping step allows to remove antibodies from PM that were not internalised within given period of time and to measure only intracellular pool of antibodies. Stripping conditions might be different for different antibodies.
6. The first spot can be labelled by printing fluorescently labelled siRNA. Alternatively, siRNA that gives a very distinct cellular phenotype could be used (i.e. siRNA targeting INCENP transcript leads to nuclei segregation defects and, as a result, multilobed nuclei, (6)). It is also possible to label the first spot from beneath with a water resistant marker as siRNA spots are visible when the LabTeks are dry and cell are not yet seeded.
7. Imaging of the surrounding areas of spots (non-transfected areas) can be used as an additional control to assess whether reverse transfection conditions are not detrimental for cells as well as to estimate whether no cross-contamination between individual spots occurs. Here, printing of labelled siRNA also could be used.
8. The Scan^R acquisition software provides images in standardised tiff format which can be imported easily by any other image processing software. This can be useful if parameters are required which are not accessible via Scan^R Analysis and/or the amount of acquired data requires algorithms for distributed computing.
9. Imaging for the primary screen should preferably be done with 10×/0.4 NA objective as it can capture the whole spot and no cells will be missed.
10. It is advisable to consider regions outside the 400 μm spot for further analysis as they can serve as an excellent internal negative control for every image. However, this approach has certain constrains. It can be applied only for the non-motile cells and for the assays, where no cell-to-cell communication and collective cell behaviour is observed.

Acknowledgments

The authors would like to acknowledge funding within the Forsys-ViroQuant consortium, Project no. 0313923, as well as by the Federal Ministry of Education and Research (BMBF). The ViroQuant-CellNetworks RNAi Screening core facility is supported by CellNetworks – Cluster of Excellence (EXC81). A.E. is supported by a Landesgraduiertenförderung fellowship from University of Heidelberg, and by the Hartmut Hoffmann-Berling International Graduate School of Molecular and Cellular Biology, University of Heidelberg.

References

1. Pellinen, T., and Ivaska, J. (2006) Integrin traffic. *J. Cell Sci.* **119**, 3723–3731.
2. Upla, P., Marjomaki, V., Kankaanpaa, P., Ivaska, J., Hyypia, T., van der Goot, F. G., and Heino, J. (2004) Clustering induces a lateral redistribution of α2β1 integrin from membrane rafts to caveolae and subsequent protein kinase C-dependent internalization. *Mol. Biol. Cell* **15**, 625–636.
3. Powelka, A. M., Sun, J., Li, J., Gao, M., Shaw, L. M., Sonnenberg, A., and Hsu, V. W. (2004) Stimulation-dependent recycling of integrin β1 regulated by ARF6 and Rab11. *Traffic* **5**, 20–36.
4. Winograd-Katz, S. E., Itzkovitz, S., Kam, Z. and Geiger, B. (2009) Multiparametric analysis of focal adhesion formation by RNAi-mediated gene knockdown. *J. Cell Biol.* **186**, 423–436.
5. Snijder, B., Sacher, R., Ramo, P., Damm, E.-M., Liberali, P., and Pelkmans, L. (2009) Population context determines cell-to-cell variability in endocytosis and virus infection. *Nature* **461**, 520–523.
6. Erfle, H., Neumann, B., Liebel, U., Rogers, P., Held, M., Walter, T., Ellengerb, J., and Pepperkok, R. (2007) Reverse transfection on cell arrays for high content screening microscopy. *Nat. Protocols* **2**, 392–399.
7. Ziauddin, J., and D.M. Sabatini. 2001. Microarrays of cells expressing defined cDNAs. *Nature* **411**, 107–110.
8. Baghdoyan, S., Y. Roupioz, A. Pitaval, D. Castel, E. Khomyakova, A. Papine, F. Soussaline, and X. Gidrol. 2004. Quantitative analysis of highly parallel transfection in cell microarrays. *Nucleic Acids Res.* **32**, e77.
9. Bartz, F., Kern, L., Erz, D., Zhu, M., Gilbert, D., Meinhof, T., Wirkner, U., Erfle, H., Muckenthaler, M., Pepperkok, R., and Runz, H. (2009) Idenitfication of cholesterol-regulating genes by targeted RNAi screening. *Cell Metab.* **10**, 1–13.
10. Neumann, B., Held, M., Liebel, U., Erfle, H., Rogers, P., Pepperkok, R., and Ellenberg, J. (2006). High-throughput RNAi screening by time-lapse imaging of live human cells. *Nat. Methods* **3**, 385–390.
11. Starkuviene, V., Seitz, A., Erfle, H. and Pepperkok, R. (2008) Measuring Secretory Membrane Traffic. *Methods in Molecular Biology* **457**, 1–9.
12. Chao, W.-T., and Kunz, J. (2009) Focal adhesion disassembly requires clathrin-dependent endocytosis of integrins. *FEBS Letters* **583**, 1337–1343.

Chapter 20

Probing Calmodulin Protein–Protein Interactions Using High-Content Protein Arrays

David J. O'Connell, Mikael Bauer, Sara Linse, and Dolores J. Cahill

Abstract

The calcium ion (Ca^{2+}) is a ubiquitous second messenger that is crucial for the regulation of a wide variety of cellular processes. The diverse transient signals transduced by Ca^{2+} are mediated by intracellular Ca^{2+}-binding proteins. Calcium ions shuttle into and out of the cytosol, transported across membranes by channels, exchangers, and pumps that regulate flux across the ER, mitochondrial and plasma membranes. Calcium regulates both rapid events, such as cytoskeleton remodelling or release of vesicle contents, and slower ones, such as transcriptional changes. Moreover, sustained cytosolic calcium elevations can lead to unwanted cellular activation or apoptosis. Calmodulin represents the most significant of the Ca^{2+}-binding proteins and is an essential regulator of intracellular processes in response to extracellular stimuli mediated by a rise in Ca^{2+} ion concentration. To profile novel protein–protein interactions that calmodulin participates in, we probed a high-content recombinant human protein array with fluorophore-labelled calmodulin in the presence of Ca^{2+}. This protein array contains 37,200 redundant proteins, incorporating over 10,000 unique human proteins expressed from a human brain cDNA library. We describe the identification of a high affinity interaction between calmodulin and the single-pass transmembrane proteins STIM1 and STIM2 that localise to the ER. Translocation of STIM1 and STIM2 from the endoplasmic reticulum to the plasma membrane is a key step in store operated calcium entry in the cell.

Key words: Calcium-binding protein, Calmodulin, Protein array, Fluorophore-labelled, STIM

1. Introduction

High-content protein arrays allow for identification of putative-binding partners over all cellular compartments. The technique is especially valuable for identifying targets of central signalling proteins that are known to regulate a large number of proteins, for example calmodulin. Array screening has several advantages over other methods, for example affinity

Ulrike Korf (ed.), *Protein Microarrays: Methods and Protocols*, Methods in Molecular Biology, vol. 785, DOI 10.1007/978-1-61779-286-1_20, © Springer Science+Business Media, LLC 2011

chromatography (1). Firstly, affinity chromatography may lead to identification of the more abundant proteins and the capture of secondary proteins that bind to primary calmodulin targets. On the protein arrays, the proteins are presented in distinct locations and secondary targets are not likely to be identified. Secondly, array screening is effective in identifying interactions with transmembrane proteins, including receptors and ion channels, which are typically not available in tissue homogenate used for identification through affinity chromatography (2). A key advantage of a protein array-based approach is the ability to directly access the protein expressing clone of an identified target protein and express it for further characterization, of major importance in defining protein–protein interaction networks.

Calmodulin is present in all eukaryotic cells and constitutes at least 0.1% of the total cellular protein. It is expressed at higher levels in brain, testes and in rapidly growing cells. It participates in signalling pathways that regulate processes, such as cell proliferation, learning and memory, growth, and movement (3–5). Regulation of these events is exerted via direct interactions of calmodulin with a large number of cytosolic proteins, including kinases, phosphatases, and cytoskeletal proteins in response to a rise in intracellular Ca^{2+} concentration. In the nucleus, calmodulin is also known to transmit Ca^{2+} signals to a number of transcription factors (5–7). Following an extracellular stimulus, Ca^{2+} moves into the cytosol either from the outside of the cell via plasma membrane Ca^{2+} channels or from intracellular stores. Recently, we have used a labelled form of calmodulin to precisely probe a high-content human protein array resulting in the identification of the high-affinity interaction of calmodulin with the ER transmembrane proteins STIM1 and STIM2 (8). This protein–protein interaction is believed to play a vital role in the homeostasis of intracellular Ca^{2+}.

2. Materials

2.1. Screening of High-Content Arrays

1. High-density protein arrays of the human brain library (hEx1) (ImaGenes, Germany).
2. Tris-buffered saline (TBS): 10 mM Tris–HCl pH 7.5, 150 mM NaCl. Store at room temperature.
3. TBS-Tween (TBS-T): 20 mM Tris–HCl pH 7.5, 500 mM NaCl, 0.05% Tween. Store at room temperature.
4. TBS-T-Triton X-100: 20 mM Tris–HCl pH 7.5, 500 mM NaCl, 0.05% Tween, 0.5% Triton X-100. Store at room temperature.
5. Blocking solution: 2% non-fat milk in TBS. Store at room temperature and use on the day of the experiment.

2.2. Fluorophore Labelling of Calcium-Binding Proteins

1. Alexa Flour 488 C5 maleimide (Invitrogen).
2. Phosphate-buffered saline (10×): 1.37 M NaCl, 27 mM KCl, 100 mM Na_2HPO_4, 18 mM KH_2PO_4. Autoclave before use and store at room temperature. Prepare working solution by dilution of one part with nine parts water.
3. Alexa Flour 488 amine reactive dye.
4. Sodium bicarbonate: Prepare a 1 M stock in deionised H_2O.
5. Sephadex G25 column (Amersham).

2.3. Expression and Purification of Recombinant Proteins from hEx1 Library

1. 2× TY medium: 16 g tryptone, 10 g yeast extract and 5 g NaCl per litre of dH_2O. Autoclave before use and store at room temperature.
2. 40% glucose: Dissolve 40 g in 100 mL of dH_2O and filter sterilise through a 0.2-micron bottle top filter. Aliquot and store at 4°C.
3. 500× ampicillin (50 mg/mL).
4. 500× kanamycin (7.5 mg/mL).
5. Overnight Express (Novagen): Dissolve 60 g in 1 L of dH_2O and autoclave.
6. Lysis buffer: (100 mM NaH_2PO_4, 10 mM Tris–HCl, 6 M guanidine HCl, adjust pH to 8.0 using NaOH).
7. Wash buffer (100 mM NaH_2PO_4, 10 mM Tris–HCl, 8 M urea, adjust pH to 6.3 using HCl).
8. Elution buffer (100 mM NaH_2PO_4, 10 mM Tris–HCl, 8 M urea, adjust pH to 4.5 using HCl).
9. Ni-NTA agarose (Qiagen), 5-mL polypropylene columns (Qiagen).

2.4. SDS-Polyacrylamide Gel Electrophoresis

1. Separating buffer (4×): 1.5 M Tris–HCl, pH 8.7, 0.4% SDS. Store at room temperature.
2. Stacking buffer (4×): 0.5 M Tris–HCl, pH 6.8, 0.4% SDS. Store at room temperature.
3. Thirty percent acrylamide/bis solution (37.5:1 with 2.6% C) (this is a neurotoxin when unpolymerised and so care should be taken to prevent exposure).
4. *N*,*N*,*N*,*N*′-tetramethylethylenediamine (TEMED, Sigma).
5. Ammonium persulfate: Prepare 10% solution in water and immediately freeze in single use (200 μL) aliquots at −20°C.
6. Water-saturated isobutanol. Shake equal volumes of water and isobutanol in a glass bottle and allow to separate. Use the top layer. Store at room temperature.
7. Running buffer (5×): 125 mM Tris–HCl, 960 mM glycine, 0.5% (w/v) SDS. Store at room temperature.

8. Prestained molecular weight markers: Kaleidoscope markers (Bio-Rad, Hercules, CA).

2.5. Surface Plasmon Resonance Measurement Validation of Binding Affinity

1. Calmodulin immobilisation: Activation of dextran matrix of CM5 chip (GE Healthcare) with 0.05 M NHS and 0.2 M EDC.
2. Reactive disulphide group addition: 100 mM PDEA, 0.1 M sodium borate, pH 8.5.
3. Immobilisation buffer: 10 mM HCO_2Na, pH 4.3.
4. Blocking buffer: 50 mM L-Cys, 100 mM HCO_2Na pH 4.3, 1 M NaCl.
5. Target protein buffer: 10 mm Tris–HCl, 150 mM KCl, 1 mM $CaCl_2$, 0.005% Tween20, pH 7.5.
6. Regeneration buffer: 350 mM EDTA, pH 8.0 and wash buffer 0.05% (w/v) SDS.
7. Ni-NTA target immobilisation: NTA sensorchips (GE Healthcare), nickel solution 500 μM $NiCl_2$, 10 mM Hepes pH 7.4, 150 mM NaCl (Sigma).

3. Methods

In these experiments, we use high-content human protein arrays on 22.2 × 22.2 cm polyvinylidene fluoride (PVDF) membranes (9). Each protein array is robotically spotted in a standard 5 × 5 spotting pattern, with each array containing 27,648 protein spots in duplicate, making a total of 55,296 protein spots per array (Fig. 1). The protein collection of 37,200 clones is spotted as follows; one with 27,648 individual clones spotted in duplicate, and one with 9,552 individual clones spotted in duplicate multiple times. Together, the two membranes hold 37,200 redundant proteins, of which over 10,000 are estimated to be unique (non-redundant) human proteins based on oligonucleotide fingerprinting and sequencing of a subset of the clones (10). Each block comprises twelve proteins arrayed in duplicate in a 5 × 5 spotting pattern around a central ink guide dot (Fig. 1). Before use, the membranes are washed clean of residual bacterial colonies leaving the expressed proteins on the membrane. The array was screened for calmodulin-binding proteins in the presence of 1 mM Ca^{2+} using calmodulin labelled at a cysteine residue with the fluorescent probe, Alexa Flour 488 (CaM-Alexa488; Fig. 2a). The cysteine was site-specifically substituted for serine 17, which is solvent exposed in all known structures of calmodulin, causing minimum perturbation of the binding interactions.

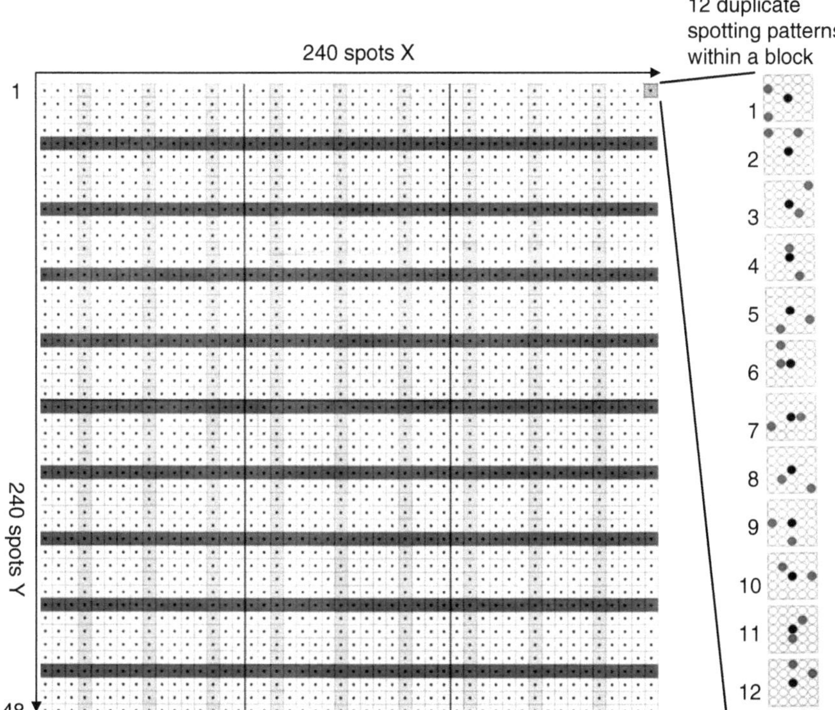

Fig. 1. Spotting format of the hEx1 human protein array. Each protein expressing clone is spotted in duplicate in a 5 × 5 pattern around a central ink dot. The 12 individual duplicate patterns are indicated in the enlarged grid patterns. Grids consisting of 25 spots, including the central ink dot, are arranged in a pattern of 48 grids across the *X*-axis and 48 grids along the *Y*-axis giving 2,304 grids that comprise 55,296 protein expressing clone spots from a total of 57,600 spots on each array.

3.1. Preparation of High-Content Arrays

1. The PVDF arrays were soaked in 95% ethanol, rinsed in deionised water and washed clean of residual bacterial colonies with 20 mM Tris/HCl, 500 mM NaCl, 0.05% (v/v) Tween20, pH 7.4 (TBS-T), with 0.5% (v/v) Triton X-100 (see Notes 1 & 2).
2. For calmodulin-binding protein profiling, the protein arrays were blocked in 2% (w/v) nonfat, dry milk powder in 20 mM Tris/HCl, 150 mM NaCl, pH 7.4 (TBS) for 2 h, washed twice in TBS-T and subsequently incubated with CaM-Alexa488 at a concentration of 1 μM in TBS with 1 mM $CaCl_2$ for 24 h (see Note 3).
3. The protein arrays were washed in TBS-T containing 1 mM $CaCl_2$ six times for 10 min each (see Note 4).
4. The arrays were illuminated with blue 460 nm epi-fluorescence, and the images were taken using a high-resolution CCD detection system (Fuji LAS3000).
5. Image analysis was performed with VisualGrid (GPC Biotech, Germany). Hits were counted as positive if both spots of a clone were significantly brighter than background (Fig. 2).

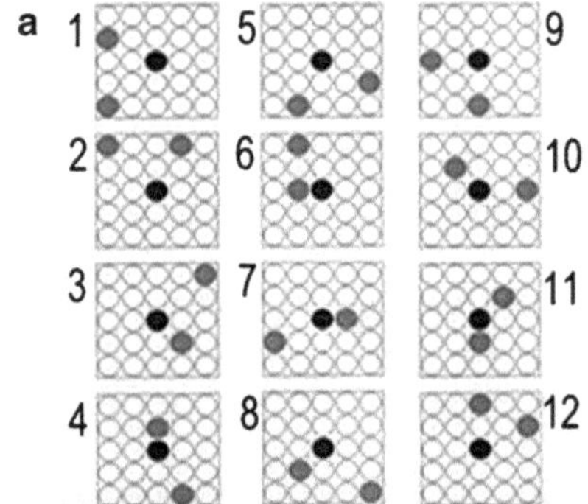

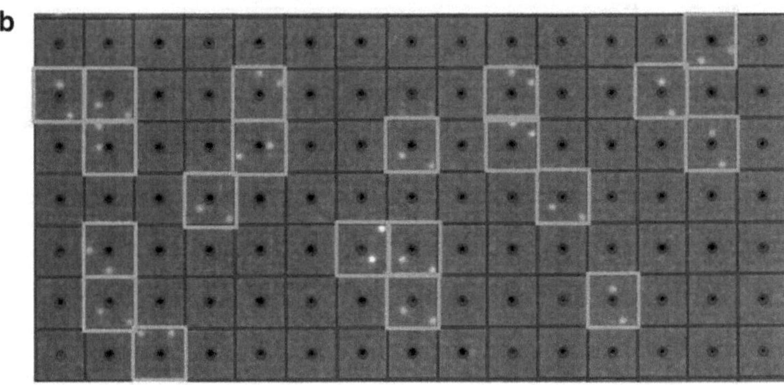

Fig. 2. Scoring positive signals on the hEx1 protein array. (**a**) 12 unique asymmetrical patterns are possible to identify a pair of positive signals reflecting binding to a unique duplicate protein pair. (**b**) Example of analysis of a field of an hEx1 array incubated with Calmodulin Alexa Flour 488 with Visual Grid Software. The array has a virtual grid superimposed allowing the scoring of positive signals that provide the specific clone coordinates and its position and clone identity in the hEx1 library.

6. Control experiments with Alexa 488 labelled secretagogin, calbindin D28k, calbindin D9k to confirm specificity of the protein–protein interactions (Fig. 3).

3.2. Fluorophore Labelling of Calcium-Binding Proteins

1. With thorough inspection of all available high-resolution structures (X-ray and NMR structures) of calmodulin in its Ca^{2+}-free, Ca^{2+}-bound and target-bound forms, Ser17 was identified as a small uncharged hydrophilic residue that is solvent-exposed in all structures.
2. Full length human calmodulin was expressed from a modified Pet3a vector ("PetSac" with NdeI- and SacI-cloning sites) containing a synthetic calmodulin gene that was built with the codons preferred by *Escherichia coli* (11).
3. The calmodulin gene with mutation Ser17→Cys was amplified by polymerase chain reaction (PCR) from this vector using primers containing the desired base change in two steps using standard procedures. The PCR product was digested by NdeI and SacI and cloned into the PetSac vector.

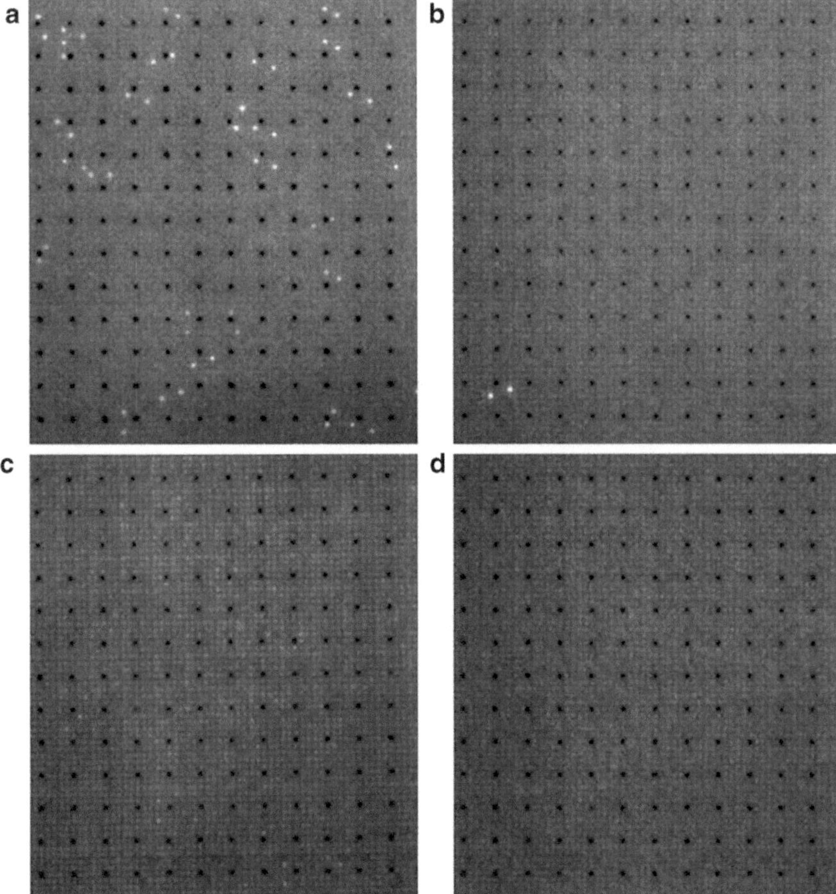

Fig. 3. Four identical fields of the hEx1 array incubated with four different calcium-binding proteins labelled with Alexa Flour 488 showing differential binding in 1 mM calcium (**a**) Calmodulin Alexa Flour 488 (**b**) Secretagogin Alexa Flour 488 (**c**) Calbindin D28k (**d**) Calbindin D9k.

4. *Apo* calmodulin with the S17C (or Ser17→Cys) mutation was dissolved in 20 mM sodium phosphate buffer, pH 8, at a concentration of 10 mg/mL. AlexaFluor488 (Invitrogen) (1.2 mol equivalents) was added from a 5 mg/mL stock in DMSO and the sample allowed to react for 1 h in the dark at room temperature.
5. The labelled protein was then separated from excess label in water on a Sephadex G25 size exclusion column, which had been pre-washed with EDTA to remove trace metal.
6. The collected protein was divided into aliquots for single use and stored frozen at −20°C and subsequently diluted for the binding experiments on the arrays (see Note 5).
7. The control proteins calbindin D9k, and calbindin D28k were labelled either by coupling to an engineered cysteine or with amine labelling for wild-type proteins, while secretagogin and anti-His antibody were labelled using amine-reactive AlexaFluor546.

3.3. Expression and Purification of Recombinant Proteins from hEx1 Library

1. Prepare single colonies of clones by streaking on selective media (2X TY, 2% glucose, 100 μg/mL ampicillin, 15 μg/mL kanamycin, 10% agar) and pick a single colony for inoculation into 1 mL of the same liquid media minus agar.
2. Inoculate 50 μL of starter culture into 10 mL of Overnight Express culture medium (see Note 13), 100 μg/mL ampicillin, 15 μg/mL kanamycin for 20 h culture. This auto-induction system allows for efficient expression of recombinant proteins without the need for IPTG induction (see Note 6) (12).
3. Cells are pelleted at 6,000 × *g*, supernatants poured off and the pellets frozen at −80°C for 1 h to disrupt the cells.
4. The pellets are resuspended in lysis buffer (100 mM NaH_2PO_4, 10 mM Tris–HCl, 6 M guanidine HCl, adjust pH to 8.0 using NaOH) and incubated for 30 min to solubilise cellular proteins. The cell debris is cleared from the lysate by centrifugation at 16,000 × *g* in a microcentrifuge (see Note 7).
5. 1 mL of Ni-NTA purification resin is added to 5 mL polypropylene columns and equilibrated with the addition of 2 × 1 mL lysis buffer.
6. The cleared cellular lysate is added to the column and allowed to purify by gravity flow. The 6 × histidine tag on the expressed recombinant proteins binds to the nickel ions immobilised on the NTA matrix.
7. The specifically bound proteins are eluted from the column in 1 mL of elution buffer (100 mM NaH2PO4, 10 mM Tris–HCl, 8 M urea, adjust pH to 4.5 using HCl) and are visualised using polyacrylamide gel electrophoresis (Fig. 4).

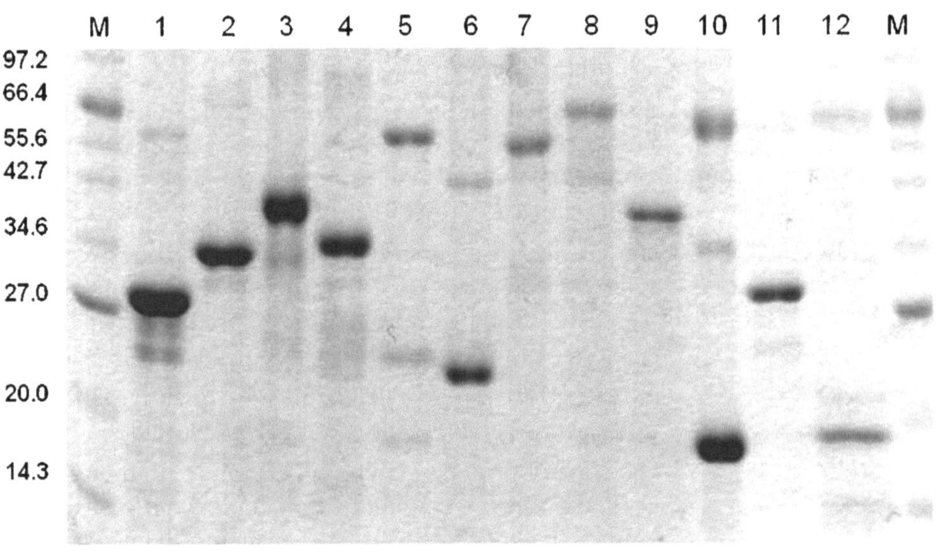

Fig. 4. SDS-PAGE of expressed target proteins after metal affinity chromatography purification on Ni-NTA resin. *Lanes M* = Molecular weight, marker, *Lanes 1–12* = examples of protein preparations.

3.4. SDS-Polyacrylamide Gel Electrophoresis

1. Prepare a 1.0-mm thick, 10% gel by mixing 7.5 mL of 4× separating buffer, with 10-mL acrylamide/bis solution, 12.5 mL water, 100 μL ammonium persulfate solution, 20 μL TEMED. Pour the gel, leaving space for a stacking gel, and overlay with water-saturated isobutanol. The gel should polymerize in about 30 min (see Note 8).
2. Pour off the isobutanol and rinse the top of the gel twice with water.
3. Prepare the stacking gel by mixing 2.5 mL of 4× stacking buffer with 1.3 mL acrylamide/bis solution, 6.1 mL water, 50 μL ammonium persulfate solution, and 10 μL TEMED. The stacking gel should polymerize within 30 min.
4. Prepare the running buffer by diluting 100 mL of the 4× running buffer with 300 mL of water in a measuring cylinder.
5. Once the stacking gel has set, carefully remove the comb and use a 3-mL syringe fitted with a 22-gauge needle to wash the wells with running buffer.
6. Add the running buffer to the upper and lower chambers of the gel unit and load the 50 μL of each sample in a well. Include one well for pre-stained molecular weight markers.
7. Complete the assembly of the gel unit and connect to a power supply. The gel can be run at a constant current of 20 mA per gel.

3.5. Surface Plasmon Resonance Measurement Validation of Binding Affinity

1. Use a Biacore 3000 instrument to carry out all the SPR experiments. S17C calmodulin is immobilised using ligand thiol disulfide exchange coupling and Ni-NTA coupling (Fig. 5).
2. Activate the dextran matrix of a CM5 chip by injecting 25 μL of a fresh mixture of 0.05 M NHS and 0.2 M EDC and introduce a reactive disulfide group on the sensorchip surface by injecting 20 μL of 100 μM PDEA, 0.1 M sodium borate, pH 8.5 (see Note 9).
3. Calmodulin is then immobilised by injecting 100 μL of 10 μg/mL calmodulin S17C in 10 mM HCO_2Na (sodium formate), pH 4.3. Finally, deactivate residual PDEA groups by injecting 40 μL of 50 mM L-cysteine, 1 M NaCl, 100 mM HCO_2Na, pH 4.3.
4. Prepare blank channels for the negative control by omitting calmodulin in the coupling step.
5. Survey the binding of different targets by injecting 150 μL of target protein solutions in 10 mM Tris/HCl, 150 mM KCl, 1 mM $CaCl_2$, 0.005% (v/v) Tween20, pH 7.5. Use the same buffer as running buffer. Monitor the dissociation of target protein from calmodulin for 90 min during buffer flow (see Note 10).
6. Regenerate the chip by injecting 100 μL of 350 mM EDTA, pH 8. A flow rate of 10 μL/min throughout the experiment is recommended (see Note 11).

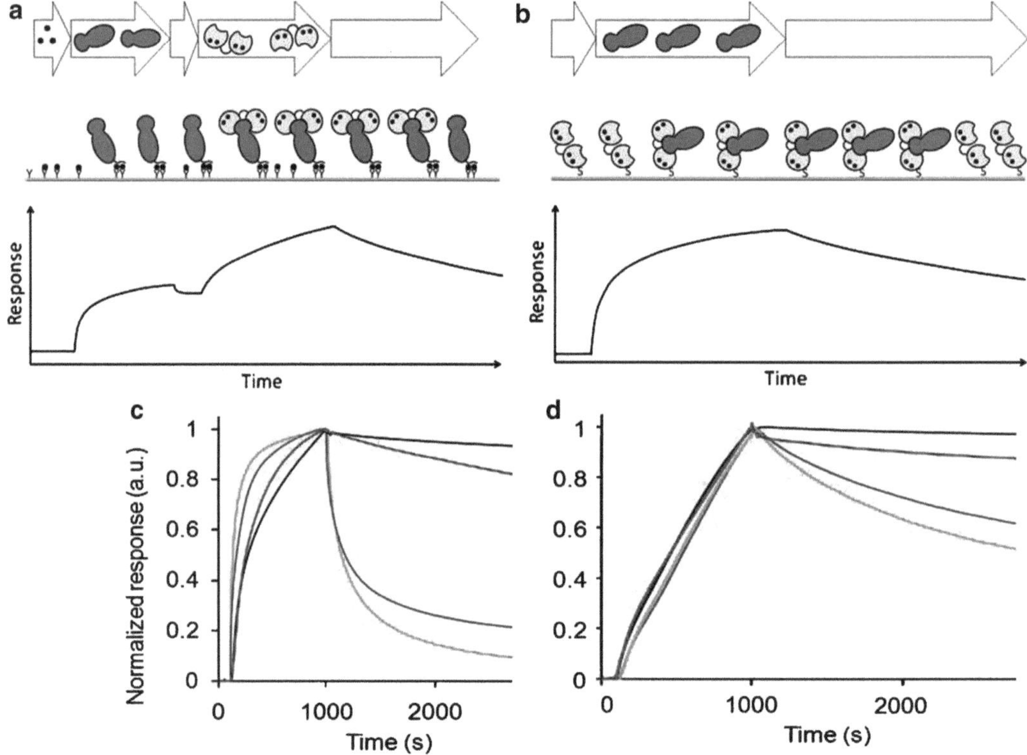

Fig. 5. Surface Plasmon Resonance analysis of the binding affinity of calmodulin with interacting proteins identified by human protein array (hEx1) array screening (**a**) schematic representation of Ni-NTA immobilisation of interacting proteins on a sensor chip with subsequent injection of free calmodulin and associated binding curve sensorgram (**b**) thiol coupling of the S17C mutant of calmodulin to the CM5 sensor chip and subsequent injection of the free interacting protein and associated binding curve sensorgram (**c**) binding curve sensorgrams showing real association and dissociation of calmodulin with immobilised interaction proteins (**d**) binding curve sensorgrams showing real association and dissociation of interacting proteins with immobilised calmodulin.

7. In an alternative approach, the purified calmodulin target proteins are immobilised on NTA sensorchips. The chip is activated by injecting 20 μL of 0.5 mM $NiCl_2$, and then each protein is immobilised by injecting 150 μL target protein solution. Calmodulin association is followed during an injection of 150 μL of 700 nM calmodulin in 10 mM Tris/HCl, 150 mM KCl, 1 mM $CaCl_2$, 0.005% (v/v) Tween20, pH 7.5. The same buffer is used as running buffer and the flow rate used is 10 μL per min. The dissociation of calmodulin is monitored for at least 90 min during buffer flow, and the chip is then regenerated by injecting first 50 μL 350 mM EDTA, pH 8, then 100 μL 0.5 % (w/v) SDS, followed by a final 50 μL injection of 350 mM EDTA (see Note 12).

3.6. Fitting of SPR Sensorgrams for Affinity Measurement

1. For the proteins that were validated using thiol-linked calmodulin, the dissociation rate constant, k_{off}, was estimated from fitting Eq. 1, $R(t) = A \times \exp(-k_{off} \times t)$.

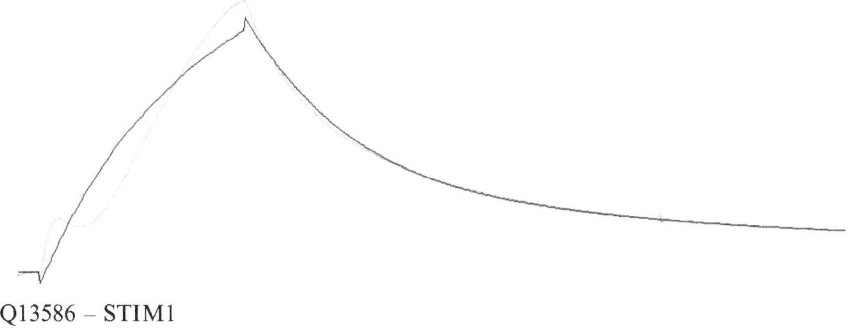

Q13586 – STIM1

Affinity: log k^{on}4 log k^{off}–3 K_D 100nM

LLVAKEGAEKIKKKRNTLFGTFHVAH – Calmodulin binding motif

Fig. 6. Surface Plasmon Resonance binding curve for calmodulin binding with purified STIM1 from the hEx1 protein array. The binding kinetics and identified calmodulin binding domain of STIM 1 are shown.

2. The concentration of each target was estimated from the intensity of gel bands on SDS-PAGE (Fig. 4), providing association rate constants by fitting Eq. 2, $R(t) = R_{max} \times (ck_{on}/(k_{off} + ck_{on})) \times (1 - \exp(-(k_{off} + ck_{on})t)) + R_0$ to the association phase data.
3. The equilibrium dissociation constant, K_D, was calculated from the estimated association and dissociation rate constants according to Eq. 3, $K_D = k_{off}/k_{on}$.
4. The binding curve for STIM1 is illustrated in Fig. 6.

4. Notes

1. Handling the protein arrays requires plenty of space as they are quite large (similar to A4 in size) and while PVDF is quite robust it is easy to tear the arrays if not handled carefully. Use a pair of forceps with flat gripping heads to grab arrays outside of the print area. There are wide margins on at least two sides of the array to facilitate the careful gripping of the membrane as it is moved around. Using two forceps reduces the risk of tearing the array.
2. It is very important to allow the membranes to equilibrate well in TBST-Triton X 100 after the membranes have been activated in ethanol and rinsed. Allow to soak at least 5 min in this buffer before wiping bacterial colonies from the surface. If the membrane is not completely equilibrated, the desiccated bacterial colonies are difficult to remove and background is significantly increased. Also ensure that the tissue paper used

to wipe the arrays clean is thoroughly wet. We have best experience using Kimwipes for this task.

3. For the blocking solution, dissolve the milk powder with stirring at high speed for enough time to ensure no milk powder aggregates are present. It is useful practice to heat this to 50°C when stirring to help dissolve the milk fully.
4. For blocking and incubation steps, it is sufficient to rock the arrays at a slow speed to ensure complete coverage of the membranes with blocking or probe solution. In washing steps, it is necessary to increase the speed of rocking to agitate the arrays and to completely remove all traces of non-specifically bound material.
5. All incubations with flourophore-labelled proteins were covered in aluminium foil to prevent photobleaching, and the foil was carefully wrapped to ensure that the arrays were level on the rocking surface to provide complete coverage of the array surface.
6. While 20 h culture for auto-induction works well for the majority of protein expressing clones and facilitates the expression of many clones simultaneously, it is also advisable to titrate for levels of protein expression for clones that have different expression characteristics. In general, high levels of expressed proteins can be obtained having been incubated for 9 h. However, the proteins expressed may suffer damage if left for incubation times past 12 h. In such cases, a titration series should be performed for individual protein expressing clones to optimise conditions.
7. For 10 mL cultures, the cell pellet is easily disrupted with pipetting and vortexing of the pellet in lysis buffer. For larger volumes, it is advisable to rock the pellet for 1 h to ensure that the cells are completely lysed before clearing the cell debris by centrifugation.
8. For polyacrylamide gel electrophoresis, it is critical that the glass plates used to cast the gels are scrubbed clean with 70% ethanol and rinsed extensively with distilled water prior to use. Failure to clean the plates properly can lead to problems in separating the plates after electrophoresis.
9. For coupling of calmodulin to the sensor chip surface to get a good yield of immobilised protein, it is of great importance to use a fresh PDEA solution. Weigh out the amount needed but wait to dissolve it until just before starting the experiment.
10. Appropriate concentrations of the target proteins to use vary depending on the affinity and kinetics of the binding reaction. Aim for a concentration, where curvature of the signal is obtained during the injection phase. To get reliable fittings of the data, use concentrations where the binding reactions reach equilibrium.

11. At the end of the binding experiment, check that the regeneration of the surface has worked well and ensure that all of the target protein has been washed off before starting a new injection. If not, make another injection of the EDTA solution.
12. The major benefit of the target protein immobilisation via the 6XHis tag is the fact that there is better control of the ligand being injected over the immobilised protein. Using a high-purity calmodulin preparation with concentration determined by amino acid quantification following acid hydrolysis, as a more accurate estimation of the fitted kinetic parameters is obtained.
13. Overnight Express Medium for auto-induction can be prepared as follows:
    (a) Separately autoclave the following four solutions
        (i) 1 M $MgSO_4$
            24.65 g $MgSO_4.7H_2O$ in 100 mL H_2O
        (ii) "50 × 5052"
            25 g glycerol
            73 mL H_2O
            2.5 g glucose
            10 g a-lactose
            (makes 100 mL)
        (iii) "20 × NPS"
            90 mL H_2O
            6.6 g $(NH_4)_2SO_4$
            13.6 g KH_2PO_4
            14.2 g Na_2HPO_4
            (makes 100 mL, check that pH of 20-fold dilution in H_2O is around pH 6.75)
        (iv) ZY
            10 g N-Z-amine AS or Tryptone (we use Tryptone)
            5 g Yeast extract
            925 mL H_2O
    (b) To make 1 L overnight express medium, mix the following:
        1 ml of 1 M $MgSO_4$
        20 ml of "50 × 5052"
        Then add
        50 ml of "20 × NPS"
        ZY up to 1 L total.
        This can be scaled for larger of smaller volumes. Add appropriate antibiotics before starting the cultures

depending on the resistance carried by the plasmid, and depending on the bacteria used, additional antibiotics may be required (e.g., chloramphenicol for BL21 DE3 PLysS Star).

(c) To grow the cultures, take a small well-isolated colony after transformation, inoculate regular LB medium (e.g., 25 or 50 mL) with appropriate antibiotics and shake during the day (6-8 hours sufficient) in baffled flasks at 37°C.

(d) Then transfer 0.5 mL of the day culture to each overnight culture of 500 mL with appropriate antibiotics (in 2.5 L baffled flask, or smaller cultures if you have smaller flasks) and grow over night at 37°C with continuous orbital shaking (120–130 rpm is perfect with baffled flasks). Harvest in the morning and purify as normal. The expression will auto-induce and, therefore, monitoring or addition of IPTG is not required. The amount of expressed protein may be optimized by varying the amount of day culture transferred (typically 0.005 to 0.5 mL to a 500 mL ON culture) and/or the culturing time for the overnight (ON) cultures (9–15 hours typical).

Acknowledgments

We gratefully acknowledge that D. J. O'Connell was supported by Science Foundation Ireland SRC BioNanoInteract (07 SRC B1155) and this work was also supported by the Science Foundation Ireland equipment grant (06/RFP/CHP031/ EC07).

References

1. Berggård, T., Arrigoni, G,, Olsson, O., Fex, M., Linse, S., and James, P. (2007) 140 mouse brain proteins identified by Ca2+-calmodulin affinity chromatography and tandem mass spectrometry. *J. Proteome. Res.* 5, 669–687
2. Larkin, D., Murphy, D., Reilly, D.F., Cahill, M., Sattler, E., Harriott, P., Cahill, D. J. and Moran, N. (2004). ICln, a Novel Integrin αIIbβ3-Associated Protein, Functionally Regulates Platelet Activation. JBC 279, 26, 27286–27293
3. Xia, Z., and Storm D. R. (2005) The role of calmodulin as a signal integrator for synaptic plasticity. *Nat. Rev. Neurosci.* 6, 267–276
4. Kahl, C. R., and Means A. R. (2003) Regulation of cell cycle progression by calcium/calmodulin-dependent pathways. *Endocr. Rev.* 24, 719–736
5. Clapham, D. E. (2007) Calcium Signaling. *Cell* 131, 1047–1058
6. Ikura, M., Osawa, M., and Ames, J.B. (2002) The role of calcium-binding proteins in the control of transcription: structure to function. *Bioessays* 24, 625–636
7. West, A. E., Chen, W. G., Dalva, M. B., Dolmetsch, R. E., Kornhauser, J. M., Shaywitz, A. J., Takasu, M. A., Tao, X., and Greenberg, M. E. (2001) Calcium regulation of neuronal gene expression. *Proc. Natl. Acad. Sci. USA* 98, 11024–11031
8. Bauer, M. C., O'Connell, D., Cahill, D. J., and Linse, S. (2008) Calmodulin binding to the polybasic C-termini of STIM proteins involved in store-operated calcium entry. *Biochemistry* 47, 6089–6091
9. Büssow, K., Cahill, D., Nietfeld, W., Bancroft, D., Scherzinger, E., Lehrach, H., and Walter, G. (1998) A method for global protein expression and antibody screening on highdensity

filters of an arrayed cDNA library. *Nucleic Acids Res.*26, 5007–5008

10. Herwig, R., Schmitt, A. O., Steinfath, M., O'Brien, J., Seidel, H., Meier-Ewert, S., Lehrach, H., and Radelof, U. (2000) Information theoretical probe selection for hybridisation experiments. *Bioinformatics* 16, 890–898

11. Yvonne Waltersson, Sara Linse, Peter Brodin and Thomas Grundstrom (1993) Mutational Effects on the Cooperativity of Ca2+ Binding in Calmodulin. Biochemistry 32, 7866–7871

12. Grabski, A, Mehler, M., and Drott, D. (2005) The Overnight Express Autoinduction System: High-density cell growth and protein expression while you sleep. *Nature Methods* 2, 233–235

Chapter 21

Protein Function Microarrays for Customised Systems-Oriented Proteome Analysis

Jonathan M. Blackburn and Aubrey Shoko

Abstract

Protein microarrays have many potential applications in the systematic, quantitative analysis of protein function. However, simple, reproducible, and robust methods for array fabrication that are compatible with the study of large, custom collections of potentially unrelated proteins are required. Here, we discuss different routes to array fabrication and describe in detail one approach in which the purification and immobilisation procedures are combined into a single step, significantly simplifying the array fabrication process. We illustrate this approach by reference to the creation of an array of human protein kinases and discuss methods for assay and data analysis on such arrays.

Key words: Protein microarray, Biotinylation, Proteomics, Functional analysis, Surface capture, Protein kinase, Biotin carboxyl carrier protein, Inhibitor specificity

1. Introduction

In the post-genomic era, attention has turned towards the systematic assignment of function to proteins encoded by genomes. Bioinformatics methods are typically now used ubiquitously as an essential first step in assigning predicted function to open reading frames (1). However, while such methods give helpful insights into possible function, there remain many examples of proteins that have closely related sequences and/or structures but which prove to have quite different functions when studied experimentally (2–4). As the number of sequenced genomes expands ever further, there is thus an ever increasing need for experimental methods that enable the determination and/or verification of protein function in high throughput. At the forefront of this monumental task, the field of proteomics can be segregated into discovery- and systems-oriented proteomics (5). Discovery-oriented proteomics

Ulrike Korf (ed.), *Protein Microarrays: Methods and Protocols*, Methods in Molecular Biology, vol. 785,
DOI 10.1007/978-1-61779-286-1_21,

is mainly concerned with documenting the abundance and localisation of individual proteins as well as building a picture of protein–protein interaction networks. This is the realm of 2-hybrid screens, 2D-gel electrophoresis and increasingly powerful more direct, isotope-labelling-based mass spectrometry methods; these latter two methods in particular are commonly used to understand the way in which expression profiles change in response to different stimuli by comparing, for example, diseased and healthy cell extracts. However, these discovery-oriented proteomics methods tell us little directly about the precise function of individual proteins or protein complexes, even when augmented by ever more sophisticated bioinformatic methods. Systems-oriented proteomics takes a different approach; rather than rediscovering each protein in each new experiment, the focus is on a pre-defined set of proteins – in principle up to an entire proteome, but in practice more typically a limited subset thereof – enabling the functionality of each member of that set to be dissected in great detail (6). However, obtaining quantitative and genuinely comparative functional data across large sets of proteins with any degree of accuracy is technically difficult, requiring isolation of each individual protein in an assayable format. We and others have chosen to focus on protein function microarray-based methods because the parallel, high-throughput nature of microarray experiments is attractive for analysing large numbers of protein interactions, while the uniform intra-array conditions both simplify and increase the accuracy of assays (6–13). Additionally, the small volumes of ligand or reaction solution required to perform assays, typically tens to hundreds of microlitres, can provide economic advantages, for example when using expensive recombinant proteins or labelled compounds. The key element to such microarray experiments is that the arrayed, immobilised proteins retain their folded structure such that meaningful functional interrogation can then be carried out. There are a number of approaches to this problem which differ fundamentally according to whether the proteins are immobilised through non-specific, poorly defined interactions or through a specific, set of known interactions. The former approach is attractive in its simplicity and is compatible with purified proteins derived from native or recombinant sources (14, 15) but suffers from a number of risks. Most notable among these is that the uncontrolled nature of the interactions between each protein and the surface might at best give rise to a heterogeneous population of proteins or at worst destroy activity altogether due to partial or complete surface-mediated unfolding of the immobilised protein. In practice, an intermediate situation probably most often occurs, where a fraction of the immobilised proteins either have undergone conformational change as a result of the non-specific interactions or have their binding/active sites occluded by surface attachment; these effects effectively reduce the specific activity of the immobilised protein, and therefore decrease the signal-to-noise ratio in any subsequent functional assay that is sensitive to conformation. It is, therefore, important to consider the possible effects of unfolding on the intended downstream assay prior to choosing an array surface: for example, an assay in which solution-phase antibodies bind to linear epitopes on the array will be unlikely to be affected by unfolding of the arrayed proteins (indeed, it may even be desirable to deliberately unfold such proteins in order to expose a greater range of potential epitopes); by comparison, an assay in which a solution phase kinase phosphorylates arrayed proteins may well be sensitive to disruption of the relative 3-dimensional arrangement of targeting and substrate domains in the arrayed proteins.

The advantages of controlling the precise mode of surface attachment are that, providing the chosen point of attachment does not directly interfere with activity, the immobilised proteins will have a homogeneous orientation resulting in a higher specific activity and higher signal-to-noise ratio in assays, with less interference from non-specific interactions (16). This may be of particular advantage when studying protein–small-molecule interactions or conformationally sensitive protein–protein interactions in an array format. The disadvantages of this approach though are that it is really only compatible with recombinant proteins or with families of proteins, such as antibodies, which have a common structural element through which they can be immobilised. However, in a systems-oriented approach the disadvantage of working with recombinant proteins is largely outweighed by the problems encountered in individually purifying large numbers of active proteins from native sources. In addition, experimental approaches that facilitate high-throughput expression and purification of many different proteins in parallel have become more generally accessible over recent years, simplifying access to larger, defined collections of recombinant proteins. An important caveat here though is that it is increasingly clear that despite its ease of use, *Escherichia coli* is not an optimal host for recombinant expression of folded, functional mammalian proteins. Furthermore, while cell-free transcription/translation-based protein microarray systems have been described (17, 18), it remains unclear how reproducible such arrays are, or what proportion of mammalian proteins produced by such approaches are properly folded and therefore functional prior to immobilisation.

In this chapter, we therefore describe the high-throughput cloning, expression, purification, array fabrication, and assay of a set of recombinant proteins in which the human proteins are expressed in insect cells and in which the mode of surface attachment is tightly controlled through the use of an appropriate affinity tag. Furthermore, we show how laborious pre-purification of the recombinant proteins prior to array fabrication can be avoided through the use of a suitable array surface combined with a suitable affinity tag, thus greatly simplifying array fabrication (7). We illustrate this approach to array fabrication with respect to a set of human protein kinases expressed in insect cells as fusions to a polypeptide tag which becomes biotinylated in vivo (7) (see Note 1). In addition, we show representative data from a number of different assays carried out on protein function arrays made in this way.

2. Materials

1. *E. coli Acc* B gene.
2. Human MAPKl cDNA.
3. *Autographa californica* baculovirus "bacmid" vector pBAC10:KO_{1629}.
4. *E. coli* strain HS996.
5. *Spodoptera frugiperda* SF21 cells.
6. Lipofectin (Invitrogen).
7. InsectXpress media (Lonza).
8. 6-well cell culture plates (Nunc).
9. Foetal bovine serum (FBS; Sigma).
10. 24-well deep well blocks (Nunc).
11. Biotin (Sigma).
12. Phosphate-buffered saline (PBS): 1.5 mM KH_2PO_4, 4.3 mM Na_2HPO_4, 137 mM NaCl, 3 mM KCl, adjust to pH 7.4 with HCl, prepare as 10× stock, and autoclave before storage at room temperature.
13. PBST: Prepare from PBS by adding 0.1% (v/v) Tween20.
14. Freezing buffer: 25 mM HEPES, 50 mM KCl, pH 7.5.
15. Lysis buffer: 25 mM HEPES pH 7.5, 20% glycerol, 50 mM KCl, 0.1% Triton X-100, 0.1% bovine serum albumin (BSA), Protease inhibitor cocktail, 1 mM DTT.
16. Streptavidin–HRP conjugate (Sigma).
17. Anti-cMyc antibody (Sigma).
18. Streptavidin (Sigma).
19. 384-well V-bottomed plate (X7022; GENETIX).
20. Nexterion Slide P (Schott).
21. Lifterslips (Nunc, USA).
22. Wash buffer: 20 mM KH_2PO_4, 0.2 mM EDTA, 5% glycerol, pH 7.4.
23. QArray II microarray robot equipped with 16 × 300 μm solid stainless steel pins (Genetix).

24. Kinase buffer: 25 mM Tris–HCl, 5 mM beta-glycerophosphate, 2 mM dithiothreitol, 0.1 mM Na_3VO_4, 10 mM $MgCl_2$, pH 7.5.
25. Adenosine triphosphate (ATP; Sigma).
26. DNA microarray scanner (Tecan LS Reloaded).
27. Ethanolamine (Sigma).
28. 150 mM Na_2HPO_4 buffer, pH 8.5: Prepared by titrating 0.2 M NaH_2PO_4 into 0.2 M Na_2HPO_4 to pH 8.5 and diluting to a final concentration of 150 mM.
29. Milk/PBST: 5% (w/v) non-fat dry milk in PBST.

3. Methods

The methods described below outline (1) the construction of a representative transfer vector in *E. coli*, (2) co-transfection of insect cells with this transfer vector and bacmid, (3) propagation of recombinant baculovirus, (4) induction of protein expression, (5) the extraction of the protein from insect cells, (6) the printing of a protein microarray, and (7) the assay of the protein microarray for protein kinase activity.

3.1. Construction of the Transfer Vector for Full Length Human MAPK1

3.1.1. Transfer Vector

The general baculoviral system used here is adapted from the work of Prof Ian Jones (Reading University, UK; (19)). The specific *E. coli* transfer vector system used is derived from pTriEx1.1 (Novagen).

3.1.2. Amplification and Cloning of the MAPK1 Gene as an N-terminal Fusion to BCCP

All DNA manipulations were carried out using standard recombinant DNA methods (20) to construct the transfer vector and are accordingly not described here in detail.

The gene encoding the *E. coli* biotin carboxyl carrier protein (BCCP) domain (amino acids 74–156 of the *E. coli accB* gene; Fig. 1) (21, 22) was amplified by PCR from an *E. coli* genomic DNA preparation and was cloned downstream of a viral polyhedrin promoter in an *E. coli* vector to create the transfer vector pJB1 (Fig. 2). Flanking this *polh*-BCCP expression cassette were the baculoviral 603 gene and the 1629 genes (19) to enable subsequent homologous recombination of the construct into a replication-deficient baculoviral genome (Fig. 2).

The full length, *MAPK1* gene was amplified by PCR from a cDNA clone (obtained from the Mammalian Gene Collection) and cloned into the pJB1 transfer vector upstream of and in-frame with the BCCP tag using ligation-independent cloning methods (24), replacing the ORF region between the *Spe* I and *Nco* I sites of pJB1 in the process, to generate pJB1-MAPK1. In the course of the PCR amplification step, the stop codon of the MAPK1 gene was

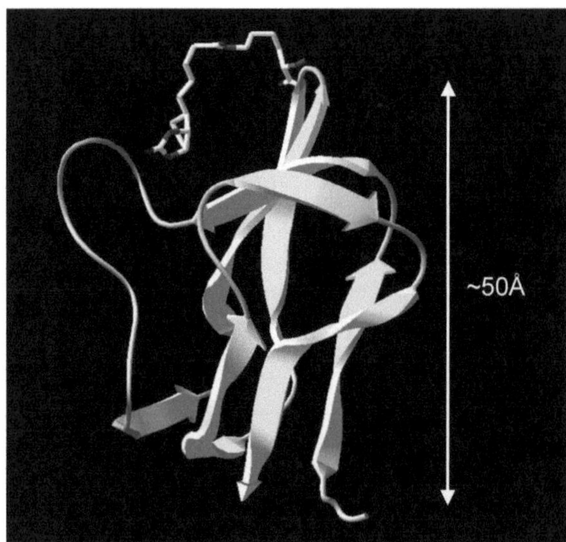

Fig. 1. Structure of the *E. coli* biotin carboxyl carrier protein (BCCP) domain. Residues 77–156 are drawn (coordinate file 1bdo), showing the N- and C-terminii and the single biotin moiety that is attached to lysine 122 in vivo by biotin ligase. Representation produced using SwissPDBViewer (23).

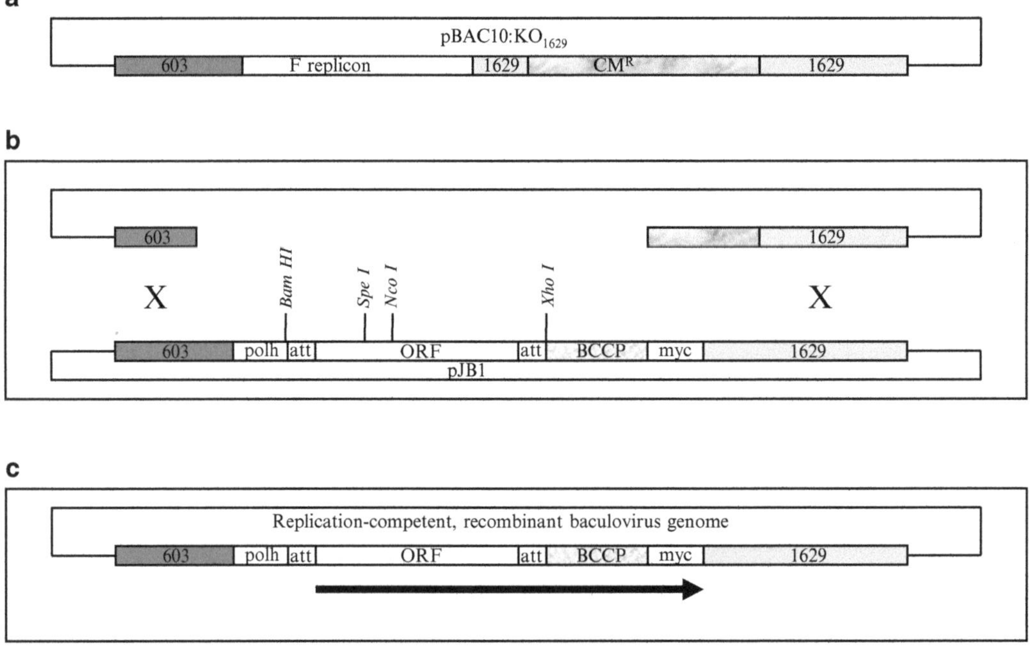

Fig. 2. Schematic of the baculoviral recombination system used in this work. (**a**) Disruption of the essential 1629 gene renders the baculoviral genome replication-deficient. (**b**) Linearised baculoviral genomic DNA and transfer vector are co-transfected into insect cells. (**c**) Homologous recombination regenerates an intact 1629 gene, enabling viral replication. Expression of the BCCP fusion protein is driven from the polh promoter.

removed such that the resulting construct encoded an in-frame MAPK1-BCCP fusion protein; the *MAPK1* gene was sequence-verified against the RefSeq database.

3.2. Insertion of the MAPK1-BCCP Expression Cassette into a Baculoviral Genome

1. Bacmid pBAC10:KO$_{1629}$ (19) was propagated in *E. coli* HS996 cells and bacmid DNA prepared according to standard procedures. pBAC10:KO$_{1629}$ was then linearised by restriction with *Bsu*361 for 5 h at 37°C, after which *Bsu*361 was heat killed at 80°C for 15 min.
2. Linearised pBAC10:KO$_{1629}$ was combined with undigested pJB1-MAPK1 and used to transfect SF21 cells according to standard protocols.
3. 500 ng of linearised pBAC10:KO$_{1629}$ was combined with 500 ng undigested pJB30-MAPK1 and the total volume made up to 12 μl with water.
4. 12 μl lipofectin (diluted 2:1 in H_2O) was then added to this DNA mix and the tube incubated at room temperature for 30 min.
5. 1 ml serum free media (InsectXpress) was added to the lipofectin/DNA mixture.
6. A 6-well plate containing 1×10^6 SF21 cells/well was prepared and incubated at 27°C for 1 h to allow the cells to adhere.
7. Excess media was aspirated from the SF21 cells and replaced with the lipofectin/DNA/serum-free mix.
8. The transfected cells were incubated at 27°C overnight.

   The media was replaced with 2 ml InsectXpress media supplemented with 2% FBS and incubated at 27°C without agitation for a further 72 h.
9. Cells were resuspended by physical agitation and then pelleted by centrifugation at $1{,}000 \times g$ for 10 min.
10. The supernatant containing recombinant baculovirus was transferred to a fresh tube and stored at 4°C; this is the P_0 stock.

3.3. Amplification of Recombinant Baculovirus

Recombinant baculoviral particles were amplified according to standard procedures. Briefly:

1. A 6-well plate was set up with 1×10^6 SF21 cells/well and incubated at 27°C for 1 h.
2. Excess media was removed and replaced with 500 μl of P_0 virus plus 500 μl InsectXpress media supplemented with 2% FBS and incubated at 27°C without agitation for a further 72 h.
3. P_1 virus was harvested as described above.
4. A 150 ml tissue culture flask was seeded with 20 ml of 1×10^6 SF21 cells/ml and incubated at 27°C for 1 h.

5. Excess media was removed and replaced with 500 µl of P_0 virus plus 3 ml InsectXpress media supplemented with 2% FBS and incubated at 27°C for 1 h, after which a further 25 ml InsectXpress media supplemented with 2% FBS were added and cells incubated without agitation for 72 h.
6. P_2 virus was harvested as described above.
7. The titre of the P_2 viral stock was determined by a SybrGreen-based quantitative PCR assay vs. a stock of known titre determined by plaque assay. The titre of the P_2 stock should be *ca.* 10^7 pfu/ml.

3.4. Protein Expression and Extraction

1. Set up a 24-well deep well plate containing 6×10^6 SF21 cells/well suspended in 3 ml InsectXpress media supplemented with 2% FBS and 50 µM biotin (see Note 3).
2. Add 200 µl of P_2 virus and incubate at 27°C for 72 h with agitation.
3. Harvest cells by centrifugation of the 24-well deep well plate prior to lysis.
4. Gently resuspend the cells in 3 ml of PBS buffer, recentrifuge the plate and discard the supernatant; repeat three times in total.
5. Gently resuspend the pellets in 350 µl of freezing buffer ensuring thorough mixing of the cells.
6. Aliquot the cells in 50 µl volumes and store at −80°C until required for cell lysis.
7. Thaw a 50 µl aliquot and add 50 µl lysis buffer plus 10 U Benzonase (Pierce) and shake on ice for 30 min.
8. Remove cell debris by centrifugation, collect the supernatant, and store on ice for up to 24 h before printing.
9. Determine the protein concentration of the soluble, crude protein extract by Bradford assay (25) to confirm that effective cell lysis has occurred (see Note 4).
10. Determine the approximate expression level of soluble BCCP fusion by SDS-PAGE together with Western blot analysis (20) using a streptavidin–HRP conjugate (Fig. 3; see Note 5).
11. To determine the extent of biotinylation of the BCCP fusion protein, carry out a supershift Western blot assay (with an anti-c-Myc antibody) in which equivalent crude lysate samples are pre-incubated with or without streptavidin (0.1 g/ml) (Fig. 4; see Note 6).

3.5. Multiplexed Cloning and Expression of Mammalian Proteins

The procedures described above (Subheadings 3.1–3.4) can clearly be applied to any mammalian cDNA. Moreover, we and others have found that the ligation-independent cloning methods, as well as insect cell transfection, baculovirus amplification, and protein

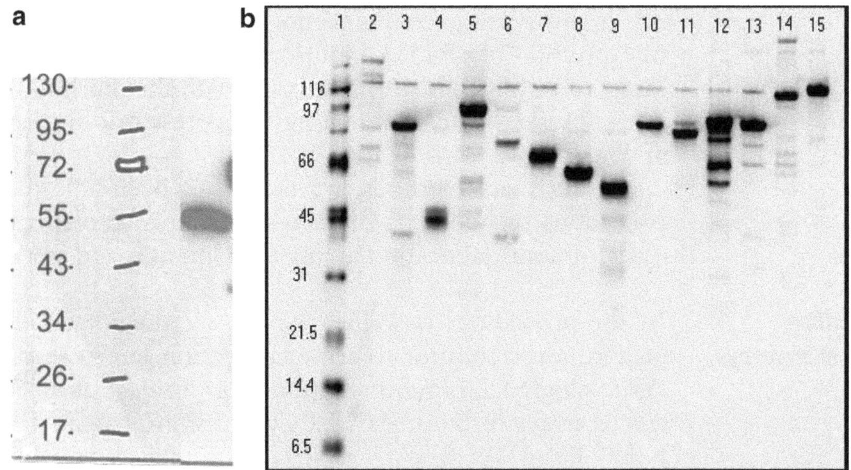

Fig. 3. Western blots of BCCP fusion proteins expressed in SF21 cells. (**a**) MAPK1-BCCP, encoded by pJB1-MAPK1. (**b**) (1) Biotinylated marker, (2) ZNF198-BCCP, (3) FUS-BCCP, (4) SDHB-BCCP, (5) STAT4-BCCP, (6) FH-BCCP, (7) MUTYH-BCCP, (8) PNUTL1-BCCP, (9) GATA1-BCCP, (10) NF2-BCCP, (11) FACL6-BCCP, (12) MSF-BCCP, (13) MSN-BCCP, (14) RAB5EP-BCCP, (15) Control. All westerns were of crude insect cell lysates and were developed using a streptavidin–hrp conjugate. Good expression levels can be observed across a range of unrelated proteins, some in excess of 100 kDa.

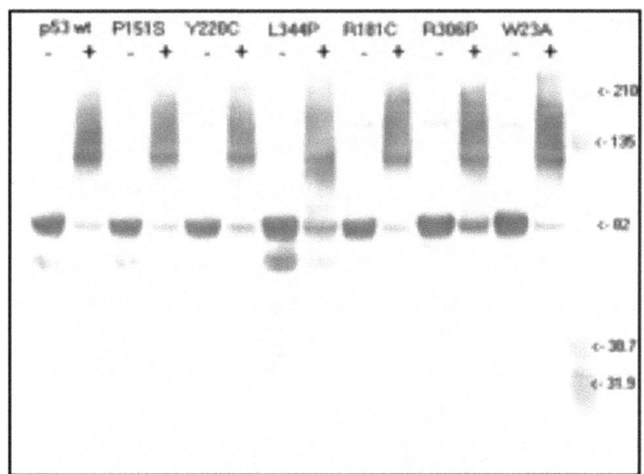

Fig. 4. Supershift assay to determine extent of biotinylation. In a supershift assay, each sample is pre-incubated either with ("+") or without ("–") streptavidin prior to separation by SDS-PAGE. The extent of biotinylation can be estimated by comparison of "+" and "–" samples. Here, we show examples using wild-type and variant p53-BCCP fusion proteins.

expression and extraction are all amenable to multiplexing through the use of appropriate multi-well plate formats. We have observed that by the use of the approach described here, we can easily achieve a ~70% success rate from starting cDNA through to expressed, folded, biotinylated protein suitable for array fabrication. Interestingly, the ~30% overall drop-out rate we observe lies almost entirely in the cloning steps and we have observed a >95% success rate in progressing from sequence-verified transfer vector through to expression of a folded, biotinylated, arrayable human protein in insect cells; this compares favourably with the much lower success rates observed when attempting to express mammalian proteins in *E. coli* (26).

Using these methods, we have thus been able to assemble collections of hundreds of expressed human proteins in a form ready for array fabrication in just a few months and at low cost.

3.6. Fabrication of Protein Microarrays

In the procedures described below, we do not employ a pre-purification step prior to array fabrication but instead rely on a rapid, single-step immobilisation and purification procedure to create arrays of biotinylated BCCP fusion proteins (Fig. 5; see Notes 1 and 2).

3.6.1. Preparation of Source Plates for Printing

1. Transfer 40 μl of crude protein extract for each BCCP-tagged protein to be arrayed into individual wells of a 384-well V-bottom plate and keep at 4°C. This is the source plate for the print runs.
2. Centrifuge the 384-well plate at 4,000 × *g* for 2 min at 4°C to pellet any cell debris that has carried over. Store plate on ice prior to print run.

3.6.2. Preparation of Streptavidin-Coated Slides for Printing

1. Equilibrate a Nexterion Slide P microarray slide to room temperature and remove from the foil package (see Note 2).
2. Make up a 1 mg/ml streptavidin solution in 150 mM Na_2HPO_4 buffer (pH 8.5).
3. Place a glass microarray "lifterslip" over the microarray surface and pipette 60 μl of the streptavidin solution along the edge of the lifterslip such that the solution is drawn under the coverslip uniformly by capillary action (see Note 2).
4. Leave for 1 h at room temperature in a humidified chamber.

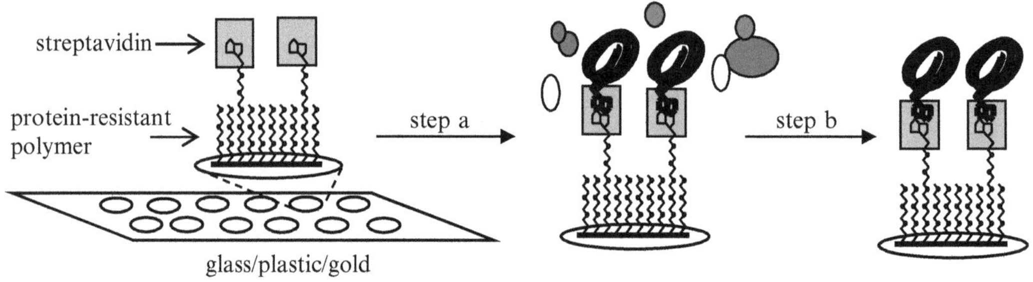

Fig. 5. Schematic of single step immobilisation/purification route to array fabrication. The array surface is intrinsically "non-stick" with respect to proteinaceous material but has a high affinity and specificity for biotinylated proteins. Crude cellular lysates containing the recombinant biotinylated proteins can then be printed onto the surface in a defined array pattern (*step a*) and all non-biotinylated proteins removed by washing (*step b*), leaving the recombinant proteins purified and specifically immobilised via the affinity tag in a single step.

5. Remove the lifterslip and wash the slide for 1 h at RT in 10 ml 150 mM Na_2HPO_4 buffer (pH 8.5) containing 50 mM ethanolamine to deactivate any remaining amine-reactive groups.
6. Wash the slide for 3×5 min in 10 ml wash buffer and then for 5 min in 10 ml water.
7. Place the slide in a 50-ml Falcon tube and centrifuge at 1,000×g for 5 min at 20°C to spin dry. Streptavidin-coated slides were placed into slide boxes, sealed in Ziploc bags and stored at −20°C.

3.6.3. Fabrication of Arrays

In general, any microarray printer could be used to print the arrays. The printing procedures can be carried out at room temperature providing the source plate is kept at 4°C and the atmosphere in the print chamber is humidified to ca. 50%. However, preferably the printing device itself should also be cooled. Here, we describe one specific set of parameters which work well on streptavidin-coated glass microarray slides at room temperature using a Genetix QArray II robot equipped with 16×300 μm tipped solid pins (Figs. 6 and 7; see Notes 7–12).

1. Load the 384-well source plate into the QArray II.
2. Load the streptavidin-coated slides onto the print bed of the QArray II.

   Print the arrays using the following key QArray II settings:

   Inking Time (ms)=500

   Microarraying pattern=7×7, 500 μm spacing

   Max stamps per ink=1

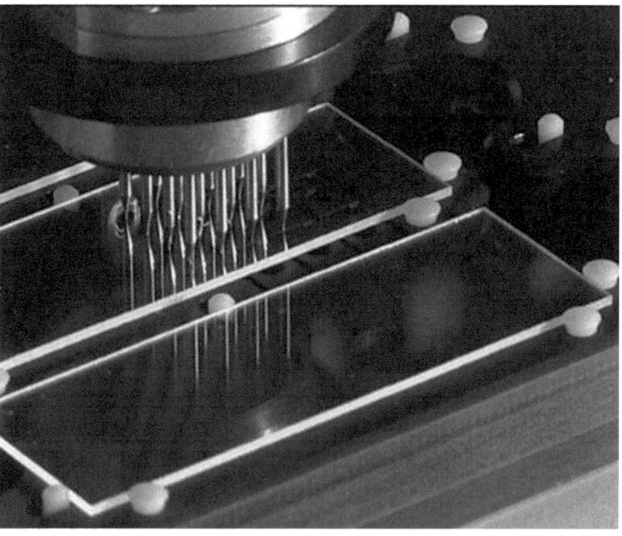

Fig. 6. Printing onto a streptavidin-coated glass microscope slide using 16 solid pins.

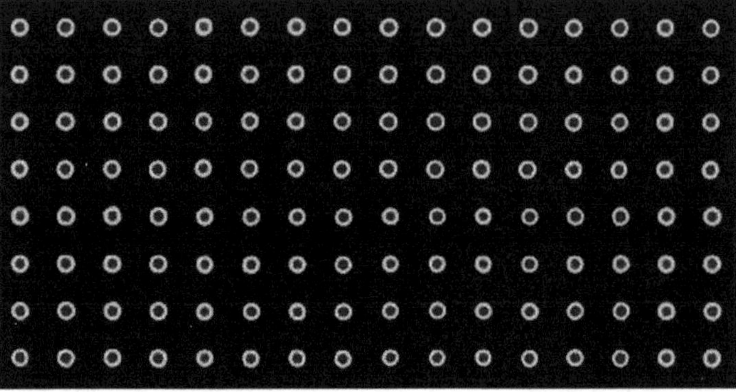

Fig. 7. Highly reproducible arrays with low intra- and inter-array variabilities are key to accurate downstream assays. Here, a single biotinylated protein was printed multiple times using all pins in a 16-pin printing head. The amount of bound protein and spot morphology was determined by imaging the array after binding of a Cy3-labelled anti-His antibody to the immobilised proteins on the array.

No. stamps per spot = 2

Printing depth = 150 μm

Water washes = 60 s wash, 0 s dry

Ethanol wash = 10 s wash, 1 s dry.

3.6.4. Post-printing Processing of Arrays

1. Remove slides from microarraying robot.
2. Wash the slides for 30 min in Falcon tubes containing 50 ml wash buffer supplemented with 100 μM biotin, 0.1 mg/ml BSA (see Note 13).
3. Store in wash buffer at 4°C until use.

3.7. Array-Based Assays for Protein Activity

Once microarrays containing sets of folded, immobilised proteins have been physically fabricated, a wide range of different, systematic, and quantitative assays can in principle be carried out on replica arrays, inter alia: protein–protein interaction assays; protein–nucleic acid interaction assays; protein–small-molecule binding assays; protein–lipid binding assays; and even enzymatic turnover reactions. Through careful experimental design, it is therefore possible now to dissect the properties of diverse collections of unrelated proteins or of specific families of proteins in order to gain a greater understanding of, among others, substrate and inhibitor selectivity. The limiting factor here thus remains the ability to devise an assay that is readable in microarray format for each of the individual arrayed proteins in parallel. Here, we exemplify such assays briefly by reference to arrays of protein kinases.

3.7.1. Verifying that the Individual BCCP-Tagged Proteins Have Indeed Become Immobilised

Following standard protocols for Western blots, it is possible, for example, to probe a replica array with an anti-cMyc antibody followed by a secondary antibody-HRP conjugate, as follows:

1. Dilute a mouse anti-cMyc antibody 1:1,000 in 1 ml milk/PBST.
2. Remove the protein array from wash buffer and equilibrate in PBST at room temperature for 5 min.
3. Drain away the PBST, add 5 ml antibody solution to the array and incubate with gentle agitation at room temperature for 30 min.
4. Wash the array for 3×5 min with 1 ml of PBST.
5. Dilute a goat anti-mouse antibody–hrp conjugate 1:1,000 in 1 ml milk/PBST.
6. Add the antibody solution to the array and incubate with gentle agitation at room temperature for 30 min.
7. Wash the array for 3×5 min with 1 ml of PBST.
8. Add 1 ml chemiluminescent detection reagents (Pierce) to the array.
9. After 1 min, remove slide to a 50-ml Falcon tube and centrifuge for 30 s to dry.
10. In a dark room, place the array against autoradiography film for varying lengths of time before developing the film (Fig. 8; see Note 14).

3.7.2. Measuring the Extent of Autophosphorylation of Each Arrayed Protein Kinase

A subset of protein kinases undergo autophosphorylation at elevated ATP concentrations; for those that do a test to demonstrate whether the arrayed kinases are active or not, based on a simple autophosphorylation assay, can be carried out as follows.

1. Dilute a Cy3-labelled broad-specificity antiphosphotyrosine antibody (Cell Signalling) 1:200 in 2 ml milk/PBST.

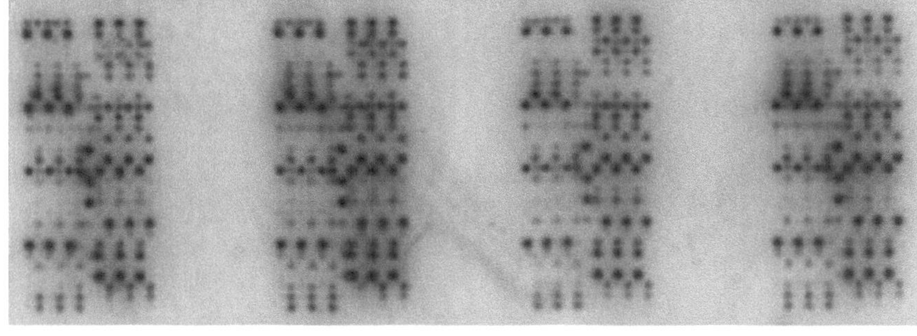

Fig. 8. Array-based qualitative analysis of immobilisation efficiencies. Here, an array of 48 diverse BCCP-tagged human proteins was printed in triplicate, with four replica arrays per slide. Western blot-style analysis, probing with an anti-c-myc antibody, reveals that all proteins are immobilised as expected; apparent gaps on the array are due to the printing pattern used here.

2. Remove two replica protein kinase arrays from wash buffer and equilibrate in PBST at room temperature for 5 min.
3. Drain away the PBST, add 1 ml of kinase buffer to the first array and 1 ml kinase buffer supplemented with 100 μM ATP to the second array. Incubate both arrays at room temperature for 1 h.
4. Wash each array for 3 × 5 min with 1 ml of PBST containing 0.1% SDS.
5. Drain away the PBST/SDS solution, add 1 ml antibody solution to each array and incubate with gentle agitation at room temperature for 30 min.
6. Wash the arrays for 3 × 5 min with 1 ml of PBST each.
7. Remove the arrays to a 50-ml Falcon tube and centrifuge for 30 s to dry.
8. Scan the arrays at 550 nm using a DNA microarray scanner and process the data using a DNA microarray data analysis software package (Fig. 9; see Note 15).

3.7.3. Phosphorylation of Arrayed Protein Kinases by Solution Phase FES Kinase

One use of protein kinase arrays is to determine and identify the subset of the kineome that are substrates for a given solution-phase kinase, as follows:

1. Remove two replica protein kinase arrays from wash buffer and equilibrate in PBST at room temperature for 5 min.

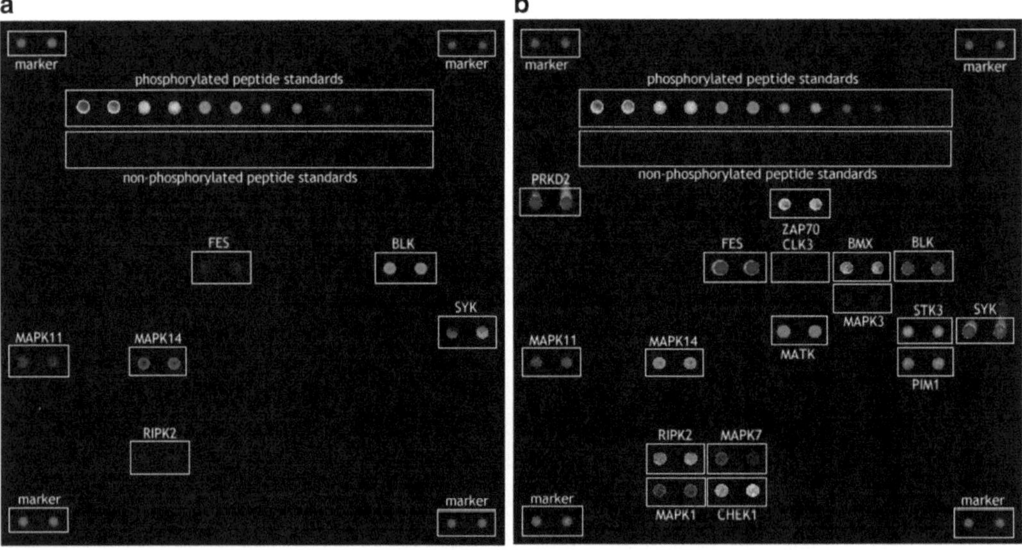

Fig. 9. Array-based autophosphorylation assays. Here, an array of 80 human protein kinases was printed in duplicate and assayed for autophosphorylation activity. (**a**) Kinase buffer only. (**b**) Kinase buffer containing 100 μM ATP. The assays were developed using a fluorescently labelled antiphosphotyrosine antibody and revealed ATP-dependent, on-array autophosphorylation for 17 human kinases, as marked.

2. Drain away the PBST, add 1 ml of kinase buffer supplemented with 100 μM ATP to each array.
3. To one of the arrays, also add 50 nM FES kinase (Upstate) and incubate both arrays at room temperature for 1 h.
4. Wash each array for 3×5 min with 1 ml of PBST containing 0.1% SDS.
5. Drain away the PBST, add 1 ml Cy3-labelled anti-phosphotyrosine antibody solution (from Subheading 3.7.2, step 1) to each array and incubate with gentle agitation at room temperature for 30 min.
6. Wash the arrays for 3×5 min with 1 ml of PBST each.
7. Remove the arrays to a 50-ml Falcon tube and centrifuge for 30 s to dry.
8. Scan the arrays at 550 nm using a DNA microarray scanner and process the data using a DNA microarray data analysis software package (Fig. 10; see Note 16).

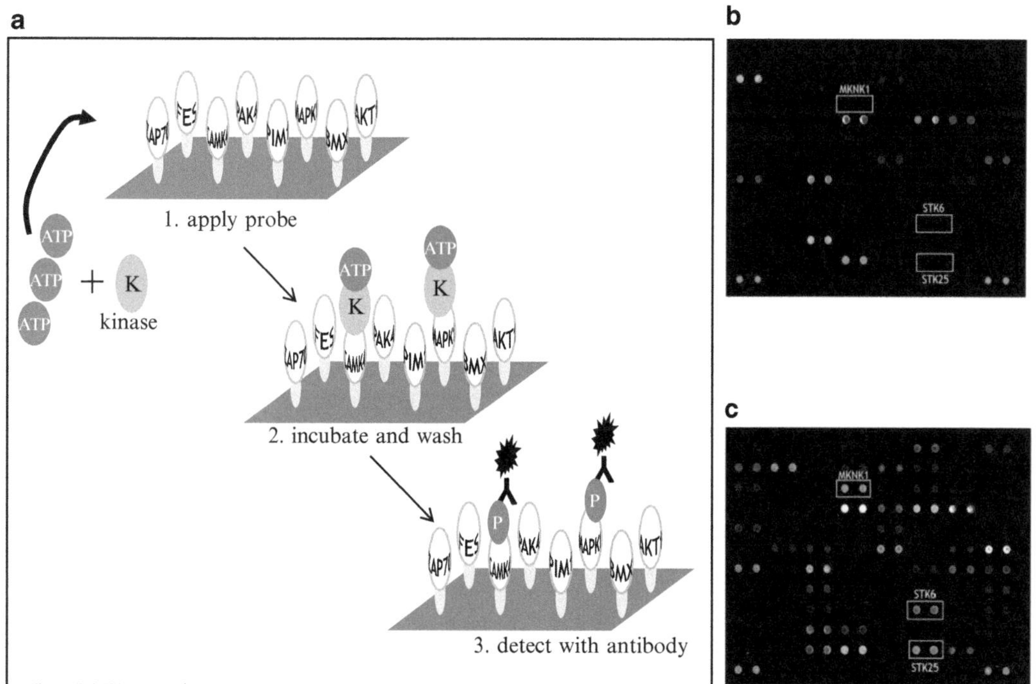

Fig. 10. Array-based phosphorylation assay using an exogenous kinase. An array of 96 human protein kinases was printed in duplicate and assayed in the presence and absence of exogenous Fes kinase. (**a**) Schematic of assay. (**b**) 100 μM ATP plus kinase buffer only. (**c**) 100 μM ATP, kinase buffer plus exogenous Fes kinase. The assays were developed using a fluorescently labelled antiphosphotyrosine antibody and revealed a number of substrates for Fes kinase, including MKNK1, STK6, and STK25, as marked.

3.7.4. Measuring the Binding of an Unlabelled Kinase Inhibitor to Each Arrayed Protein Kinase

Protein kinases are currently regarded as tractable drug targets in many different disease states. Most drug-like current kinase inhibitors target the ATP binding pocket to a greater or lesser degree, yet data from structural biology suggests that the ATP-binding pocket of protein kinases is strongly conserved. Cross-reactivity of protein kinase inhibitors is, therefore, likely to be an issue so low cost, simple, quantitative, yet high-throughput approaches to assess the selectivity of such inhibitors across large panels of human kinases have considerable potential. Protein kinase arrays can be used to provide such data as follows:

1. Remove eight replica protein kinase arrays from wash buffer and equilibrate in PBST at room temperature for 5 min.
2. Drain away the PBST, add 1 ml of kinase buffer to each array.
3. To each array, add an increasing concentration of a fluorescently labelled broad-spectrum kinase inhibitor (e.g. Cy3-labelled staurosporine; a fluorescent-ATP analogue; or fluorescently labelled polycyclic heteroaromatic compounds (27); concentration range from 0.5–50 nM) and incubate all arrays at room temperature for 1 h.
4. Wash each array for 3×5 min with 1 ml of PBST containing 0.1% SDS.
5. Remove the arrays to a 50-ml Falcon tube and centrifuge for 30 s to dry.
6. Scan the arrays at 550 nm using a DNA microarray scanner and process the data using a DNA microarray data analysis software package to determine the K_d for the binding of the fluorescent ligand to each arrayed kinase.
7. Remove a further eight replica protein kinase arrays from wash buffer and equilibrate in PBST at room temperature for 5 min.
8. Drain away the PBST, add 1 ml of kinase buffer to each array.
9. To each array, add fluorescently labelled broad-spectrum kinase inhibitor (10 nM final concentration) plus an increasing concentration of an unlabelled kinase inhibitor (e.g. Geftinib; concentration range from 0.5–50 nM) and incubate all arrays at room temperature for 1 h.
10. Wash each array for 3×5 min with 1 ml of PBST containing 0.1% SDS.
11. Remove the arrays to a 50-ml Falcon tube and centrifuge for 30 s to dry.
12. Scan the arrays at 550 nm using a DNA microarray scanner and process the data using a DNA microarray data analysis software package to determine the IC_{50} for the binding of the unlabelled inhibitor to each arrayed kinase (Figs. 11 and 12).

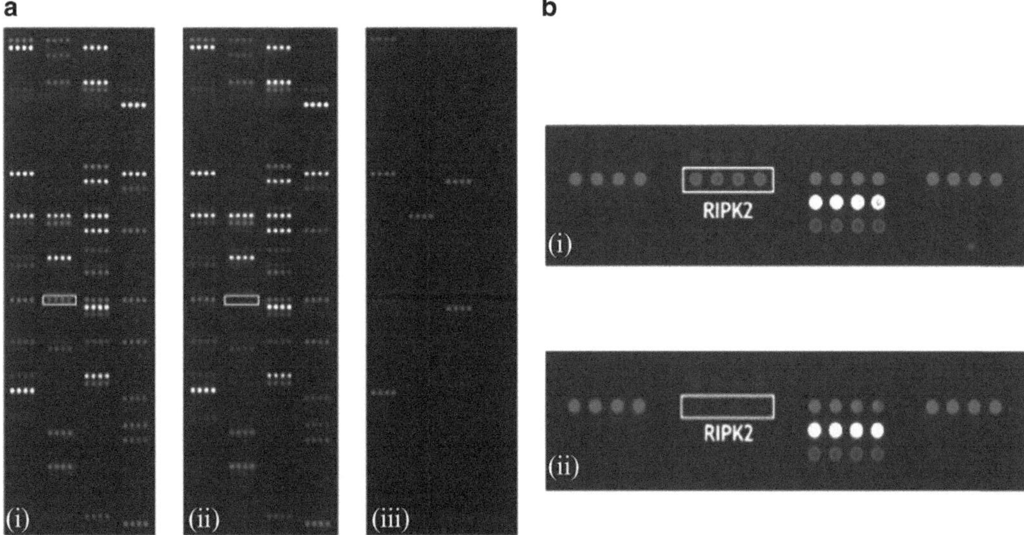

Fig. 11. Array-based inhibitor-binding assays. An array of 150 human protein kinases was printed in quadruplicate, one array per slide, and assayed for binding to a universal fluorescent kinase ligand in the presence and absence of small-molecule kinase inhibitors. (**a**) (i) Universal fluorescent ligand only; (ii) Universal fluorescent ligand plus Iressa, a specific kinase inhibitor; (iii) Universal fluorescent ligand plus staurosporine, a broad-spectrum kinase inhibitor. (**b**) RIPK2 was identified as a target for Iressa because the drug successfully competed with the universal fluorescent ligand for binding to this kinase. Note that the specific "universal" fluorescent ligand used in this assay only bound to 48 out of 150 kinases on the array; by using combinations of such ligands, higher coverage can be achieved.

NB. From the IC_{50} and K_d values, it is possible to derive the K_i value for the unlabelled inhibitor binding to each kinase by standard manipulations.

3.8. Data Analysis: General Principles

3.8.1. Raw Data Extraction

There are in principle numerous different ways in which the raw data can be pre-processed prior to analysis, depending on the precise microarray scanner software package used. We find that the most reliable method is as follows:

1. Use the highest gain setting on the microarray scanner that does not cause any signals to saturate.
2. Set a "pixel inclusion" threshold to exclude any dark pixels (i.e., pixels that strictly form part of the background) from within the analysis area.
3. Extract the raw data and determine the median foreground pixel intensity for each spot, as well as the median local background pixel intensity for each spot.
4. Subtract the median local background pixel intensity from the median foreground pixel intensity for each spot to give the net pixel intensity for each spot.
5. Using the technical replicate data, determine the mean of the median net pixel intensities for each arrayed protein; this is then the data to use in subsequent analyses.

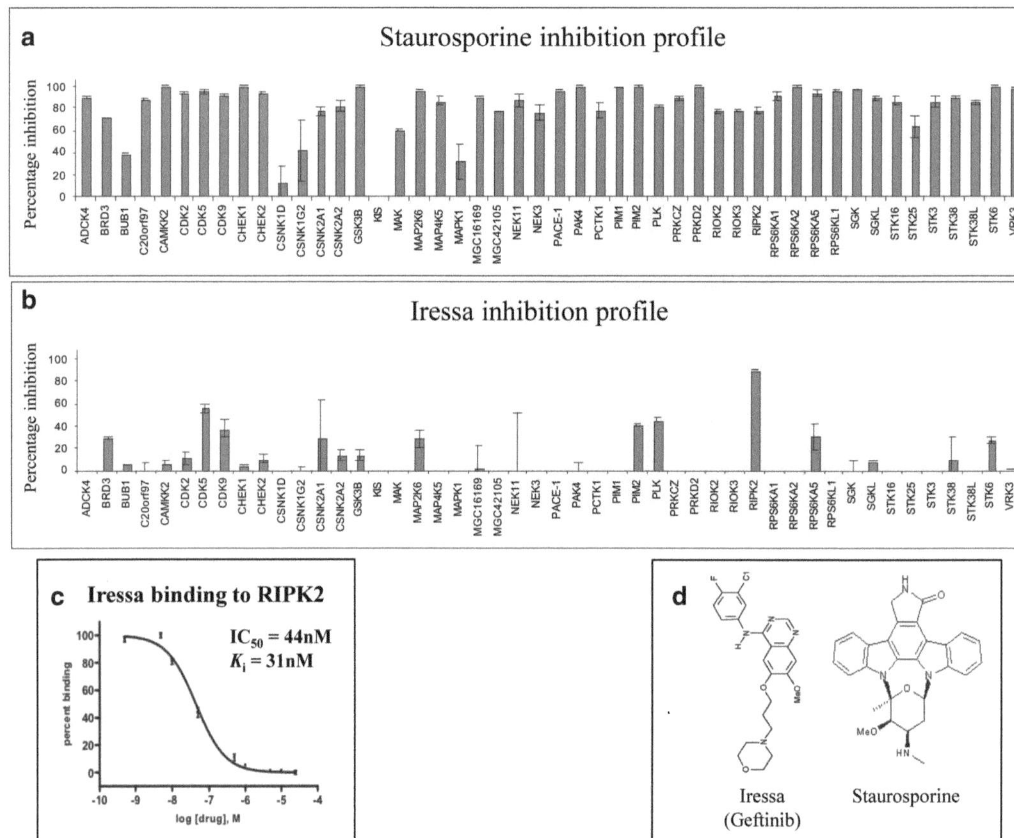

Fig. 12. Data analysis. From the primary data in Fig. 11, the ability of staurosporine and Iressa to compete with the universal fluorescent ligand for binding to each arrayed kinase can be quantified. (**a**) Percentage inhibition of arrayed kinases by 1 μM unlabelled staurosporine. (**b**) Percentage inhibition of arrayed kinases by 1 μM unlabelled Iressa. (**c**) Repeating the assay at different concentrations of Iressa enabled an IC_{50} value of 44 nM to be determined for the Iressa/RIPK2 interaction, in good agreement with literature values. From the IC_{50} value, we calculated that K_i (Iressa/RIPK2) = 31 nM. (**d**) Structures of Iressa and staurosporine, which both bind in the ATP-binding pocket of their target kinases. As expected, at 1 μM concentration, staurosporine competed effectively with the universal fluorescent ligand across the majority of arrayed kinases while Iressa was much more selective in its binding profile.

3.8.2. Data Analysis

Fluorescent ligand-binding data can be analysed in a number of different ways, including the use of global fitting algorithms. Manual analysis can be carried out by the use of Scatchard-type plots as follows:

1. Plot the relative amount of bound fluorescent ligand (FL) as a function of "ligand concentration in solution" for each protein in the array (7) and fit to a simple hyperbolic concentration–response curve according to:

$$R = \frac{B_{\max}[\mathrm{FL}]}{\{K_d + [\mathrm{FL}]\}},$$

where R is the response in relative fluorescence units and [FL] is the ligand concentration. From this, the ligand-binding constants (K_d) and maximum ligand-binding capacities (B_{max}) for each arrayed protein can be determined (7).

2. In competition-binding assays of the type described above (Subheading 3.7.4), the IC_{50} values for a specific protein–ligand interaction are dependent on the intrinsic K_i of the unlabelled ligand for the protein and on the K_d of the fluorescent ligand competitor, as well as on the concentration of the fluorescent ligand in the assay. The relationship is:

$$K_i = \frac{IC_{50}}{\left\{1 + \left(\frac{[FL]}{K_d}\right)\right\}}$$

4. Notes

1. In order to avoid laborious pre-purification of each recombinant protein prior to array fabrication, we have developed a procedure in which we combine the immobilisation and purification stages into a single-step (7). For this approach to work, the array surface itself must have a low capacity for non-specific binding of proteins, yet must have a high specificity and high-binding affinity for the proteins to be arrayed. In principle, a range of different affinity tags could be used here but in practice few actually offer sufficiently high-specificity interactions with the surface affinity matrix. For example, His-tags are not well-suited to such an approach since the specificity of the interaction with Ni^{2+} ligands is too low (particularly, in the context of expression in eukaryotic cells where there are numerous Ni^{2+}-binding proteins) and the intrinsic affinity of the interaction is also relatively low, resulting in leaching of protein from the array even in the absence of imidazole. To circumvent these problems, we have chosen to use an affinity tag that becomes biotinylated in vivo at a single specific residue, allowing us to make use of the very high affinity ($K_d \sim 10^{-15}$M) and specificity of the streptavidin–biotin interaction. Array fabrication thus becomes the simple process of printing crude lysates containing the recombinant biotinylated proteins onto streptavidin-coated surfaces which have a low non-specific protein-binding capacity (see Note 2), followed by washing to remove all non-biotinylated proteins from the array surface (Fig. 5).

In the absence of any pre-purification step, it might be expected intuitively that host cell proteins which are endogenously biotinylated would compete with the biotinylated recombinant proteins for the available streptavidin-binding sites on the array surface. Insect cells have five such endogenous proteins but under native conditions we have observed that these proteins do not compete efficiently with biotinylated recombinant proteins for binding to streptavidin. This perhaps reflects low natural expression levels and the fact that in endogenous biotinylated protein, the biotin is typically less solvent accessible in the native protein (28) meaning that the biotin may not be physically accessible to bind to streptavidin.

We have also observed that using the streptavidin–biotin interaction as the basis for array fabrication confers an additional major advantage: the very high affinity of the streptavidin–biotin interaction means that we quickly start to saturate the available biotin-binding sites on the surface so a crude normalisation of protein loading can be achieved without pre-adjusting the concentrations of the crude lysates to compensate for differences in the individual expression levels of the different recombinant proteins (7).

As with many affinity tags, biotinylated affinity tags can be positioned at either the N- or C-terminus of the protein to be arrayed, dependent on the structural and functional characteristics of the protein. In the context of oriented protein microarray fabrication, "biochemical" biotinylation seems preferable to "chemical" biotinylation since the latter offers little control over the site of biotinylation and still requires pre-purification of each protein to remove excess biotinylation reagent. There are currently three main alternatives for biochemical introduction of a biotin moiety into a recombinant protein: two involve affinity tags that can be biotinylated in vivo or in vitro while the third involves an intein-mediated introduction of biotinylated cysteine.

The AviTag (Avidity, Colorado USA) is an in vitro evolved 15 residue peptide that is specifically biotinylated exclusively by the *E. coli* biotin ligase (29). Alternatively, a single biotinylated cysteine moiety can be added to the C-terminus of a recombinant protein during intein-mediated protein splicing of fusion proteins (30). However, it is not entirely clear what advantages this latter route offers over.

We have chosen however to use a compact, folded, biotinylated, ~80 residue domain derived from the *E. coli* BCCP (Fig. 1) (21, 22) since this affords two significant advantages over the AviTag and intein-based tag. Firstly, the BCCP domain is cross-recognised by eukaryotic biotin ligases enabling it to be biotinylated efficiently in yeast, insect, and mammalian cells without the need to co-express the *E. coli* biotin ligase (31–33). Secondly, the N- and C-termini of BCCP are physically separated from the site of biotinylation by ~50 Å (Fig. 1) (21), so the BCCP domain can be thought of as a stalk which presents the recombinant proteins away from the surface, thus minimising any deleterious effects due to immobilisation.

Vectors for expressing proteins as fusions to a BCCP domain derived from *Klebsiella pneumoniae* are now available from Invitrogen (pET104 DEST Bioease). This *K. pneumoniae* domain is highly homologous to the *E. coli* BCCP protein and confers the same properties.

2. We and others have found that organic polymer coatings, such as those based on dextran or polyethylene glycol (PEG) (either in the form of long chain PEGs or of short chain PEGs supported on a self-assembled monolayer), are considerably superior to proteinaceous blocking agents, such as BSA or powdered milk, in reducing the non-specific-binding background in surface-based assays (34); a number of commercially available surfaces conform to this specification.

   One such commercially available PEG-based streptavidin-coated surface that works well in resisting non-specific protein binding is the Nexterion Slide S from Schott, although we have also used other streptavidin-coated glass or plastic surfaces with similar "non-stick" properties. However, most if not all pre-coated streptavidin slides must be shipped on dry-ice in order to preserve the integrity of the streptavidin layer and the shipping costs can therefore be substantial, depending on location. In order to circumvent this cost issue, we have found that we can custom coat amine-reactive slides (for example, the PEG-based, NHS-activated Nexterion Slide P from Schott, which ships at room temperature and therefore at significantly lower cost) with streptavidin at point of use. Importantly, we have found that our home-made slides show as good uniformity of the streptavidin coating as commercial, pre-made streptavidin surfaces (data not shown); furthermore, we found that utilising an automated hybridisation station in an effort to gain greater control over the streptavidin-coating process gave no obvious advantage over the manual process described here (data not shown).

3. We observed that the addition of free biotin to the growth medium increases the extent of biotinylation of the recombinant BCCP fusion protein. Importantly, we have not found it necessary to also overexpress the *E. coli* biotin ligase in any expression host when using the BCCP tag. The wash step prior to cell lysis is needed to remove free biotin in the media before array fabrication; if the protein is purified before array fabrication, this is then not necessary.

4. When expressing a large number of clones in parallel for array fabrication, the Bradford assay can conveniently be done on all clones in parallel using a microtitre plate format. However, it

would be laborious to carry out SDS PAGE analysis on all clones for each and every repeat expression run, so typically we only assess a selection of clones in this way since the absolute expression level is not critical for array fabrication. For the majority of clones, we have found that a simpler dot-blot assay can form a reliable qualitative indicator that biotinylated recombinant protein has been expressed in the specific set of insect cell cultures in question (data not shown).

5. The growth temperatures and times described here have proven to be directly applicable to all BCCP fusions we have sought to express in insect cells, which number in the hundreds. Since our approach to downstream array fabrication does not require pre-purification of recombinant proteins prior to printing, we have not found it necessary to try to maximise expression levels of any one protein. To the contrary, we have found heuristically that as long as we can observe an expressed, biotinylated recombinant protein by Western blot, there will then be enough recombinant protein to array and assay.
6. The biotinylated component of the sample is supershifted by streptavidin even under denaturing conditions, enabling a simple side-by-side comparison to be carried out. To save time, this assay need not be done on all clones since we have observed that if the recombinant BCCP proteins are expressed and folded, the BCCP domain is efficiently biotinylated.
7. We have found solid pins the easiest to clean rapidly and have observed no carry over between samples using such pins. In addition, we have found that solid pins also work well printing directly onto many different surface types.
8. Each spotting event delivered ~10 nl liquid. We typically use multiple stamps per spot to increase the protein loading at each position in the array, and we have found that these general printing parameters work well with other surfaces, although care must be taken in calibrating the z-height and touch-down velocities on the robot when using fragile surfaces. In addition, by using the same printing parameters under the same conditions of printing (i.e. in a glycerol-containing buffer to reduce evaporation rates), pre-purified proteins can be spotted onto non-selective surfaces which bind proteins by chemical cross-linking (e.g. epoxide or aldehyde-coated glass (14, 15)) or by non-covalent adsorption (e.g. supported nitrocellulose, agarose, or polyacrylamide (15)).
9. Each 50 μl aliquot of harvested expression cells provides enough recombinant protein to print 25 replica slides in 4-plex format, with each protein printed in triplicate in each sub-array. Thus, one 3 ml baculovirus culture yields enough expressed

protein to fabricate 700 replica sub-arrays, or >2,000 replica spots of each protein, which is an important consideration when seeking to minimise array-to-array variability. It is important to note, though, that not all the sample volume loaded into a 96-well V-bottomed source plate can be used for printing because care needs to be taken to ensure that the uniformity of the spot size is not affected by any change in the volume remaining in the source plate.

10. We routinely print each protein in triplicate within each array so that we have three "technical replicates" during downstream data analysis. Furthermore, depending on the desired downstream assay format, it is simple with robotic arrayers to print 1, 2, 4, 8, or 16 replica sub-arrays on each 7.5 × 2.5-cm glass slide. Obviously, the higher the number of sub-arrays per slide, the lower the number of discrete protein spots that can be accommodated in each sub-array. By way of example, using a "4-plex" format, with 8 × 300 μm solid pins it is possible to print eight 8 × 8 panels (=512 discrete protein spots) in each of the four replica sub-arrays with a spot-to-spot spacing of 500 μm; this would enable, for example, 170 individual protein types to be printed in triplicate in each of the four sub-arrays.

11. Importantly, with the ready availability of reusable gasket systems today (e.g., Gentel Biosciences SIM*plex*™ system; Whatman FAST Frame system; etc.), it is possible now to assay each sub-array on any one slide under independent conditions, thus minimising the impact of the cost of the base slides on the "cost per assay" and "cost per data" point to much more reasonable levels. This capability also allows a number of parallel assays to be carried out on a single slide, thus reducing the possible impact of slide-to-slide variability on experimental error.

12. By printing, processing, and assaying arrays using these protocols, we have been able to achieve spot-to-spot CVs of 10% (Fig. 7).

13. The blocking and wash steps should remove all non-biotinylated proteins from the array surface while the biotin in the milk powder blocks any remaining biotin-binding sites on the streptavidin surface such that any excess biotinylated proteins which have not bound within the specific spot cannot then rebind elsewhere on the array. We have found that under the simple storage conditions described here, our arrays remain viable for around 3 weeks, after which the loss of activity of the arrayed proteins starts to be observed. We have made no effort to address this issue though since our interest does not lie in trying to make protein array products here.

14. We have found better performance in this assay when using a chemiluminescent end point than when using a fluorescent end point. The reasons for this are not entirely clear, but may relate to factors surrounding the physical accessibility of the c-Myc epitope in a non-denatured protein array (the c-Myc epitope is on the C-terminus of the BCCP tag, which itself forms the link to the surface, so may be occluded by the fusion protein unless deliberately denatured prior to assay); the signal amplification obtained via chemiluminescence appears to be helpful, but this also suggests that the resultant data from this assay may not be quantifiable in a meaningful sense.
15. We routinely observe that some human protein kinases expressed in insect cells are already phosphorylated prior to assay. However, many other kinases become newly phosphorylated during this assay. There are two obvious explanations for this: firstly, the autophosphorylation of these latter kinases might simply require the higher ATP concentration used in the assay (compared to in vivo concentrations); or secondly, that the autophosphorylation event requires dimerisation of the kinases in question and that such dimerisation is promoted by the surface attachment in the arrays.
16. It is also possible to carry out the same assay using radioactive ATP if desired (we have found 100 μM ATP at 60 Ci/mmol to work well, in which case the detection is by direct autoradiography. This assay format has the immediate advantage of not being restricted by the specificity of the anti-phosphopeptide antibodies. Furthermore, in such cases, it is also then possible to mask competing autophosphorylation events by carrying out a pre-incubation with 100 μM cold ATP.

Acknowledgments

The authors thank Nashied Peton, Natasha Beeton-Kempen, Sarah Joyce, Colin Wheeler, Jens Koopman, Nick Workman, Steve Parham & Mike McAndrew for their help in generating the data detailed herein. We thank Procognia Ltd. (UK) for provision of arrays and also thank the Centre for Proteomic & Genomic Research, Cape Town, for access to equipment. JMB thanks the National Research Foundation (South Africa) for a Research Chair. The research was supported by grants from the National Research Foundation, Procognia Ltd. and Genetix PLC (UK).

References

1. Hunter S., Apweiler R. *et al.* (2009) InterPro: the integrative protein signature database (2009). Nucleic Acids Res. **37**, D224–228.
2. Wise, E.Y., Yew, W.S., Babbitt, P.C., Gerlt, J.A., & Rayment, I. (2002) Homologous $(b/a)_8$-Barrel Enzymes That Catalyze Unrelated Reactions: Orotidine 5'-Monophosphate Decarboxylase and 3-Keto-L-Gulonate 6-Phosphate Decarboxylase. *Biochemistry* **41**, 3861–3869.
3. Schmidt, D.M.Z, Mundorff, E.C., Dojka, M., Bermudez, E. *et al.* (2003) Evolutionary potential of $(b/a)_8$-Barrels: Functional promiscuity produced by single substitutions in the enolase superfamily. *Biochemistry* **42**, 8387–8393.
4. The genome international sequencing consortium (2001) Initial sequencing and analysis of the human genome. *Nature* **409**, 860–921.
5. MacBeath, G. (2002) Protein microarrays and proteomics. *Nature Genetics* **32**, 526–532.
6. Wolf-Yadlin, A., Sevecka, M. & MacBeath, G. (2009) Dissecting protein function and signaling using protein microarrays. *Current Op. Chem. Biol.* **13**, 398–405.
7. Boutell, J.M., Hart, D.J., Godber, B.L.J., Kozlowski, R.Z. & Blackburn, J.M. (2004) Analysis of the effect of clinically-relevant mutations on p53 function using protein microarray technology. *Proteomics* **4**, 1950–1958.
8. Kodadek, T. (2001) Protein microarrays: prospects and problems. *Chem. Biol.* **8**, 105–115.
9. Predki, P. (2004) Functional protein microarrays: Ripe for discovery. *Curr. Opin. Chem. Biol.* **8**, 8–13.
10. Zhu, H., Klemic, J.F., Chang, S., Bertone, P. *et al.* (2000) Analysis of yeast protein kinases using protein chips. *Nat. Genet.* **26**, 283–289.
11. Zhu, H., Bilgin, M., Bangham, R., Hall, D. *et al.* (2001) Global analysis of protein activities using proteome chips. *Science* **293**, 2101–2105.
12. Michaud, G.A., Salcius, M. Zhou, F. Bangham, R. et al. (2003) Analyzing antibody specificity with whole proteome microarrays. *Nature Biotechnology* **21**, 1509–12.
13. Fang, Y., Lahiri, J. & Picard, L. (2003) G-Protein-coupled receptor microarrays for drug discovery. *Drug Discovery Today* **8**, 755–761.
14. MacBeath, G. & Schreiber, S.L. (2000) Printing proteins as microarrays for high-throughput function determination. *Science* **289**, 1760–1763.
15. Angenendt, P., Glokler, J., Sobek, J., Lehrach, H. & Cahill, D.J. (2003) Next generation of protein microarray support materials: Evaluation for protein and antibody microarray applications. *J. Chromatogr. A* **1009**, 97–104.
16. Koopmann, J.-O. & Blackburn, J.M. (2003) High Affinity Capture Surface for MALDI compatible Protein Microarrays. *Rapid Communication in Mass Spectrometry* **17**, 1–8.
17. He, M. & Taussig, M. (2001) Single step generation of protein arrays from DNA by cell-free expression and *in situ* immobilization (PISA method). *Nucleic Acids Res.* **29**, E73.
18. Ramachandran, N., Hainsworth, E. et al. (2004) Self-Assembling Protein Microarrays. *Science* **305**, 86–90.
19. Zhao, Y. Chapman, D.A.G. & Jones, I.M. (2003) Improving baculovirus recombination. *Nucleic Acids Res.* **31**, e6.
20. Sambrook, J., *MacCallum, P. & Russell*, D. (2001) *Molecular Cloning, A Laboratory Manual*, Third ed. Cold Spring Harbor, New York: Cold Spring Harbor Laboratory Press.
21. Athappilly, F.K. & Hendrickson, W.A. (1995) Structure of the biotinyl domain of acetyl-coenzymeA carboxylase determined by MAD phasing. *Structure* **3**, 1407–19.
22. Chapman-Smith, A. & Cronan, J.E. (1999) The enzymatic biotinylation of proteins: a post-translational modification of exceptional specificity. *Trends Biochem. Sci.* **24**, 359–363.
23. Guex, N. & Peitsch, M.C. (1997) SWISS-MODEL and the Swiss-PdbViewer: An environment for comparative protein modeling. *Electrophoresis* **18**, 2714–2723.
24. Yang,Y-S., Watson, W.J., Tucker, P.W. & and Capra, J.D. (1993) Construction of recombinant DNA by exonuclease recession. Nucleic Acids Res. **21**, 1889–1893.
25. Bradford, M.M. (1976) A rapid and sensitive method for the quantitation of microgram quantities of protein utilizing the principle of protein-dye binding. *Anal Biochem* **72**, 248–254.
26. Terwilliger, T.C., Stuart, D. & Yokoyama, S. (2009) Lessons from Structural Genomics. *Ann. Rev. Biophys.* **38**, 371–383.
27. Brown, M. (2007) Novel fluorescent kinase ligands and assays employing the same. Patent application no. WO2008071937.
28. Choi-Rhee, E. & Cronan, J.E. (2003) The biotin carboxylase-biotin carboxyl carrier

protein complex of *Escherichia coli* acetyl-CoA carboxylase. *J. Biol. Chem.* **278**, 30806–30812.

29. Cull, M.G. & Schatz, P.J. (2000) Biotinylation of proteins in vivo and in vitro using small peptide tags. *Methods Enzymol.* **326**, 430–40.

30. Lue, R.Y., Chen, G.Y., Hu, Y., Zhu, Q. & Yao, S.Q. (2004) Versatile protein biotinylation strategies for potential high-throughput proteomics. *J. Am. Chem. Soc.* **126**, 1055–62.

31. Berliner, E., Mahtani, H.K., Karki, S., Chu, L.F., *et al.* (1994) Microtubule movement by a biotinated kinesin bound to streptavidin-coated surface. *J. Biol. Chem.* **269**, 8610–5.

32. Lerner, C.G. & Saiki, A.Y. (1996) Scintillation proximity assay for human DNA topoisomerase I using recombinant biotinyl-fusion protein produced in baculovirus-infected insect cells. *Anal. Biochem.* **240**, 185–96.

33. Parrott M.B. & Barry M.A. (2001) Metabolic biotinylation of secreted and cell surface proteins from mammalian cells. *Biochem. Biophys. Res. Commun.* **281**, 993–1000.

34. Zheng, J. Li, L. *et al.* (2005) Strong Repulsive Forces between Protein and Oligo (ethylene glycol) Self- Assembled Monolayers: A Molecular Simulation Study. *Biophys. J.* **89**, 158–166.

Chapter 22

Optimized Autoantibody Profiling on Protein Arrays

Sara L. O'Kane, John K. O'Brien, and Dolores J. Cahill

Abstract

Profiling the autoantibody (AAb) repertoire in serum has been routinely used for many years for the diagnosis of autoimmune diseases, including rheumatoid arthritis, scleroderma, and lupus. In recent years, AAb profiling of cancers has become a prominent field in oncology research. Protein arrays enable high-throughput screening of clinical samples, characterising the serum profile using low volumes of samples. This chapter describes the use of a protein array comprising 37,200 redundant proteins (containing over 10,000 non-redundant human recombinant proteins) for identification of the proteins bound by the antibodies in human sera using a test set of serum samples. The proteins identified have the potential to be candidate biomarkers. These recombinant proteins are expressed, purified, and robotically spotted on microarrays or chips to facilitate the screening of additional serum samples with the aim of identifying a candidate biomarker or panel of potential biomarkers for applications in disease diagnosis, stage, progression, or response to therapy.

Key words: Autoantibody profiling, Protein array, Chip, Serum screening, Biomarker, Immunoproteomics

1. Introduction

All human blood, sera, or plasma, including from healthy individuals and those with disease, contain antibodies (1). Some of these antibodies recognise self-antigens (known as autoantibodies, AAbs). These self-antigens include intracellular and cell surface components, proteins, and extracellular molecules (2, 3). While the pathogenic role for most AAbs in various diseases is not clear, the specificity of the AAb response and its appearance early in the disease may facilitate improved disease diagnosis. The use of protein array technology facilitates the identification of potential biomarkers of disease (or health) from blood (4, 5). This non-invasive sample collection diminishes the need for biopsies and facilitates more rapid and simple tests.

Ulrike Korf (ed.), *Protein Microarrays: Methods and Protocols*, Methods in Molecular Biology, vol. 785,
DOI 10.1007/978-1-61779-286-1_22,

The AAb repertoire is profiled using protein arrays to determine the antigens bound by these antibodies (i.e. the antibodies in the blood bind to the proteins which are present on the protein array). In this chapter, we describe the method of profiling the antibody repertoire in serum using high-density protein arrays (6). The protein arrays used are generated by in situ expression of 37,200 human proteins from the human foetal brain cDNA library (hEx1) on polyvinylidene fluoride (PVDF) membrane (7), containing a set of over 10,000 non-redundant human proteins (8). These arrays are used to screen serum samples from controls and disease samples of clinically non-remarkable subjects ("healthy"), disease-related control samples (see Note 1), in addition to background incubations with anti-human IgG. Following bioinformatics and statistical analysis of the results, proteins identified in the disease subjects can be identified. These proteins are expressed, purified, and spotted in nitrocelluose-coated microarrays (FAST slides) enabling the screening of larger numbers of serum samples at a lower cost to establish a candidate biomarker or panel of potential biomarkers that require further validation.

Advantages of this method for applications in diagnosis include that as the immune systems amplify the response to the disease-associated proteins, a simple dilution of serum (e.g., 1:100 dilution) is required for testing in either the laboratory or discovery phase of the research or in the clinical assay format. In addition, no pretreatment or sub-fractionation of the serum or plasma is required. The signal amplification facilitated by the immune system means the assays can be easily reproduced between laboratories. An important advantage when protein arrays are used to profile the AAb repertoire is that the assay format is compatible with ELISA resulting in an assay that can be easily incorporated into the clinical setting.

2. Materials

Unless otherwise stated, reagents were obtained from Sigma Aldrich.

2.1. Screening High-Density Protein Arrays

1. High-density protein arrays of the human brain library (hEx1) (ImaGenes, Berlin, Germany): Tris-buffered saline (TBS) (10 mM Tris–HCl, pH 7.5, 150 mM NaCl).
2. TBS-Tween (TBST) [20 mM Tris–HCl, pH 7.5, 500 mM NaCl, 0.05% (v/v)Tween-20 (Reidel de Hein)].
3. TBST-Triton X-100 (TBSTT) (20 mM Tris–HCl, pH 7.5, 500 mM NaCl, 0.05% (v/v) Tween-20, 0.5% (v/v) Triton X-100).

4. Blocking solution 1 (2% (w/v) non-fat milk, TBS).
5. Blocking solution 2 (2% (w/v) BSA, TBS).
6. Human serum (see Note 1).
7. Primary antibody monoclonal mouse anti-human IgG (Fc specific).
8. Secondary antibody anti-mouse IgG (Fc specific), alkaline phosphatase conjugate.
9. Attophos buffer (100 mM Tris–HCL, pH 9.5, 1 mM $MgCl_2$).
10. Diluted Attophos Substrate (25 mM Attophos Substrate (Roche), 100 mM Tris–HCL, pH 9.5, 1 mM $MgCl_2$).
11. Fujifilm LAS 3000 imaging system.

2.2. Expression and Purification of Denatured Human Recombinant Proteins from hEx1 Library

1. 2× TY medium, 0.8% (w/v) glucose, 100 μg/ml ampicillin, 15 μg/ml kanamycin.
2. Overnight Express (Novagen), 10% (v/v) glycerol.
3. Lysis buffer (100 mM NaH_2PO_4, 10 mM Tris–HCl, 6 M guanidine HCl, pH 8.0).
4. Wash buffer (100 mM NaH_2PO_4, 10 mM Tris–HCl, 8 M urea, pH 6.3).
5. Elution buffer (100 mM NaH_2PO_4, 10 mM Tris–HCl, 8 M urea, pH 4.5).
6. Ni-NTA agarose (Qiagen).
7. 1 ml polypropylene columns (Qiagen).
8. BCA Protein Assay Kit (Thermo Scientific).

2.3. SDS-Polyacrylamide Gel Electrophoresis

1. Separating buffer (4×): 1.5 M Tris–HCl, pH 8.7, 0.4% SDS. Store at room temperature.
2. Stacking buffer (4×): 0.5 M Tris–HCl, pH 6.8, 0.4% SDS. Store at room temperature.
3. Thirty percent acrylamide/bis solution (37.5:1 with 2.6% C).
4. *N*, *N*, *N*, *N*'-Tetramethyl-ethylenediamine (TEMED).
5. Ammonium persulfate: 10% (v/v) ammonium persulfate (store in aliquots at −20°C).
6. Water-saturated isobutanol: Shake equal volumes of water and isobutanol in a glass bottle and allow to separate. Use the top layer. Store at room temperature.
7. Running buffer (5×): 125 mM Tris–HCl, 960 mM glycine, 0.5% (w/v) SDS. Store at room temperature.
8. Pre-stained molecular weight markers: Precision Plus Protein dual colour marker (BioRad). Aliquots stored at −20°C.

9. Coomassie: 0.1% (w/v) Coomassie R-250, 5% (v/v) acetic acid, 50% (v/v) methanol.
10. Destain: 10% (v/v) acetic acid, 30% (v/v) methanol.

2.4. Microarray Production and Screening

1. Protein Arrayer: QArray (Genetix, UK).
2. Phosphate buffered saline (PBS) (137 mM NaCl, 2.68 mM KCl, 10.1 mM NaH_2PO_4, 1.76 mM KH_2PO_4).
3. Secondary antibody: Cy3-labelled goat anti-human IgG (Jackson ImmunoResearch).
4. Array scanner: ScanArray 4000B (Axon, UK).
5. 384-well microtitre plates (Genetix, UK).

3. Methods

All experiments were conducted at room temperature unless otherwise stated.

3.1. High-Density Protein Arrays

3.1.1. Serum Screening with High-Density Protein Arrays

The high-density protein arrays purchased from ImaGenes consist of two PVDF membranes each with 27,648 clones spotted, in duplicate, around a central ink guide dot in a 5 × 5 spotting pattern. Internal reproducibility is calculated from the second membrane which contains 9,552 clones (in duplicate) spotted multiple times throughout the membrane. All washing steps involved agitation at 30 rpm on a rocker.

1. The PVDF membranes were soaked in ethanol, rinsed in deionised water, and residual bacterial colonies removed by washing with TBSTT. The arrays were subsequently rinsed with 2× 5-min washes of TBS and 1× 5-min wash with TBST.
2. The membranes were incubated in blocking solution 1 for 2 h at room temperature and agitated at 8 rpm. Blocking solution 1 was discarded after use.
3. Defrosted serum was diluted 1:100 in blocking solution 2 to a total volume of 20 ml and incubated with the membranes (see Note 2). The membranes were incubated in serum for 16 h and agitated at 8 rpm. Diluted serum was collected and stored at −20°C for possible future use.
4. Residual serum was removed by 3× 30-min washes in TBST.
5. The primary antibody was diluted 1:5,000 in blocking solution 2 to a total volume of 50 ml. The membranes were incubated in diluted primary antibody for 1 h and agitated at 8 rpm.
6. After incubation, diluted primary antibody was discarded, and residual antibody was removed by 3× 30-min washes with TBST.

7. The secondary antibody (AP-labelled) was diluted 1:5,000 in blocking solution 2, with a total volume of 50 ml. The membranes were incubated in diluted secondary antibody for 1 h and agitated at 8 rpm.
8. After incubation, diluted secondary antibody was discarded, and residual antibody was removed by 2× 30-min washes with TBST and 1× 20-min wash with TBS.
9. Membranes were incubated for 10 min in Attophos buffer and then incubated (in dark) for 10 min in diluted Attophos Substrate (the substrate for the AP-labelled secondary antibody).
10. The Fujifilm LAS 3000 CCD detection system was used to illuminate the membranes and record the images.

3.1.2. Analysis of High-Density Protein Arrays

1. Positive proteins were identified using VisualGrid (GPC Biotech, Germany). A positive result was only recorded when both of the duplicate spots were identified (see Fig. 1).
2. The clones in the libraries are identified by 5′ tag DNA sequencing. These tags are processed to trim vector sequences, polyA tails, and stretches of ambiguous bases from the ends and then examined for in-frame open reading frames which are translated to protein sequence. For identification, the sequences

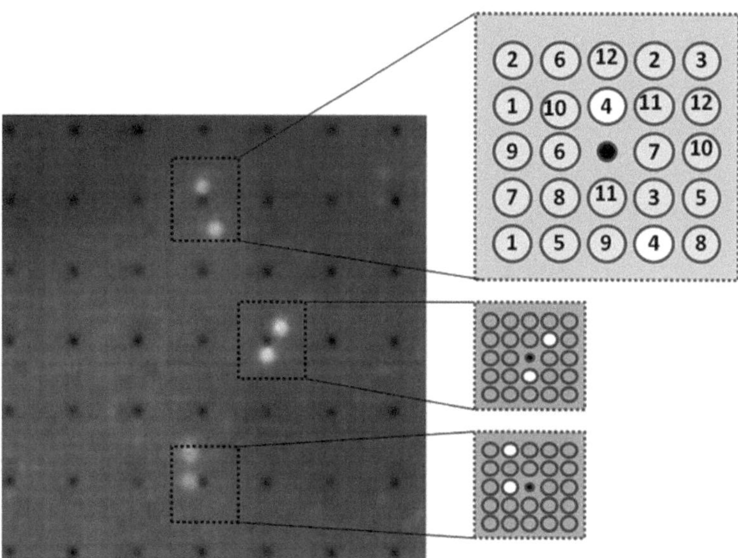

Fig. 1. Section of human protein macroarray (hEx1) following incubation with healthy control serum. The protein-expressing clones on the high-density protein macroarray are spotted in duplicate around a central ink guide dot in a 5×5 pattern. The Visual Grid analysis software produces a grid with the identity of each clone on the macroarray. Proteins are selected as positive when both duplicate spots are detected. The image shows that spots are visible in differing intensities with strong signals detected with three sets of duplicate clones.

are compared to public databases. A blast database of each sequence type (DNA and translated ORF) is constructed so that the library can be searched for the presence of genes or proteins of interest or to compare the library with other clone libraries.

3. Experimental results are managed in a PostgreSQL database. Experimental subjects, samples derived from them, protein arrays, and the list of positively scored clones, along with various experimental details, are all tracked along with the blast homologies.
4. Perl scripts, utilising the CPAN Rose::DB modules, are used to interrogate the database and to collate and compare the data (see Note 3).
5. The initial result required from these experiments is to identify the clones that are positive in the greatest proportion of one group while being negative in the greatest proportion of another group.
6. Further scripts interrogate the database as to the identity of these clones based on the best available match to known genes or proteins. Clones, where no match has been found, are also of interest as these may represent novel or uncharacterised proteins (see Note 4).
7. Further analysis can be performed, for example, by the clustering of subjects based on the profiles of positive clones, searching for common pathways, or ontological groupings among groups of positive proteins or the identification of common sequence motifs. These analyses are performed using the R statistical environment, public databases, and open-source bioinformatic tools, such as BioPerl or EMBOSS.

3.2. Protein Expression and Purification

1. 1 ml of culture was grown for 8 h from a frozen glycerol stock in 2× YT (supplemented with 100 μg/ml ampicillin and 15 μg/ml kanamycin).
2. Single colonies were prepared by streaking the culture onto plates of 1.5% (w/v) agar in 2× YT with kanamycin and ampicillin. The colonies were incubated at 37°C overnight.
3. A single colony was selected and inoculated into 1 ml of supplemented 2× YT and grown for 12 h at 37°C, agitated at 300 rpm (starter culture).
4. A 10 ml aliquot of Overnight Express culture medium (supplemented with ampicillin and kanaymcin) was inoculated with 50 μl of the starter culture. This was agitated at 300 rpm at 37°C for 16 h. The remaining starter culture is used to create a new glycerol stock of the clone. Overnight Express culture medium enables efficient expression of recombinant proteins without IPTG induction (9).

5. The culture was centrifuged at 2,500 × *g* for 15 min and the supernatant was discarded (waste is autoclaved before being disposed). The cell pellets were frozen at −80°C for 1 h to aid cell lysis.
6. The pellets were resuspended in lysis buffer, gently agitated for 30 min, and then centrifuged at 4,000 × *g* for 30 min at 4°C.
7. The supernatant was transferred to a tube containing 500 μl of nickel-nitrilotriacetic acid (Ni-NTA) beads and agitated gently for 1 h. The Ni-NTA and supernatant mix were transferred into 1 ml polypropylene columns (Qiagen) and the sample was purified using affinity chromatography.
8. The 6-HIS tag on the proteins binds to the Ni-NTA beads allowing 2× 4 ml washes with wash buffer to wash away any non-specific binding.
9. Elution of the proteins was achieved by altering the pH, 500 ul of elution buffer was added to the column and the eluate is collected in a low protein-binding tube.
10. The proteins were quantified using a Thermo (Pierce) BCA protein quantification kit and stored in aliquots at −20°C (see Note 5).
11. Proteins were run on a 12% (w/v) polyacrylamide gel at a constant current of 20 mA per gel. The gel was stained in warm Coomassie stain for 15 min and incubated overnight in destain solution (see Note 6).

3.3. Protein Microarray Production

1. The programming of the layout for the microtitre plate for the preparation of microarrays is dependent on the type of robotic microarrayer being used. Using the software provided by the QArray, the density of the pattern is calculated based on the number of proteins to be spotted. The software produces a GAL file which indicates where each protein needs to be positioned in the microtitre plate for the array spotting (see Note 7).
2. The proteins are diluted (see Note 8) to approximately similar concentrations (see Note 9).
3. Nitrocellulose coated FAST slides (Schleicher & Schuell, Whatman) were placed in a Q-Array System (Genetix) at 60–65% humidity.
4. Sixteen blunt-ended stainless steel print tips with a tip diameter of 150 μm were used to generate the protein arrays.
5. The antigens were spotted in quadruplicate onto two identical fields in three concentrations (see Fig. 2).
6. Spotted microarrays were stored at 4°C for at least 8 h before use (see Note 10).

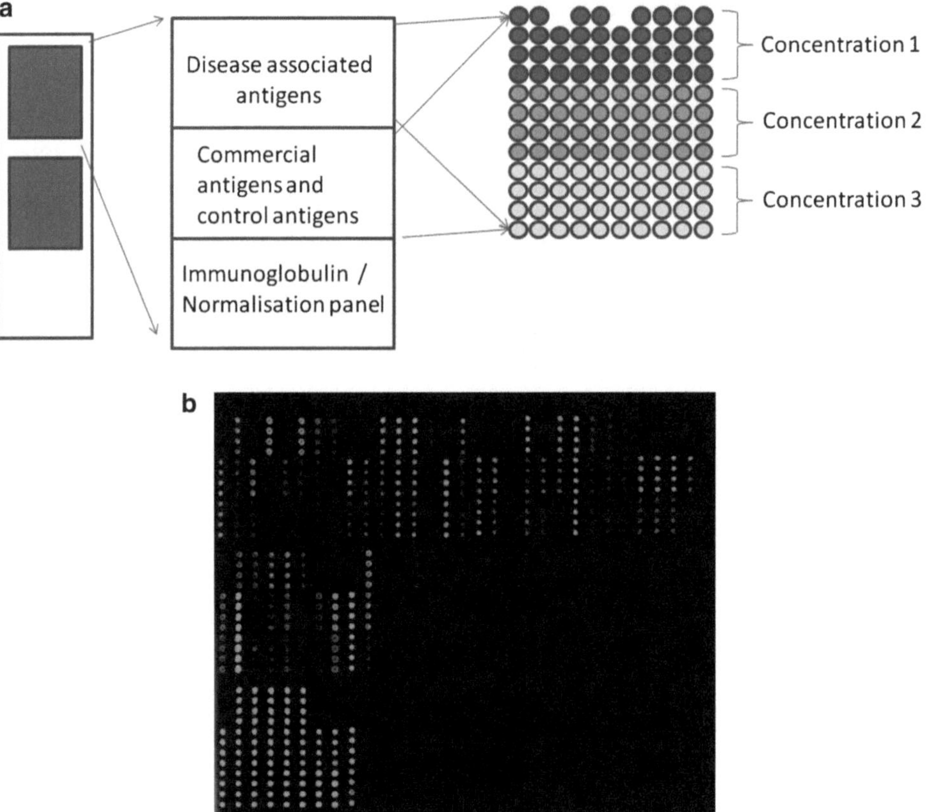

Fig. 2. (**a**) Pictorial layout of the protein microarray used for validation. The microarray consists of two identical fields (Fields 1 and 2) with each field containing three sections. The panels on the microarray consist of antigens relevant to the experiment, a panel of commercially available autoimmune antigens, and within the final panel the "normalisation panel" containing immunoglobulins to enable normalisation for analysis of the microarrays. Each antigen in the top two panels is spotted in three concentrations (where the concentration is available). The immunoglobulins are spotted in four concentrations. Each antigen concentration is spotted a total of eight times on the microarray (four times in each field). (**b**) Indicates a representative image of protein microarray screened with a human serum sample.

3.4. Serum Profiling of Protein Microarrays

All antibody incubation steps were completed in a 150 μl volume under a cover slide in a humidified atmosphere at room temperature.

1. Microarrays were processed in batches of ten.
2. The protein microarrays were blocked in 2% (w/v) BSA/PBS for 1 h at room temperature.
3. Diluted serum samples (1:100 in 2% (w/v) BSA/TBST) were incubated on the microarrays for 2 h and subsequently washed three times in PBST (0.1% (v/v) Tween 20).
4. The arrays were incubated in the dark with the secondary antibody (Cy3-labelled goat anti-IgG 1:300 dilution in 2% (w/v) BSA/PBS for 1 h). The arrays were then washed twice for 30 min in PBS.

6. The slides were centrifuged dry at 1,000 × *g* for 2 min.

7. The arrays were scanned using a confocal microarray reader (Genepix 4000B) and analysed using Genepix Pro 5.1 software (Axon Instruments).

4. Notes

1. All controls should be age and gender matched to the diseased subjects, where possible. When studying a disease, a cohort of non-remarkable (healthy) controls (either certified healthy or as a selection of the general population), a cohort of samples from similar diseases, and a cohort of samples from other disease-related conditions should be included. For example, when studying a specific cancer, subjects with the same cancer type/sub-type at different disease stages should be included. In addition, these results should be compared with results from subjects with benign disease, patients with inflammatory, autoimmune, and/or other diseases of that tissue, and from patients with other types of cancer to ensure that you can accurately describe the identified candidate biomarkers for a specific histological type/sub-type of cancer, and to distinguish them from cancer-related biomarkers, from biomarkers for inflammation, etc.
2. A dilution of serum to 1:100 in blocking solution 2 is used as this represents the standard serum dilution carried out in the clinical setting. Other dilutions can be optimised as this increases/decreases the stringency of the screen. Alternatively, the screening can be carried out by calculation of the immunoglobulin class, such as IgG or IgM, concentration in each sample and screening using a defined IgG or IgM antibody concentration for all the serum samples screened.
3. Typically, the subjects are collected into various groups according primarily to the research programme they are part of, but also stratified according to disease state, age, gender, etc.
4. This approach of profiling the antibody repertoire against high-content proteins arrays identifies the proteins bound by the repertoire of human antibodies in serum or plasma. However, it is recognised that the signal seen by the binding of the antibody (such as IgG) to the human protein on the array could be due to homology (or cross-reactivity). For example, the original antibody response could have been generated against viral or bacterial epitope(s) or protein(s) or as a result of previous vaccination, immunisation, or environmental exposure. If the protein or an epitope had homology to the human protein, the signal seen on the array could be as a result of this homology or cross-reactivity. This can be confirmed

by more extensive sequence analysis, by peptide mapping, or by incorporating known (relevant) control proteins into the microarray content.

5. For protein quantification, it is essential to select an appropriate protein quantification kit that is compatible with the reagents of the protein eluate. As the elution buffer used contains a high concentration of urea, the sample is diluted in PBS to bring the salt concentration within the accepted range of the kit.
6. This allows the mass of the proteins to be estimated when comparing the bands to the bands produced from the molecular weight marker. As the sequence of each of the clones is available, the expected mass of the purified proteins can be estimated. If many additional background bands are visible, the protein expression and purification stages may have not been successful and may need to be repeated.
7. When planning the layout of the microarray, it is possible to design the microarray layout in such a way that all the candidate biomarkers of interest can be represented in one field which can aid the analysis and presentation of the results.
8. The protein spotting buffer comprises elution buffer and PBS (containing 0.1% (w/v) sodium azide). Glycerol is not included as it decreases the binding efficiency when spotting onto nitrocellulose-coated slides as previously reported (10).
9. Although it is not possible to exactly calculate the molar concentration of the protein solutions used for spotting on the microarray, it is assumed that the levels of background proteins are negligible.
10. As the proteins used are HIS-tagged, the efficiency of the robot print run is checked by incubating a slide with anti-HIS antibody before using the batch of arrays for a screening project (6).

Acknowledgments

We gratefully acknowledge that S. L. O'Kane was supported by Enterprise Ireland grant PC/2009/085 and is supported by Health Research Board grant HRA/2009/137 and this work was supported also by the Science Foundation Ireland equipment grant 06/RFP/CHP031/EC07.

References

1. Silverman G. J., Gronwall C., Vas J., and Chen Y. (2009) Natural Autoantibodies to Apoptotic Cell Membranes Regulate Fundamental Innate Immune Functions and Suppress Inflammation. *Discovery Medicine* **8**, 151–156
2. Guilbert B., Dighiero G., and Avrameas S. (1982) Naturally occurring antibodies against nine common antigens in human sera. I. Detection, isolation and characterization. *J Immunol* **128**, 2779–2787
3. Pfueller S. L., Logan D., Tran T. T., and Bilston RA (1990) Naturally occurring human IgG antibodies to intracellular and cytoskeletal components of human platelets. *Clin Exp Immunol.* **79**, 367–373
4. Leuking A., Huber O., Wirths C., Schulte K., Stieler K. M., Blume-Peytavi U., Koward A., Hensel-Wiegel K., Tauber R., Lehrach H., Meyer H. E., and Cahill D. J. (2005) Profiling of Alopecia Areata Autoantigens Based on Protein Microarray Technology. *Molecular & Cellular Proteomics* **4**, 1382–1390
5. Anderson K. S., Ramachandran N, Wong J., Raphael J. V., Hainsworth E., Demirkan G, Cramer D, Aronzon D, Hodi FS, Harris L, Logvinenko T, LaBaer J. (2008) Application of protein microarrays for multiplexed detection of antibodies to tumor antigens in breast cancer. *J Proteome Res* 7, 1490–1499
6. Lueking a., Possling A., Huber O., Beveridge A., Horn M., Eickhoff H., Schuchardt J., Lebrach H., Cahill D. J. (2003). A nonredundant human protein chip for antibody screening and serum profiling *Mol Cell Proteomics* **2**, 1342–1349
7. Büssow, K., Cahill, D., Nietfeld, W., Bancroft, D., Scherzinger, E., Lehrach, H., and Walter, G. (1998) A method for global protein expression and antibody screening on high density filters of an arrayed cDNA library. *Nucleic Acids Res.***26**, 5007–5008
8. Herwig, R., Schmitt, A. O., Steinfath, M., O'Brien, J., Seidel, H., Meier-Ewert, S., Lehrach, H., and Radelof, U. (2000) Information theoretical probe selection for hybridisation experiments. *Bioinformatics* **16**, 890 –898
9. Grabski, A, Mehler, M., and Drott, D. (2005) The Overnight Express Autoinduction System: High-density cell growth and protein expression while you sleep. *Nature Methods* **2**, 233–235
10. Espejo A., and Bedford M. T. (2004) Protein-domain microarrays. *Methods Mol Biol* **264**, 173–181

Part IV

Sample Immobilization Strategies

Chapter 23

Inkjet Printing for the Production of Protein Microarrays

Iain McWilliam, Marisa Chong Kwan, and Duncan Hall

Abstract

A significant proportion of protein microarray researchers would like the arrays they develop to become widely used research, screening, validation or diagnostic devices. For this to be achievable the arrays must be compatible with high-throughput techniques that allow manufacturing scale production. In order to simplify the transition from laboratory bench to market, Arrayjet have developed a range of inkjet microarray printers, which, at one end of the scale, are suitable for R&D and, at the other end, are capable of true high-throughput array output. To maintain scalability, all Arrayjet microarray printers utilise identical core technology comprising a JetSpyder™ liquid handling adaptor, which enables automated loading of an industry standard inkjet printhead compatible with non-contact on-the-fly printing.

This chapter contains a detailed explanation of Arrayjet technology followed by a historical look at the development of inkjet technologies for protein microarray production. The method described subsequently is a simple example of an antibody array printed onto nitrocellulose-coated slides with specific detection with fluorescently labelled IgG. The method is linked to a notes section with advice on best practice and sources of useful information for protein microarray production using inkjet technology.

Key words: Inkjet, Microarrayer, Protein microarray, Printing buffer, Diagnostic microarray, Non-contact, JetSpyder, JetGuard, Microarray production, Antibody microarray

1. Introduction

Since their inception in 1995, microarrays have been the icon of the "omics revolution". Whereas DNA can be printed onto a wide range of surface chemistries with a wide range of buffers, proteins must be printed in buffers which protect them and onto substrates which maintain their structural integrity, binding sites and activity. Common buffers for protein storage contain cryoprotectorants (e.g., glycerol (1) or ethylene glycol) which add to the often viscous

Ulrike Korf (ed.), *Protein Microarrays: Methods and Protocols*, Methods in Molecular Biology, vol. 785,
DOI 10.1007/978-1-61779-286-1_23,

nature of protein solutions, so the ability of a microarrayer to print such solutions without extensive modification is a distinct advantage when producing protein microarrays. The most popular substrates for protein microarrays, designed specifically with structural protection in mind, are made with thin nitrocellulose or hydrogel coatings and are fragile themselves, which encourages the use of a non-contact method of printing. When addressing production throughput requirements, the ability to handle multiple samples simultaneously and to print those samples quickly on-the-fly is important to minimise production timescales. The technology should also be scalable and robust enough to accommodate a shift from R+D level production to full-scale, manufacturing levels of production. Arrayjet has used the key factors of being able to print potentially viscous protein solutions with a non-contact printing method to develop a range of microarrayers, all centred around an industrially proven inkjet printhead, which are highly suited to printing high-quality protein microarrays at all levels of production.

This chapter contains a detailed explanation of Arrayjet technology followed by a historical look at the development of inkjet technologies for protein microarray production. The method described subsequently is a simple example of an antibody array printed onto nitrocellulose-coated slides and subsequent detection with fluorescently labelled IgG. The method is linked to a notes section offering advice on best practice and further sources of useful information for protein microarray production.

1.1. Inkjet Printing

Inkjet printing is the ejection, from a nozzle, of liquid droplets (e.g., protein solution) which travel a short distance (1–5 mm) through the air to land on a substrate in a predetermined pattern. The different methods of inkjet droplet ejection are well-reviewed (2, 3), and there are three modes of action as follows:

1. Piezo actuation uses a volumetric change to induce the pressure required for droplet ejection.
2. Valve-jet uses a continuous pressure stream in conjunction with a valve which opens and closes to eject a stream of droplets.
3. Thermal inkjet, also known as bubble-jet, uses the rapid heating of samples to create a pocket of gas to induce the required pressure for droplet ejection (4, 5).

Arrayjet printheads use piezo actuation; the interior walls of the printhead channels are made from ceramic, piezoelectric material called lead zirconate titanate (PZT). When PZT is subjected to an electrical charge, it changes shape causing a volumetric change and a subsequent acoustic wave which ejects a droplet of sample from the nozzle (Fig. 1). Unlike thermal inkjet, with which Arrayjet

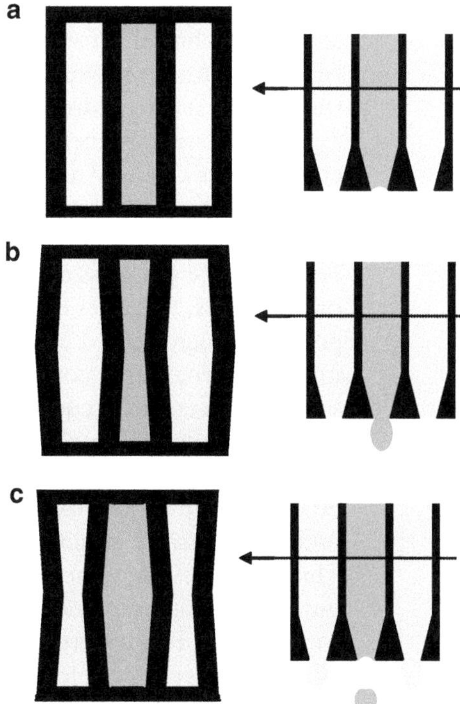

Fig. 1. Simplified diagram of the mechanism of sample droplet ejection from a small section of a shared wall printhead shown in two cross-sectional planes. (**a**) Three channels of the printhead, each terminating in a nozzle, in the non-actuated state. The *dashed line* indicates a cross-section part way up the channels. (**b**) Actuation of the piezoelectric walls of the central channel; the walls change form momentarily to create a pressure wave to eject a droplet of sample from the nozzle. Capillary action replenishes the sample to make up for the volume ejected so ensure that the channel is ready for the next ejection. (**c**) Neighbouring channels undergo the same process while the droplet from the central nozzle travels through the air to the substrate.

was incorrectly associated (6), there is no heating of the sample with piezo-actuated printing. Piezo actuation and acoustic wave drop formation impart almost no mechanical stress on the printhead; nozzles consistently print even after actuations exceeding 10^{13} per nozzle (7) making this technology ideal for printing reliable microarrays.

Arrayjet technology is comparatively new, so it is not surprising that a number of review articles published since its emergence have inadvertently omitted to mention it. There has also been some confusion as to the nature and speed of Arrayjet technology (6, 8), particularly with reference to printing speeds of inkjet microarraying in the literature published before 2004 (9, 10). Use of Arrayjet technology is gaining momentum and while many users print different types of nucleic acid arrays (11–15), the very same microarrayers are being used increasingly for protein microarray production (16), including label-free protein techniques (17–19) and nanotechnology applications (20). There is, however, limited accurate technical information in the scientific literature on how Arrayjet technology works, best practice, and troubleshooting tips; the present article addresses this need.

Traditional commercially available piezoelectric microarrayers employ glass capillary dispenser(s) with a ceramic piezo-collar (21) while Arrayjet utilises the Xaar XJ126, an industrially proven, multi-nozzle, shared wall printhead which has 126 linearly arranged nozzles. Whereas glass capillary spotters use a low number (1–16) of dispensers, this does not apply, as was suggested (6), to Arrayjet printers which have 126 nozzles. The Arrayjet printhead has a unique internal surge control mechanism, ensuring consistent drop size and placement accuracy from each nozzle. The systems handle 12 or 32 samples simultaneously, and printing is performed on-the-fly, with a head speed of 0.2 m/s. Arrayjet systems currently print at a rate of approximately 475 features per second which is understood to be significantly faster than competitors (22, 23). This translates into the ability to print an entire 384-well plate onto all 100 slides in under 25 min; tests performed in conjunction with DKFZ, Heidelberg indicated that the same task would take peer instruments between 6 and 12 h to complete. This represents a paradigm shift in throughput when compared to both pin spotters and competitor non-contact microarrayers. As a real-life comparison with a pin-printer, one Arrayjet user prints a 576 feature array onto 100 Schott Nexterion 16 pad slides in fewer than 6 h; a run which takes their MicroGrid II (BioRobotics), utilising four pins, 58 h to print. Suggestions that printing rates of commercially available piezoelectric printers are less than (9) or typically similar to single-pin contact printing due to limitations of the number of tips on the spotter (21) are not applicable to Arrayjet technology.

1.2. Arrayjet's Contribution to Inkjet Microarrayers

Arrayjet offers four microarrayers to address the different levels of throughput required by users. All four instruments are built around the same core components, discussed below, which allow users to increase throughput as their requirements for production grow. The Sprint is an entry level benchtop research and development system which prints two microplates (four with reloading) onto 20 slides. The other three systems are the Marathon, Super Marathon, and Ultra Marathon which are built around a modular, upgradeable design. The Marathon prints six microplates (96 with reloading) onto 100 slides. With the addition of a microplate stacker module, the Marathon becomes the Super Marathon; enabling 48 microplates to be loaded at once with a further 156 by reloading. The Ultra Marathon, which can also accommodate the microplate stacker, critically has a slide stacking module which transforms it into a 1,000 slide production platform for truly industrial-scale microarray printing; this level of throughput remains unique in the industry.

1.3. The JetSpyder™ Concept for Automated Sample Loading

The key components of Arrayjet systems are a 126 nozzle inkjet printhead, a liquid handling device called a JetSpyder™ and a stepper motor-driven syringe. The printhead is located on a robotic gantry and is connected to a hydraulic system which incorporates the syringe. The JetSpyder™ is used during the automated loading of samples into the printhead and the syringe drives bulk liquid movements, such as system purging and aspiration of samples.

Samples are aspirated into the printhead via the patented liquid handling adapter, the JetSpyder™ (Fig. 2). The JetSpyder™ has a rubber seal on its upper surface comprising 12 or 32 diamond

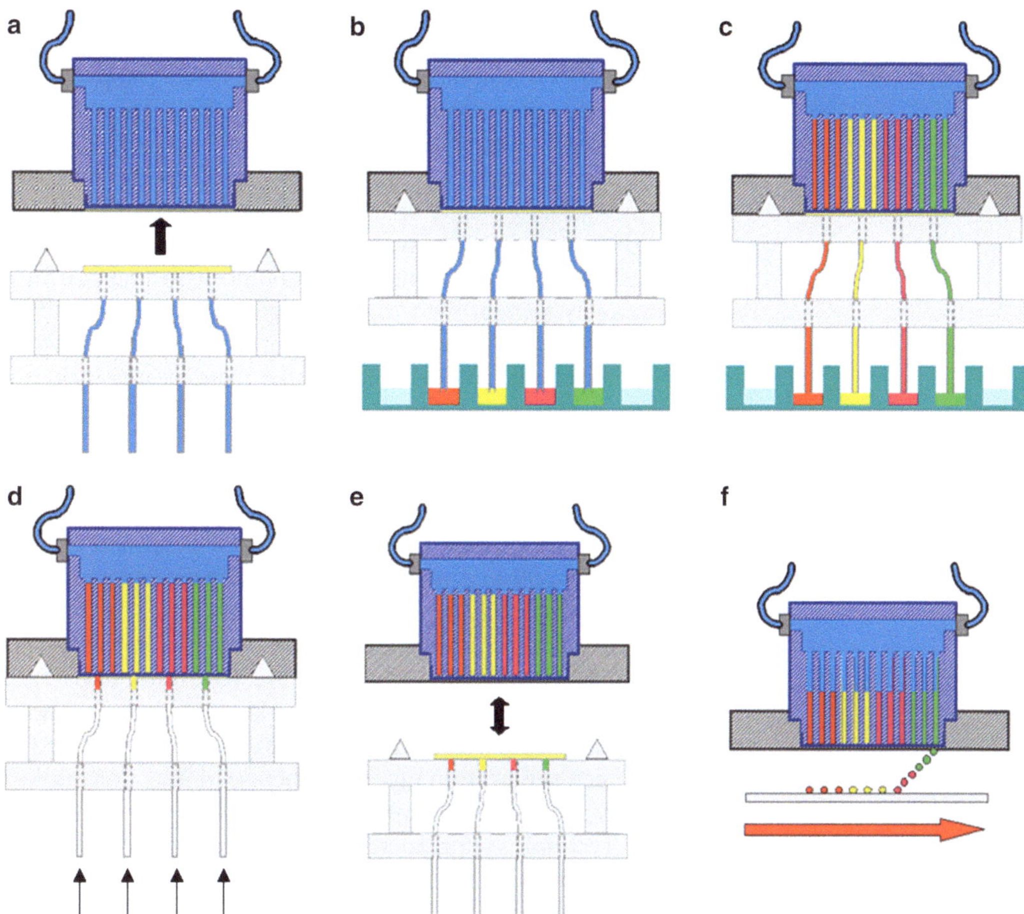

Fig. 2. (**a**) The JetSpyder™ is docked to the printhead and held in place by vacuum. System buffer is pushed through the printhead and JetSpyder™ to prime them before sample aspiration. (**b**) The docked JetSpyder™ is positioned over a 96- or 384-microplate, and the capillaries are lowered into the samples. (**c**) A servo-driven syringe (not shown) is used to aspirate 12 or 32 separate samples simultaneously into the printhead. (**d**) The JetSpyder™ is removed from the samples and air is aspirated (*arrows*) forcing the sample deeper into the printhead and reducing the dead volume. (**e**) The printhead is undocked, depositing the JetSpyder™ in its cleaning station. (**f**) The printhead performs on-the-fly, non-contact printing onto the substrates.

shaped enclosures arranged linearly. These enclosures correspond to the line of nozzles on the underside of the printhead that lead to the channels which hold samples during printing. The printhead can be docked to the JetSpyder™ which causes the upper surface of the rubber seal to compress around groups of nozzles. The walls of each enclosure isolate a group of nozzles which will ultimately share a common sample. The JetSpyder™ has 12 or 32 stainless steel capillaries which extend from the underside of the seal enclosures and are manipulated into a 4×8 or 3×4 Society for Biomolecular Screening (SBS) footprint. When the JetSpyder™ is lowered into a 96 or 384 microplate, 12 or 32 samples are simultaneously aspirated into the capillaries. Each sample travels through a separate capillary to the printhead, where it enters an enclosure and continues into up to six adjacent nozzles. Air is then aspirated into the capillaries to displace the samples still in the JetSpyder™ to ensure that dead volumes are minimised (Fig. 2d). The printhead is now loaded with samples and the JetSpyder™ is replaced into its cleaning station. Printing is performed on-the-fly to the substrates by the printhead alone and up to 500 drops can be printed per nozzle, with each droplet having a volume of 100 pL. This arrangement of separate sampling and printing components make Arrayjet systems uniquely compatible with JetGuard™, a disposable rubber cover for microplates with a valved septum for each well which protects samples held in 384 microplates from evaporation and contamination at all times during microarraying (Fig. 3).

1.4. A Brief History of Non-contact Printing Technology in Microarray Production

Early microarray fabrication technologies employed photolithography, as used by Affymetrix, or pin technology championed by their academic inventors and proponents, both described in now-famous articles published in Science by Steven Fodor et al. (24) and Mark Schena et al. (25), respectively. Non-contact technology had previously been demonstrated to have potential in this area by Schober et al. (26), who showed the precise deposition of bacterial colonies onto agar plates in frame-like layouts that resembled printed microarrays. These publications were swiftly followed by arguably the first publication involving the use of inkjet technology for microarray production, adapted by Leroy Hood at the California Institute of Technology for microarray production via in situ synthesis of nucleotides on a solid substrate (27). Later publications also followed, including a chapter in the widely selling DNA Microarrays – A Practical Approach, which described a non-contact spotting instrument called the GeneJet (28). Hood, meanwhile, together with Leland Hartwell and Steven Friend founded Rosetta Inpharmatics in 1996. A strategic partnership agreement with Agilent Technologies to commercialise Rosetta's DNA microarrays followed in 1999 and the Gene Expression Solution, which used inkjet printed nucleotides for in situ oligo synthesis.

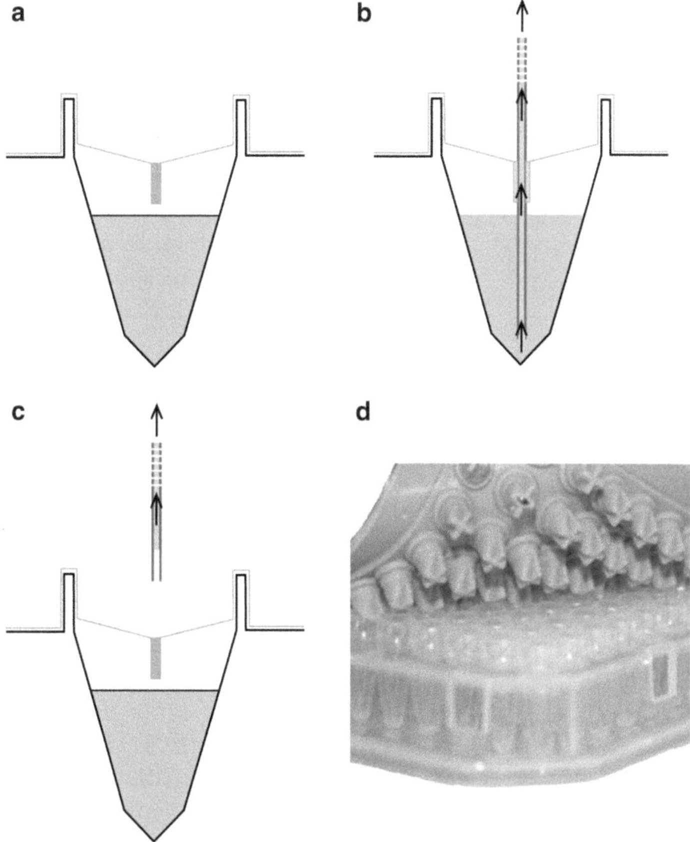

Fig. 3. (**a**) The JetGuard™ seal creates a microclimate with 100% relative humidity around the sample, thus minimising sample evaporation. The sample is also protected from foreign body contamination. (**b**) A JetSpyder™ capillary has entered the well through the duckbill valve and sample is aspirated; the arrows indicate the direction of flow towards the printhead and the dashes through top of the capillary indicate its extension towards the printhead. (**c**) The duckbill valve closes as the JetSpyder™ capillary is removed from the well. Air is aspirated into the capillary to displace the sample further into the printhead. (**d**) A photograph of a JetGuard™ seal being pulled peeled away from a 384-well microplate.

1.5. Commercial Non-contact Microarrayers: In the Beginning

As the microarray boom continued in the latter part of the 1990s and on into the early part of this century, technology companies were attracted to the possibilities it opened up. There has been a general trend towards systems which are increasingly robust, easy to use, and offer printing that is faster and more consistent, resulting in smaller spots with better morphology and higher density. Home-spotting technologies were initially thought to be reserved to contact spotters, though non-contact technology to produce microarrays, such as BioDot's Microdoser was available. In 1997, BioDot spun out Cartesian Technologies to focus on the life sciences market and microarrays in particular. The company was very successful, reporting an installed base of over 170 microarrayers worldwide at the time of its acquisition in 2001 by Genomic Solutions. A close rival to

Cartesian in this regard was Packard Biosciences who developed the BioChip Arrayer 1 (BCA1) based on technology owned and developed by Microdrop, and which launched in 1999 after 5 years of development. Packard Biosciences was acquired by Perkin Elmer Life Sciences in 2001 prompting a significant revision of the BCA1 (with improvements to the linear motions and operating system) before its relaunch as the Piezoarray in 2003. The revisions made the Piezoarray more user-friendly but it still required high levels of maintenance and frequent recalibration. Ninety Piezoarray units were shipped during the product's life time before being withdrawn from the market. Concurrently, a number of start-up companies launched new products – specifically two German companies, GeSiM (Gesellschaft für Silizium-Mikrosysteme mbH, translated as Silicon Microsystems Company mbh) and Scienion AG. GeSiM's Nanoplotter range of instruments was launched at the end of the 1990s. The first SciFLEXARRAYER of Scienion, which followed shortly after, was an instrument of an unusual design in that the substrate tray and motions were based on a turntable.

Therefore, by 2003 a number of commercial possibilities existed for scientists interested in using non-contact technology for microarray production. All of these microarrayers were, however, based largely on the same technology: the use of microdosing piezoelectric tip dispensers. The tips are dipped into the wells holding the probe sample material to be arrayed, in order to aspirate the samples into the dispensing device. They are then located over the substrates to allow the probe samples to be dispensed in droplets during microarray production.

1.6. The Need for Speed: Accelerating Non-contact Microarray Production

While Agilent Technologies were able to produce very large numbers of high-quality DNA microarrays with its proprietary inkjet technology, many scientists were frustrated that they could not acquire the technology themselves for use in their own laboratories. The Agilent array writers, as they were often described, were complex machines which required intensive maintenance and engineering support to obtain maximum performance. This level of support is not available to most scientists and the prospect of owning and running such a device was beyond their means. In its early years, non-contact technology was not especially reliable or robust and it was certainly not fast. Only Packard Biosciences had been able to develop and bring to market a high-throughput version of its microarray technology. The Spot Array Enterprise appeared in 2001 and was based on a new dispensing head which held eight tips, as opposed to the usual four on the BCA1 and piezoarray. While the Spot Array product was never officially launched, fifteen were shipped between 2001 and 2003, when the product was discontinued.

In 2005, following successful trials at three locations beginning in 2003, Arrayjet officially launched its first product: the Aj100 – an inkjet microarrayer which uniquely employed modern piezoelectric printhead technology from Xaar as described above. The launch of the Aj100 was followed by the Aj120 (now rebranded the Super Marathon) which had a microplate stacker to increase its walk-away printing capacity from 6 to 48 microtitre plates, making it a true core facility scale microarrayer. Shortly afterward, Arrayjet announced that it had developed a new liquid handling device which increased sample handling from 12 to 32 samples simultaneously, thereby reducing the time required to print 384 samples to 100 slides from 48 to only 25 min. In 2007, Arrayjet launched the Sprint Inkjet Microarrayer, an R&D-scale instrument capable of printing microarrays on up to 20 slides from a maximum of four microtitre plates, and simultaneously rebranded the entire productline. Then, in 2008, the Ultra-Marathon Inkjet Microarrayer was unveiled at the Advances in Microarray Technology conference in Barcelona. The Ultra-Marathon represented a paradigm shift in microarray production, taking it to a new level of throughput. As Arrayjet launched its new products, the world of microarray was being reinvigorated. While many scientists had ceased making their own DNA microarrays, the technology was constantly maturing, and many had turned their attention to protein microarray technology as a biomarker discovery and validation platform.

1.7. Protein Microarrays: A Renaissance for Microarrayers

Since the beginning of proteomics, the number of protein array applications has increased greatly. Protein arrays are being used as a crucial tool in major research projects which have the ability to change the world of diagnostics and, therefore, therapeutics. Areas, such as biomarker discovery, drug interactions, and expression profiling, require high-throughput technology compatible with proteins. Unlike DNA applications, where commercially available chips are widely used, the protein field often requires a customised spotting for both research and manufacturing for diagnostic purposes. Many researchers have found themselves forced to manufacture their own protein microarrays and have found their old DNA microarrayer unsuited to the task. Diagnostic companies, with greater demands for speed and quality, find themselves in a position where the chosen technology needs to fulfil more specific challenges, including high throughput. Antibody arrays are especially powerful as a proteomic technology. Multiplexed and ultra-sensitive assays, specifically targeting several analytes in a single experiment, can be performed, while consuming only minute amounts of the sample (29). Many antibody applications mimic the Enzyme-Linked Immunosorbent Assay (ELISA) technique in that they either employ a capture reagent and a detector reagent, often both IgG molecules, as per a sandwich ELISA. Alternatively, a target is adsorbed to a suitable substrate prior to detection with either a primary, or a primary and a secondary antibody, as in a direct ELISA. Colorimetric and fluorescent-labelling approaches are both widely used. Many routine diagnostic tests are ELISA-based, and the prospect of migrating these tests to protein microarray is attractive not only for the savings in reagents, sample and time, but also due to the potential advantages in sensitivity and parallel ability known as multiplexing. For these reasons, a method to effectively print antibodies has been the chosen subject for this manuscript.

Non-contact microarrayers offer benefits for the production of protein microarrays. This has stimulated the replacement market in microarrayers, and further validates recent product offerings not only from Arrayjet, but from its non-contact competitors GeSiM and Scienion. The latter has recently updated its existing product offering and added new products in the form of an entry-level system, an updated mid-scale system, the SciFLEXARRAYER SX, and a high-throughput version, the S100. Furthermore, during 2009 two new arrivals in non-contact microarrayer space have emerged in the shape of M2 Automation and Olivetti. M2's microarrayer – iONE – is without doubt the fastest moving microarrayer on the market, though its limited input and output capacity, and the number of dispensers makes it suitable only for low-throughput, low-density microarrays. Moreover, the dispenser type employed by M2 is very similar to those known to have problems of reproducibility and robustness – only time will tell if M2 has been able to address these issues with its new products. Olivetti's thermal BioJet technology is clearly aimed at the high-throughput market though to date it has only been seen at the proof of concept stage.

2. Materials

1. Arrayjet Sprint, Marathon, Super-Marathon, or Ultra-Marathon Inkjet Microarrayer.
2. JetMosphere environmental control system.
3. 2× JetStar™ Optimum Protein Printing Buffer (Arrayjet, UK).
4. Arrayjet system buffer: 47% (v/v) glycerol (G6279, Sigma Aldrich, UK) and 0.06% (v/v) Triton X-100 (T8787, Sigma Aldrich, UK).
5. Command Centre™ GUI microarray printing software provided with the microarrayer.
6. Mouse monoclonal IgG_1 antibody P-Tyr (PY20), 200 μg/ml (sc-508, Santa Cruz Biotechnology, USA).
7. Mouse monoclonal IgG_{2b} antibody p-Tyr (PY99), 200 μg/ml (sc-7020, Santa Cruz Biotechnology, USA).
8. Mouse monoclonal I gG_{2a} antibody EGFR (528), 200 μg/ml (sc-120, Santa Cruz Biotechnology, USA).

9. Mouse monoclonal IgG_1 antibody β-Actin (C4), 200 μg/ml (sc-47778, Santa Cruz Biotechnology, USA).
10. Blocking solution: 2% bovine serum albumin (A9647, Sigma Aldrich, UK) in PBS-T (P3563, Sigma Aldrich, UK).
11. Incubation solution: Atto 550 goat anti-Mouse IgG (43394, Lot No. 1267477, Sigma Aldrich) diluted to 2.8 μg/ml in PBS-T (as before).
12. Wash buffers: 1× PBS-T and 1× PBS (both as before).
13. Whatman® FAST® nitrocellulose-coated slides (10484182, Sigma Aldrich, UK) (see Note 1).
14. Lifterslips (22×25l-2-4816, Erie Scientific Company, USA).
15. JetGuard™ Probe Protector as shown in Fig. 3 (AJPP20, Arrayjet, UK).
16. JetGuard™ compatible 384-well microplate (PCR-384-55-C, Axygen, UK).
17. Innoscan 700 microarray scanner (Innopsys, France).
18. Mapix image analysis software v2.9.5 (Innopsys, France).

3. Methods

1. Create a serial dilution of each antibody with 2× JetStar™ Optimum Protein Printing Buffer (100, 20, 8, 4, and 2 μg/ml) (see Note 2).
2. Overlay JetGuard™ probe protector onto the 384-well microplate.
3. Populate a 384-well microplate with >5 μl of sample per well.
4. Centrifuge the microplate at 3,000×g for 5 min to remove any bubbles from the sample.
5. Program a printrun in Command Centre™ graphical user interface software to print six slides, each with three miniarrays per slide, three replicate spots per miniarray (nine replicate spots per slide), and one drop per (100 pL) spot volume (see Note 3).
6. Mount the slides into the spring-loaded holders.
7. Set the JetMosphere™ environmental control system to 50% relative humidity and 5 °C below ambient (see Note 4).
8. Initialise the microarrayer; this will take about 1 min.
9. Print a test slide, an automated procedure taking about 20 s, and verify that all nozzles are functioning correctly by visually inspecting the slide.
10. If test slide is not satisfactory, request automatic printhead purging and reprint a test slide.

11. If any nozzles are performing sub-optimally, they can be switched off in the software. There is inbuilt redundancy in the printhead allowing automatic reallocation of nozzles without affecting the printrun.
12. Start the printrun (the above array takes under 10 min to print in its entirety).
13. Once printing is complete, incubate the slides 37 °C and 50% relative humidity for 3 h (see Note 5).
14. Immerse the printed slides in the blocking solution for 1 h at room temperature with constant, gentle agitation.
15. Wash the slides twice in PBS-T for 5 min.
16. Overlay the array with sufficient incubation solution, cover with a lifterslip and incubate at room temperature for 1 h. Protect the slides from the light during the incubation and for the remainder of the process until they have been scanned.
17. Wash the slides in PBS-T for 10 min with gentle agitation to remove unbound detection antibody.
18. Wash the slides twice in PBS for 5 min with gentle agitation.
19. Centrifuge the slides for 30 s or until dry.
20. Scan the slides using the 532 nm channel (see Note 6). A scanned image is shown in Fig. 4. The PMT was set at 1 and laser power was low.
21. Extract the signal intensity data from the images with Mapix analysis software using local background correction. The data extracted from this experiment is shown in Fig. 5.

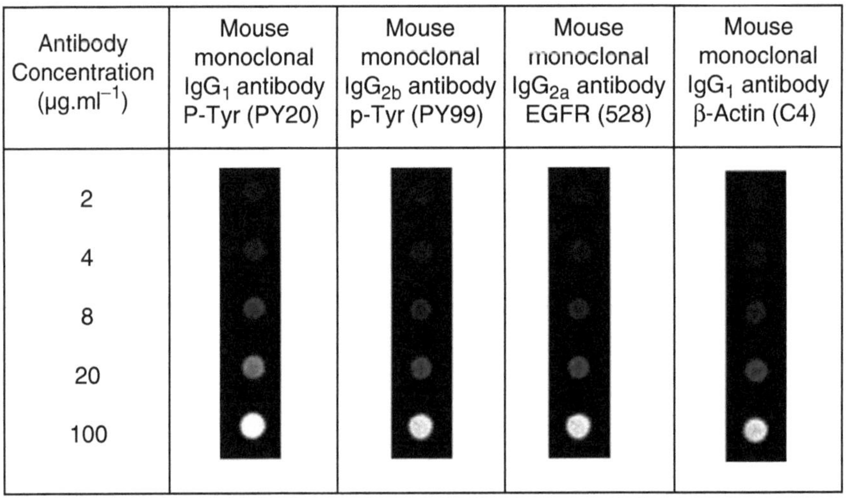

Antibody Concentration (μg.ml^{-1})	Mouse monoclonal IgG_1 antibody P-Tyr (PY20)	Mouse monoclonal IgG_{2b} antibody p-Tyr (PY99)	Mouse monoclonal IgG_{2a} antibody EGFR (528)	Mouse monoclonal IgG_1 antibody β-Actin (C4)
2				
4				
8				
20				
100				

Fig. 4. Scanned images showing one set of replicates for each antibody at each of five dilution points. The spot morphology is consistently good across all spots.

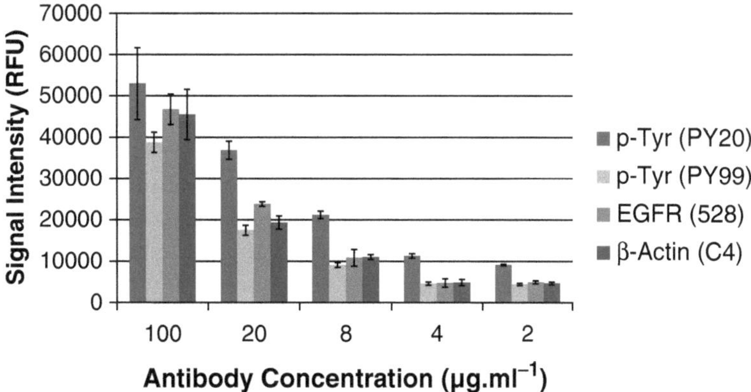

Fig. 5. Mean signal intensity (relative fluorescence units) corrected for local background of four different antibodies printed at 100, 20, 8, 4, and 20 μg/ml onto six nitrocellulose slides with nine replicate spots per slide. Error bars are ± standard deviation and the mean coefficient of variation was 8.7% (±5.4%).

4. Notes

1. In our experience, protein microarrays are most commonly printed onto nitrocellulose membranes bonded to glass slides (Whatman FAST, Gentel PATH, JetStar™ Nitrocellulose, Schott Nexterion NC), nitrocellulose or nylon membrane sheets; they are polysorbant and hold the proteins in a fibrous matrix. Other coated slides employ aldehyde, epoxy or NHS ester chemistry either as a dry coating or hydrogel matrix to chemically bind proteins to the substrate randomly via functional groups on the protein surface. Oriented protein binding can be achieved by introducing affinity tags, such as His-6 (30) or biotin (31) groups at specific sites in the protein and then arraying them onto nickel-NTA or streptavidin-coated substrates, respectively. Protein microarrays are also finding their way onto novel surfaces for label free signal detection on slides compatible with surface plasmon resonance spectroscopy (SPR) or pre-patterned substrates for electrical detection of analytes, both covered in a review by Yu et al. (32). Each surface type has advantages but consideration must be given to aspects, such as membrane pore size, which will affect spot size and adsorption efficiency, or whether chemical binding of the protein may disrupt the protein conformation and function detrimentally. The reader is directed to detailed reviews of available slide substrates (33–35).
2. For optimal printing and best spot morphology, the printed samples should have a fluid viscosity between 4 and 20 centipoises (cP, a measure of fluid viscosity). As a guide, water has a viscosity

of 1 cP and a 67.5% (v/v) glycerol solution has a viscosity of 20 cP. The correct viscosity ensures that each droplet of sample remains intact during its flight between printhead and substrate. Protein solutions are often inherently viscous but where modulation is required, a viscosifier can be added to increase the Newtonian viscosity of the samples. Recommendations include glycerol or ethylene glycol at concentrations up to 50%; both of which confer the added benefit of protein stabilization and cryoprotection. Sets of samples aspirated together should have a viscosity range within ±40% of the mean viscosity to ensure print consistency. Other hydroxylated printing buffer additives (e.g., polyvinyl alcohol 9,000, 0.05–0.5% (36)) or carrier protein additives (e.g., BSA, 0.1% (w/v) (37)) can be used to further enhance spot morphology and signal intensity. There is a detailed review of the printability of proteins by Joseph Delaney et al. (38). The JetStar™ range of printing buffers has been developed and optimised by Arrayjet for a wide range of sample types, including proteins and peptides.

3. The assay performed here gave sufficient signal from features of ~100 μm in diameter but users may wish to amplify the signal or achieve larger spots. This can be achieved by printing multiple drops of the same sample to each spot location. Every time the printhead passes over a slide, up to six drops (600 pL) per spot can be printed and repeat passes allow the spot volume to be increased to a maximum of 10 nL. Spot size will increase with spot volume; 100 pL spots are ~100 μm in diameter, 600 pL spots will be ~300 μm; but the precise ratio between spot volume and diameter will depend on sample surface tension and the substrate in use; a comprehensive explanation is given by Frits Dijksman and Anke Pierik (39).

4. Consistent environmental conditions are critical in protein microarray production to ensure that spots on every batch of slides have optimal morphology, and also to maintain samples in their optimal state to maximise stability and limit evaporation from microplates. Ideally, the printing environment should be HEPA filtered to prevent foreign bodies from contaminating source microplates and printed substrates. Different assays and substrates will have different optimal operating conditions. The environmental operating range of Arrayjet systems is 6–30 °C, and a relative humidity between 40 and 60%. An ideal printing environment for protein array printing can be created with JetMosphere™ environmental control system, which can be installed on all Arrayjet systems. It allows the user to define, create, and maintain a stable printing environment as low as 6° C with a relative humidity between 40 and 60%. JetMosphere™ comprises a humidifier, dehumidifier, refrigerating element, and

an air pump with built-in HEPA filter which circulates the air within the arrayer. Either the refrigerating element can be mounted directly onto the arrayer to cool up to 5°C below ambient or the arrayer can be placed within an insulated unit and cooled as low as 6°C. There are no consumable elements in JetMosphere™ and the only user intervention is to empty or fill water tanks before the start of a printrun.

5. Post-printing environmental conditions can be as important as printing conditions for the highest quality arrays as they affect the evaporation rates of spots and hence their morphology and distribution of immobilised protein. An alternative post-printing incubation would be at room temperature overnight and 50% relative humidity. Spots should be dry before proceeding to the blocking step to avoid "comet tail" artefacts. JetMosphere™ can also be used to maintain appropriate post-printing conditions for consistent protein immobilisation between different production batches.
6. Slides should be scanned with PMT and laser power set to get maximum signal intensity without pixel saturation to maximise the dynamic range of the assay.

References

1. Olle, E. W., Messamore, J., Deogracias, M. P., McClintock, S. D., Anderson, T. D., and Johnson, K. J. (2005) Comparison of antibody array substrates and the use of glycerol to normalize spot morphology, *Exp Mol Pathol* **79**, 206–209.
2. Derby, B. (2008) Bioprinting: inkjet printing proteins and hybrid cell-containing materials and structures, *Journal of Materials Chemistry* **18**, 5717–5721.
3. Cooley, P., Wallace, D, Antohe, B. (2001) Applications of ink-jet printing technology to BioMEMS and microfluidic systems. *J Assoc Lab Autom* 7, 33–39.
4. Okamoto, T., Suzuki, T., and Yamamoto, N. (2000) Microarray fabrication with covalent attachment of DNA using Bubble Jet technology, *Nat Biotech* **18**, 438–441.
5. Roda, A., Guardigli, M., Russo, C., Pasini, P., and Baraldini, M. (2000) Protein Microdeposition Using a Conventional Ink-Jet Printer *Biotechniques* **28**, 492–496.
6. Lausted, C. G., Warren, C. B., Hood, L. E., and Lasky, S. R. (2006) Printing your own inkjet microarrays. In Kimmel, A. and Oliver B. (eds.) *DNA Microarrays,* Academic Press, California.
7. Alexander, M. (2008) The Xaar Guide to Single Pass Printing, http://www.xaar.com/reg-download.aspx?file=uploads/xaar_guide_to_single_pass_printing_white_paper.pdf. Accessed 15 August 2010.
8. Dufva, M. (2009) Fabrication of DNA Microarray. In Dufva M. (ed) *DNA Microarrays for Biomedical Research: Methods and Protocols,* Humana Press Inc, New York.
9. Martinsky, T. (2003) Printing technologies and microarray manufacturing techniques: making the perfect microarray. In Blalock, E. (ed) *A beginner's guide to microarrays,* Kluwer Academic Publishers, Norwell.
10. Heller, M. J. (2002) DNA microarray technology: devices, systems, and applications, *Annu Rev Biomed Eng* **4**, 129–153.
11. Sun, A., Devi-Rao, G. V., Rice, M. K., Gary, L. W., Bloom, D. C., Sandri-Goldin, R. M., Wagner, P., and Wager, E. K. (2004) The TATGARAT box of the HSV-1 ICP27 gene is essential for immediate early expression but not critical for efficient replication in vitro or in vivo, *Virus Genes* **29**, 335–343.
12. Evans, H., Mello, L.V., Yongxiang, F., Wit, E., Thompson, F.J., Viney, M.E., Paterson, S. (2008) Microarray analysis of gender-and

parasite-specific gene transcription in Strongyloides ratti, *Int J Parasitol* **38**, Elsevier, Oxford, UK.

13. Grinfeld, E., Ross, A., Forster, T., Ghazal, P., Kennedy, P. (2009) Genome-wide reduction in transcriptomal profiles of varicella-zoster virus vaccine strains compared with Parental Oka strain using long oligonucleotide microarrays, *Virus Genes* **38**, 19–29.
14. Williams, D. R., Li, W., Hughes, M. A., Gonzalez, S. F., Vernon, C., Vidal, M. C., Jeney, Z., Jeney, G., Dixon, P., McAndrew, B., Bartfai, R., Orban, L., Trudeau, V., Rogers, J., Matthews, L., Fraser, E. J., Gracey, A. Y., and Cossins, A. R. (2008) Genomic resources and microarrays for the common carp *Cyprinus carpio* L. *Journal of Fish Biology* **72**, 2095–2117.
15. Taggart, J. B., Bron, J. E., Martin, S. A., Seear, P. J., Hoyheim, B., Talbot, R., Carmichael, S. N., Villeneuve, L. A., Sweeney, G. E., Houlihan, D. F., Secombes, C. J., Tocher, D. R., and Teale, A. J. (2008) A description of the origins, design and performance of the TRAITS-SGP Atlantic salmon Salmo salar L. cDNA microarray, *J Fish Biol* **72**, 2071–2094.
16. Korf, U., Henjes, F., Schmidt, C., Tresch, A., Mannsperger, H., Löbke, C., Beissbarth, T., and Poustka, A. (2008) Antibody Microarrays as an Experimental Platform for the Analysis of Signal Transduction Networks, *Adv Biochem Eng Biotechnol* **110,** 153–175.
17. Olkhov, R. V., and Shaw, A. M. (2009) Quantitative label-free screening for antibodies using scattering biophotonic microarray imaging, *Anal Biochem* **396**, 30–35.
18. Olkhov, R. V., and Shaw, A. M. (2008) Whole serum BSA antibody screening using a label-free biophotonic nanoparticle array, *Anal Biochem* **385**, 234–241.
19. Olkhov, R. V. and Shaw, A. M. (2008) Label-free antibody-antigen binding detection by optical sensor array based on surface-synthesized gold nanoparticles, *Biosens Bioelectron* **23**, 1298–1302.
20. Tan, C. P., Cipriany, B. R., Lin, D. M., and Craighead, H. G. (2010) Nanoscale Resolution, Multicomponent Biomolecular Arrays Generated By Aligned Printing With Parylene Peel-Off, *Nano Lett* **10,** 719–725.
21. Clarke, S. (2005) Protein and Peptide Microarray- Based Assay Technology, in Walker J. M. (ed) *The Proteomics Protocols Handbook*, Humana Press Inc., Totowa, New Jersey.
22. Scienion AG, (2010) sciFLEXARRAYER S5 and S11 - versatile tool for ultra-low volume applications in array manufacturing. http://www.scienion.com/index.php?mid=42&vid=&lang=en. Accessed 15 August 2010.
23. Leung, F. L., Pang C. P. and Browne K. (2001) Guide to microarray hardware- a researcher perspective. http://www.images.technologynetworks.net/resources/comptab.asp. Accessed 15 August 2010.
24. Fodor, S. P., Read, J. L., Pirrung, M. C., Stryer, L., Lu, A. T., and Solas, D. (1991) Light-directed, spatially addressable parallel chemical synthesis, *Science* **251**, 767–773.
25. Schena, M., Shalon, D., Davis, R. W., and Brown, P. O. (1995) Quantitative monitoring of gene expression patterns with a complementary DNA microarray, *Science* **270**, 467–470.
26. Schober, A., Gunther, R., Schwienhorst, A., Doring, M., and Lindemann, B. F. (1993) Accurate high-speed liquid handling of very small biological samples, *Biotechniques* **15**, 324–329.
27. Blanchard, A. P., Kaiser, R.J. and Hood, L.E. (1998) High-density oligonucleotide arrays *Biosens Bioelectron* **11**, 687–690
28. Theriault, T. P., Winder, S.C. and Gamble, R.C. (1999) Application of ink-jet printing technology to the manufacture of molecular arrays. In Schena, M (ed) *DNA Microarrays - A Practical Approach*, Oxford University Press, Oxford.
29. Wingren, C., and Borrebaeck, C. A. (2009) Antibody-based microarrays. In Bilitewski U. (ed) *Microchip Methods in Diagnostics*, Humana Press, New York.
30. Cornelia, S., Christer, W., Farid, K., Mingyue, H., Michael, J. T., and Carl, A. K. B. (2006) Improved affinity coupling for antibody microarrays: Engineering of double-(His) 6-tagged single framework recombinant antibody fragments, *Proteomics* **6**, 4227–4234.
31. Peluso, P., Wilson, D. S., Do, D., Tran, H., Venkatasubbaiah, M., Quincy, D., Heidecker, B., Poindexter, K., Tolani, N., Phelan, M., Witte, K., Jung, L. S., Wagner, P., and Nock, S. (2003) Optimizing antibody immobilization strategies for the construction of protein microarrays, *Anal Biochem* **312**, 113–124.
32. Yu, X., Xu, D., and Cheng, Q. (2006) Label-free detection methods for protein microarrays, *Proteomics* **6**, 5493–5503.
33. Angenendt, P., Glokler, J., Sobek, J., Lehrach, H., and Cahill, D. J. (2003) Next generation of protein microarray support materials: evaluation for protein and antibody microarray applications, *J Chromatogr A* **1009**, 97–104.
34. Seurynck-Servoss, S. L., White, A. M., Baird, C. L., Rodland, K. D., and Zangar, R. C. (2007) Evaluation of surface chemistries for antibody microarrays, *Anal Biochem* **371**, 105–115.
35. Grainger, D. W., Greef, C. H., Gong, P., and Lochhead, M. J. (2007) Current microarray

surface chemistries. In Rampal J.B. (ed) *Microarrays: Volume 1: Synthesis Methods*, Humana Press, New York.

36. Wu, P., and Grainger, D. W. (2006) Comparison of Hydroxylated Print Additives on Antibody Microarray Performance, *J Proteome Res* **5**, 2956–2965.
37. Delehanty, J. B. (2004) Printing Functional Protein Microarrays Using Piezoelectric Capillaries. In Fung E.T. (ed) *Protein Arrays: Methods and Protocols*, Humana Press, New York.
38. Delaney, J. T., Smith, P.J., Schubert, U.S. (2009) Inkjet printing of proteins, *Roy Soc Chem* **5**, 4866–4877.
39. Dijksman, J. F., and Pierik, A. (2008) Fluid dynamical analysis of the distribution of ink jet printed biomolecules in microarray substrates for genotyping applications, *Biomicrofluidics 2*, 044101–044122.

Chapter 24

Impact of Substrates for Probe Immobilization

Ursula Sauer

Abstract

Protein chips are becoming a key technology in proteomic research and medical diagnostics. Surface chemistry for immobilization of proteins forms the basis for assay design and determines the properties of protein microarrays. Optimal substrates provide a homogeneous environment for probes, preventing loss of biological activity and unspecific adsorption. Numerous immobilization approaches, based on covalent binding, affinity, or adsorption, have been proposed thus far, and these represent the toolbox for choosing optimized strategies for each individual application.

Key words: Protein patterning, Coating, Surface chemistry, Affinity binding, Adsorption, Covalent binding, Hydrogel

1. Introduction

Solid supports and immobilization strategies for probe attachment play a central role in the development of protein biochips by determining sensitivity, specificity, and reproducibility. The immobilization of proteins to solid phase surfaces has been of interest since immunological techniques emerged. With the emergence of biosensor arrays (also referred to as protein microarrays or protein biochips), the demand for suitable immobilization strategies has grown. In contrast to immobilization of proteins for immunosensors in general, parallelization as a main feature of microarrays requires a patterning of probes as opposed to simple coating. Nonetheless, when applicable for patterning, knowledge about immobilization derived from ELISA techniques, affinity chromatography or biosensors can often be employed. This "patterning" of protein chips, providing regions of specific binding of ligands and nonadhesive regions, is primarily done by robotic printing,

Ulrike Korf (ed.), *Protein Microarrays: Methods and Protocols*, Methods in Molecular Biology, vol. 785,
DOI 10.1007/978-1-61779-286-1_24,

yet arrays may also be created by means of self-assembling, photolithography, photochemistry, or plasma polymerization.

The second resource for potential protein biochip substrates is DNA microarray technology, in which consideration must be given to the differing chemical and physical properties of proteins and nucleic acids. While DNA, being negatively charged, provides a uniform chemistry, proteins exhibit a vast chemical and structural diversity; they differ in size, charge, and reactive groups on the surface. Protein purification is complicated, and the lack of an amplification method, such as PCR causes sensitivity problems. Furthermore, proteins are less stable than DNA and more prone to lose biological activity when immobilized. The role of substrates in the dynamics of rapidly drying protein spots after printing is not yet fully elucidated.

Immobilization matrices for protein microarrays can be classified according to their coupling chemistry (adsorption, affinity binding, covalent binding) or their dimensionality, namely, one-dimensional (monolayers), two-dimensional (2D), more or less planar surfaces or three-dimensional (3D) surfaces, e.g., membranes or hydrogels (Fig. 1).

Reports on characterization and in-depth comparisons of the performance of such substrates are rather sparse and little has been published about the underlying biophysical mechanisms of protein binding to surfaces (1–10).

No general recommendation or recipe for an immobilization method can be given here; an optimal protocol in fact needs to be chosen for each application.

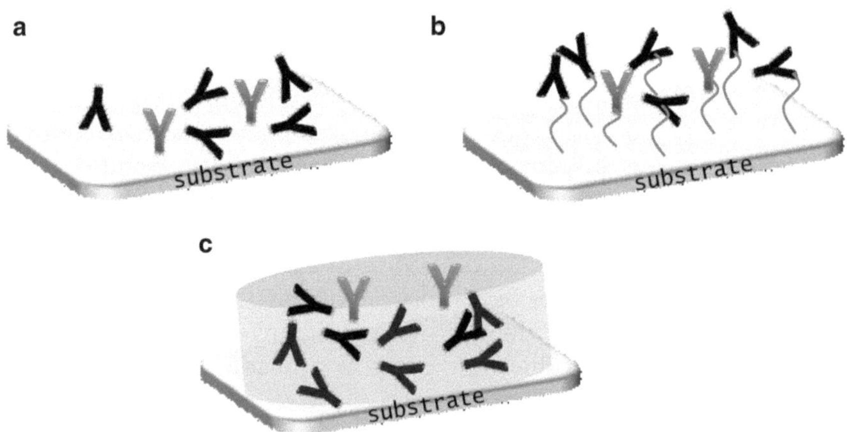

Fig. 1. Scheme of randomly immobilized antibodies in (**a**) one-dimensional (monolayers), (**b**) 2D (silanes, cross-linkers), and (**c**) 3D (hydrogels, membranes, dendrimers) coatings.

Decision criteria for an immobilization method

Solid support
Required sensitivity
Assay format/targeted analytes
Stability of the probes
Detection system
Possible costs/economic mass production

The most common sources of solid support for microarray systems are fused silica and glass substrate materials due to their good optical properties (low autofluorescence at excitation wavelengths), mechanical and thermal stability and chemical inertness. For the microfabrication of lab-on chip systems, alternative materials had to be introduced, e.g., cycloolefin copolymers (COC), poly(methyl methacrylate) (PMMA), and polycarbonate; these substances are suitable for high-throughput processing, such as molding, hot embossing, or laser welding. Either immobilization matrices are coated onto these substrates or the substrates themselves are chemically modified in order to create appropriate binding sites for biomolecules.

Choice of surface chemistry is also driven by the detection system. Planar waveguide-based detection only works for coatings with a layer thickness of less than 100 nm and surface plasmon resonance (SPR) requires immobilization onto gold surfaces while fluorescence-based approaches call for substrates of low autofluorescence at the excitation wavelengths.

2. General Requirements for Immobilization Matrices

An important criterion of biomolecule immobilization is the high functionality of the chip surface. A proper density of binding sites consistent over the entire slide surface is a prerequisite for effective biosensing. Increasing the solid phase concentration of antibodies results in increased sensitivity and extended working range (11, 12). Otherwise, when capture molecules are bound too densely, steric hindrance and decreased target-binding efficiency may result. Furthermore, attached molecules have to be presented in such a way that epitopes/binding sites are lifted away from the surface and hence well-accessible for the target.

One of the major issues in microarray development is fabricating a surface that, in addition to excellent signals, results in as little background noise as possible. Figure 2 shows a typical high-quality scan image of a biomarker chip using an epoxy resin as

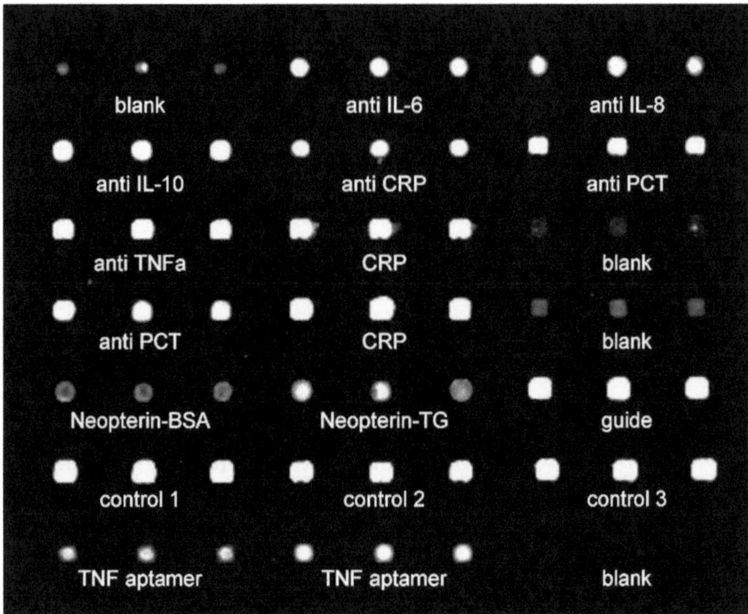

Fig. 2. Array of capture antibodies, antigens, and antigen conjugates printed on ARChip Epoxy. Detail of a chip for the quantification of biomarkers, after an assay with a mixture of spiked proteins and fluorescently labeled antibodies.

immobilization matrix. One source of noise is unspecific protein adsorption, which is controlled by the choice of surface chemistry and blocking protocols. For optical read-out based on fluorescence, the intrinsic autofluorescence of the surface can be a major contributor to noise, especially for nitrocellulose. Fluorescence background is usually computed for each individual spot based on a local background subtraction technique.

Kusnezow et al. point out the importance of antibody microspot kinetics, namely, for the analyte to migrate in solution as well as across the immobilization surface (13). Protein receptors in solution display homogeneous-binding affinities and kinetics for their ligand; while upon immobilization, they display heterogeneous-binding characteristics. Vijayendran and co-workers evaluated this heterogeneity of five different immobilization strategies (14). The most homogeneous behavior was found with antibodies immobilized oriented via their carbohydrate moiety: the amount of heterogeneity with respect to affinity to the ligand was closely related to heterogeneity in analyte-antibody kinetics.

High-quality spots with uniform pixel intensities are a key requirement for meaningful data analysis (15). Spots must be of the same shape and size throughout a slide and from one slide to another. This is achieved by the optimized interplay of surface chemistry, probe, printing technique, and print buffer. Hydrophobic surfaces tend to produce small but inhomogeneous spots, whereas most hydrophilic surfaces yield homogeneous spots, which, however, are often irregular in shape (9). High reproducibility is even more important for quantitative analyses of biomarker detection, since the standard deviation of replicate spots is part of sensitivity measures, such as the calculation of limit of detection (LOD) and limit of quantification (LOQ).

In order to be applicable as routine analytical tools, microarray substrates have to be affordable, suitable for mass production, easy to handle and provide reasonable shelf life.

Requirements for immobilization matrices
Availability of protein-binding sites
Accessibility of the binding sites for capturing the target molecules
Maintained capability of specifically capturing antigen, nondenaturing conditions
Reproducibility
Low nonspecific binding of the surface (background)
Excellent spot morphology for reproducible image analysis
Reasonable shelf life

3. Sensing Molecules to Be Immobilized (Probes)

For sandwich type and competitive on-chip immunoassays, as well as for protein expression profiling, antibodies or their fragments have to be attached to a solid support while proteins, recombinant proteins, or peptides are used as probes for binding inhibition assays. In the first case, probes are more or less of similar nature, namely, monoclonal or polyclonal antibodies; consequently, the requirements for the immobilization chemistry are comparable. In the second case, proper probe immobilization may be more sophisticated, as molecules of different chemistry, size, quaternary structure, and loading have to be immobilized on one common solid support, without biological activity being negatively affected by denaturation and conformational changes.

Further designs posing high demands on surface chemistries include antigen arrays for studying autoimmune diseases [http://proteomics.stanford.edu/robinson/antigen.html], gylcan-microarrays for the interrogation of glycan–protein interactions to study cell communication (16) (e.g., see the Consortium for Functional Glyomics; http://www.functionalglycomics.org/static/index.shtml), protein kinase assays (17), ATP and GTP binding assays, and studies of protein–protein interactions.

4. Physical Adsorption

A simple and affordable immobilization method is the adsorption of proteins via intermolecular forces as employed previously in microtiter ELISAs. Adsorption is based on nonspecific electrostatic, hydrophobic, or Van der Waals forces. Local dipoles in the participating molecules are stationary, forming strong hydrogen bonds, or alternating dipoles in nonpolar regions of the reagents, forming weaker hydrophobic interactions (18). Proteins are oriented randomly upon adsorption. Drawbacks are desorption of the proteins during assays, structural deformation, and the denaturation of biomolecules commonly observed. High-bulk concentrations lead to less contact with the surface per molecule and hence to less unfolding. Butler et al. (19) report on high losses of protein function upon adsorption on polystyrene. Only 5–10% of polyclonal antibodies were capable of capturing antigen while a streptavidin-mediated immobilization of biotinylated IgG resulted in up to 70% preservation of the antigen-binding sites.

Widely used materials for protein adsorption are polystyrene (http://www.nuncbrand.com) (20), poly-L-lysine, aminosilane, and nitrocellulose (21–23). Figure 3 shows the microporous 3D structure, responsible for the high-binding capacity of nitrocellulose, imaged by means of scanning electron microscopy (http://www.whatman.com).

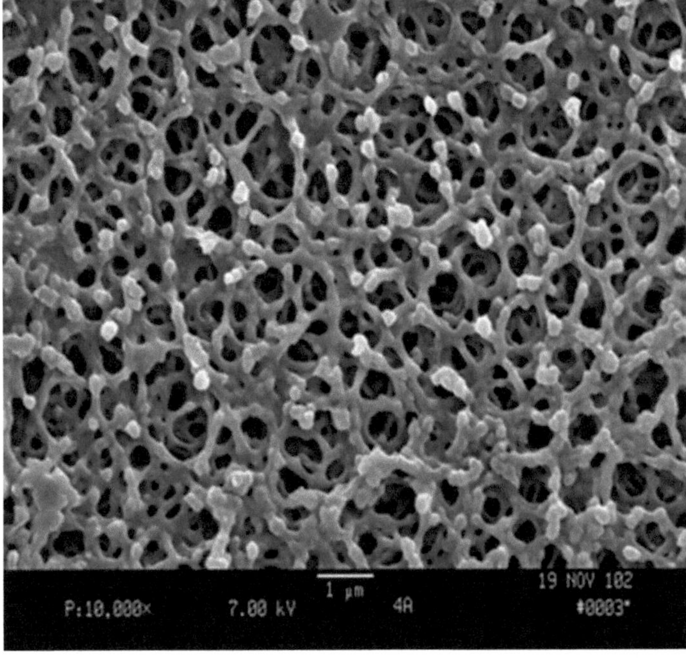

Fig. 3. 3D Structure of Whatman nitrocellulose, scanning electron microscopy, magnification ×10,000 (http://www.whatman.com: The FAST Guide to Protein Microarrays).

5. Hydrogels

Hydrogels are considered especially suitable for protein immobilization, providing a controlled nano-environment that can keep the protein hydrated and stabilize the structure. Binding principles on hydrogels are adsorption or, where reactive groups are available, covalent attachment, e.g., the hydrophilic polymer of Nexterion Slide H (http://www.schott.com) is activated with *N*-hydroxysuccinimide (NHS) ester, which reacts with primary amino groups of proteins covalently (Fig. 4).

Widely used as 3D-immobilization matrices are agarose (24, 25), poly(acrylamide) (26), polyurethane (27), poly(vinyl alcohol) (28), dextran (29, 30), and polyethyleneglycol (31). Hydrogels can be tuned to a certain extent in order to mimic the biological environment of proteins. For example, Moorthy et al. (32) found the binding interactions between IgG and protein A enhanced as the pore size of polyacrylamide decreased.

Hydrogel coatings are produced by spin-coating, dip-coating onto solid supports or by covalent binding of the gel on silanized substrates. Gels are not only produced as coatings but directly co-spotted with probes. Rubina et al. (26) used a polymerization-mediated immobilization method to produce hydrogel protein chips for the detection of biotoxins, in which case the polymethacrylamide hydrogel containing the proteins is spotted onto the slides. Dominguez et al. (33) fabricated antibody-entrapped hydrogel chambers by arraying solutions of both tetra- or octa-amine functionalized peptide-based branch macromolecules and IgG on aldehyde glass slides. These methods are single-step, rapid and keep the antibody hydrated and in its original conformation, since no modification of the antibody is necessary.

Clearly, diffusion coefficients for proteins should be lower in gels, slowing down assay times. Kinetic curves for binding Cy3-labeled

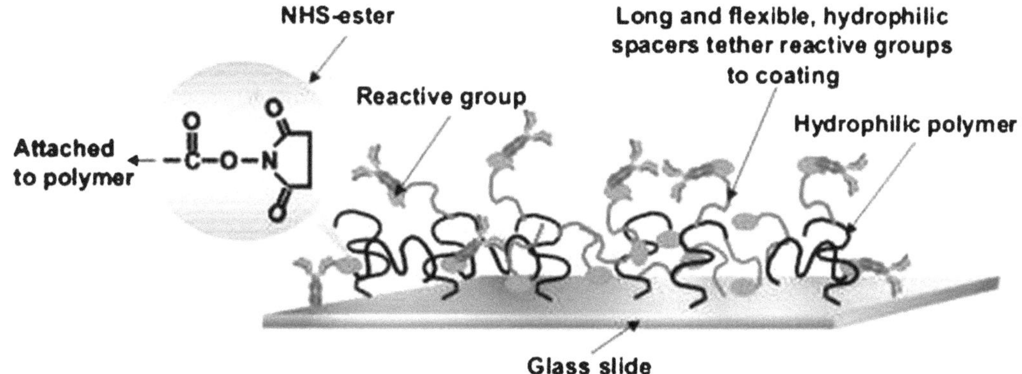

Fig. 4. Scheme of immobilization chemistry of Nexterion Slide H (http://www.schott.com).

ricin with immobilized antibodies in acrylamide gel-pads reached equilibrium only after 15 h (26). Several methods for accelerating diffusion in microarray experiments have been suggested, such as peristaltic pumps (34), creating a constant flow, and ultrasonic mixing.

6. Covalent Binding of Probes

Functional groups of amino acids exposed to the protein surface can be employed for direct covalent attachment, which results in "statistically oriented" immobilization. Lysines are numerous on a protein surface for example while cysteines are less abundant.

Commonly used 2D surfaces for covalent immobilization provide aldehyde, epoxy, amino, mercapto, isothiocyanate groups (35) or NHS ester. Proteins are coupled to amine reactive surfaces on the formation of a Schiff's base linkage, primary amines may be provided by lysines on the protein surface. Epoxy functionalities bind via nucleophilic substitution to highly abundant groups on protein surfaces as amino, thiol, and hydroxyl groups. McBeath et al. (36) attached proteins covalently to aldehyde-derivated glass slides as well as to BSA-NHS slides for three applications: screening for protein–protein interactions, identifying substrates of protein kinases and finding protein targets of small molecules. Other chemical functionalities of amino acids are: –SH (cysteine), –COOH (aspartic acid, glutamic acid), –OH (serine). Carbohydrate or carboxyl groups can be activated with (1-ethyl-3-(3-dimethylaminopropyl)carbodiimide) EDC and bind to an amine reactive surface (1) while thiol groups may be used for covalent coupling to epoxy and maleimide surfaces. Numerous coupling strategies have been developed for immobilizing antibodies on different solid surfaces through the formation of defined linkages in which glutaraldehyde, carbodiimide, and other reagents, such as succinimide ester, maleinimide, and periodate, are employed. However, problems can be seen in many cases, associated with the loss of the native functional state upon immobilization of antibodies (37). Covalent attachment is less denaturing than adsorption while nonetheless only a certain proportion of the antibodies stay biologically active and accessible.

Seong (38) compared IgG immobilization on commercial silylated slides and epoxy slides, both of which are amine-reactive, and reported a superior binding capacity on epoxy-coated slides (http://www.xenopore.com). According to the findings of Olle (39) as well, epoxysilane (http://www.eriesci.com) was superior to Hydrogel™ (Perkin Elmer) and SuperAldehyde (Telechem) in IgG binding with respect to signal intensity and low background.

The covalent immobilization of histone proteins onto NHS ester and aminoated surface modified with maleic anhydride-*alt*-methyl vinyl ether (MAMVE) copolymer was investigated (40). The immunoassay on MAMVE-functionalized surfaces displayed an LOD 50 times lower than that of the ELISA assay in polystyrene plates.

The accessibility of antibodies immobilized via a long and flexible spacer, such as poly(ethylene glycol) was investigated by AFM (41). An AFM tip was coated with *Escherichia coli* in order to analyze by means of force–distance curves the interaction between bacteria and the specific antibody and evaluate optimal surface coverage and spacer length.

Dendrimers are highly branched macromolecules that form a three-dimensional structure with a variety of possible chemical functionalities, maximizing the density of binding sites (for examples, see http://www.dendritech.com; http://www.dendrimercenter.org; http://www.sigmaaldrich.com). High-density protein chips were prepared by activation of Si or glass wafers and poly (propyleneimine) dendrimers modified with a sulfosuccinimide ester, providing a fixed number of functionalities for covalent protein binding (42). Yam et al. (43) prepared poly(amidoamine) dendrimers (PAMAM) functionalized with biotinylated oligo (ethylene glycol) (OEG) derivatives consisting of self-assembled monolayers (SAMs) of 11-mercaptoundecanoic acid (MUA) on gold substrates, which minimized nonspecific protein adsorption and at the same time provided a high density of avidin-binding sites.

7. Affinity Binding of Probes

Immobilization via biochemical affinity ideally results in the oriented attachment of probes. The avidin/biotin system is especially widely used since it offers several advantages, particularly the strong affinity and specificity of the interaction (44). Biotinylated proteins are attached to streptavidin-coated surfaces (Fig. 5); several are commercially available (e.g., http://www.xenopore.com; http://www.arrayit.com). Bathia et al. describe silanization and treatment with succinimide ester for subsequent coating with Neutravidin, ready for binding biotinylated probes (45). Provided that biotinylation takes places in a nonbinding region of the protein, this approach is more likely to maintain the native function of a protein. For biotinylation, amino groups of the proteins are often used, resulting in random attachment of biotin and consequently random immobilization of the biotinylated probes. Site-directed biotinylation at the hinge region of $F(ab')_2$, on the other hand, was demonstrated to allow controlled oriented antibody immobilization

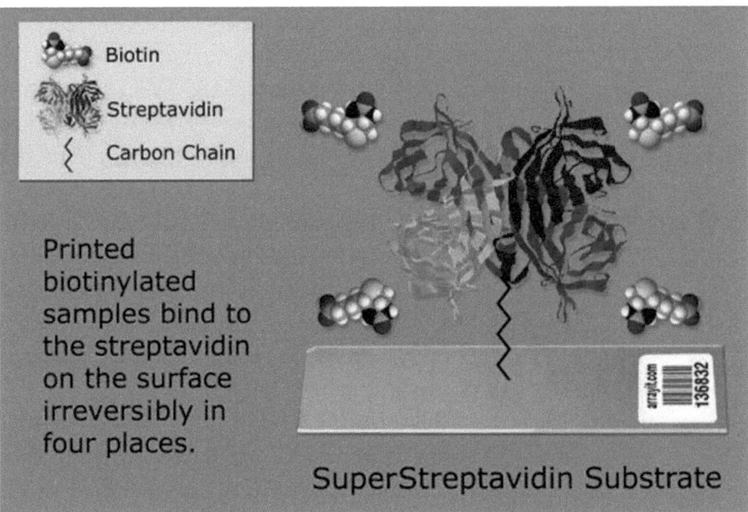

Fig. 5. ArrayIt Super Streptavidin Substrate. Image provided by http://www.arrayit.com. Copyright 2010, Arrayit Corporation. All Rights Reserved. World Wide.

with detection capabilities up to 20 times greater compared to random biotinylation (46). Peluso et al. (47) studied the effect of four different methods of binding biotinylated antibodies or fragments onto streptavidin surfaces with respect to surface density and binding activity. The study involved comparing random biotinylation of monoclonal IgG and Fab′ fragments to biotinylation with the biotin-aminooxy compound ARP after oxidizing the glycosylation site at the Fc portion. Oriented immobilization outperformed random coupling, with up to a tenfold increase in analyte-binding capacity.

Metal complexes had been employed for affinity binding of proteins in affinity chromatography prior to being used in microarray technology. Histidine -tagged recombinant proteins are captured with high affinity by metal ions, retaining the native conformation (http://www.xenopore.com). Nitriloacetic acid (NTA) forms a tetradentate chelate with the Ni^{2+} ion, although other transition metal ions with a coordination number of 6 can be used (e.g., Co^{2+}, Cu^{2+}, Zn^{2+}). In order to overcome the drawback of low affinity, which reduces the yield of immobilized protein on Ni-NTA surfaces, double his-tags were introduced by Khan and coworkers (48).

A screening of libraries of polymer-chelating surfaces containing different metal ions for efficient antibody binding was done by Muir et al. (49), and the secondary amine, the metal counter ion, and chelating ligand were identified as main variables.

8. Oriented Versus Random Immobilization

Non-oriented immobilization of antibodies does not discriminate binding sites in or near the Fab fragment of an antibody, therefore the antigen-binding sites may not be accessible, which either entirely blocks or at least hinders the ability to specifically bind antigen. In order to avoid such problems, several strategies for oriented antibody immobilization have been developed (37). The advantages of the oriented immobilization of proteins are a good steric accessibility of the active binding site and increased stability. Danczyk et al. (1) found improved antigen capture capabilities of antibodies attached using protein A, although the amount of immobilized antibodies is smaller than the number directly bound or adsorbed, suggesting that a higher proportion stays functional.

Oriented antibody immobilization
Via antibody receptors
Chemical or enzymatic oxidation
Disulfide bond reduction
Site-specific biotinylation
Recombinant antibodies with tags
DNA-directed

Orientation is achieved by coupling via antibody receptors, such as protein A, protein G, or recombinant protein A/G, which bind in the Fc region of the antibody. Coupling via antibody receptors may cause problems with IgG from serum that may also bind to the receptors if they are not saturated with capture antibodies or if the blocking of remaining receptors is insufficient. Affinity of protein A for IgG subclasses differs, and the same is true for protein G; an alternative product combining binding sites from protein A and G is the recombinant protein A/G (http://www.arrayit.com).

Chemical or enzymatic oxidation of the carbohydrate moiety located in the Fc fragment to aldehyde groups goes without significantly impairing the active sites of the antibody. The oxidized antibodies can then be immobilized to hydrazide-activated supports by forming covalent hydrazone bonds. Periodate oxidized antibodies were also first used for immunoaffinity gels. The method was successful with polyclonal antibodies while for monoclonal antibodies milder oxidation conditions have to be employed. Another approach uses the sulfhydryl group of the Fab region between the light and heavy chain to create an oriented antibody fragment.

SAMs consist of a single layer of molecules on a substrate. In self-assembly techniques, thiols and disulfides are mostly used on metal substrates, such as gold and silver, while silanes are used on nonmetallic surfaces, such as SiO_2 and TiO_2 (8).

Surface preparations of SAMs composed of ssDNA thiols and OEG-terminated thiols were introduced for DNA-directed protein immobilization (50). The mixed SAM allows rational control over the DNA probe surface density. Antibodies conjugated to ssDNA with a sequence complementary to the surface-bound ssDNA are hybridized on the biosensor and convert the DNA surface into a protein surface in a single step. The surface can be completely regenerated with NaOH to dehybridize the DNA. Alternatively, DNA-streptavidin conjugates were used to immobilize biotinylated antibodies onto DNA surfaces (51). DNA-directed immobilization was compared to direct spotting on activated glass and strepatividin–biotin attachment with regard to signal intensity, assay sensitivity, and reproducibility (52). All three methods allowed the detection of 150 pg/mL IgG in a sandwich immunoassay while DNA-directed immobilization was superior with regard to very low antibody consumption, spot homogeneity, and reproducibility.

Shriver-Lake et al. (53) compared nine heterobifunctional cross-linkers as to their ability to bind antibodies, and tested the immunological activity with a fiber-optic biosensor. One approach used thiol-terminal silanes and heterobifunctional cross-linker with a succinimide moiety, reacting with the primary amines of the antibody (non-oriented). The other groups were cross-linkers containing hydrazide, reacting with the carbohydrate moiety in the Fc region of the antibody, and therefore providing orientation. Immobilization via the carbohydrate region resulted in higher packing density and higher levels of antigen-binding capacity (over 30% of the antibodies being active), which is explained by the distance between cross-linker reaction site and antigen-binding site. Disadvantages were high loss of antibody (up to 50%) in the multistep immobilization procedure and reports of decreased antibody activity after periodate treatment.

Some methods for oriented immobilization listed above require chemical treatments of the probes, which may result in a significant loss of material, and hence neutralize the positive effect of oriented immobilization. Kusnezow et al. reported a loss of up to 40% of antibodies as a result of activation and purification, leading to similar signal/noise ratios as for nonactivated probes (54).

9. Advanced Materials

Protein biochips have to detect low target concentrations. In addition to optical techniques, such as planar waveguide, evanescent resonator platforms, integration of micro-optical elements, mirror slides, and optical interference coatings, strategies for enhancing the sensitivity of protein microarrays include substrates with increased surface area, allowing high density of probes in a highly ordered manner.

Brush polymeric coating based on a copolymer of *N*, *N*-dimethylacrylamide (DMA) and *N*, *N*-acryloyloxysuccinimide (NAS) were produced for the detection of allergen-specific immunoglobulins (55). Since IgE-binding epitopes are mostly conformational, it is imperative to maintain the native conformation of the immobilized allergens.

Nijdam et al. (56) used modified silicon as a substrate for reverse phase protein microarrays (RPMA), yielding protein binding comparable to nitrocellulose. Mixtures of proteins from cellular lysate were directly spotted onto silicon that was roughened by reactive ion etching and chemically functionalized using 3-aminopropyltriethoxysilane (APTES) and mercaptopropyltrimethoxysilane (MPTMS).

Coatings with calixarene derivatives for amine glass or gold were demonstrated to bind proteins in an oriented manner, yielding excellent sensitivity as low as 1–10 fg/mL of analyte (57). The authors proposed that the calixarenes, bifunctional affinity linkers, form a SAM which binds antibodies in the Fc region, stretching the antigen-binding sites to the solution phase. Oh and coworkers (58) found a Calixcrown chip to be 10–100 times as sensitive than aldehyde and carboxyl chip in a sandwich immunoassay for PSA.

Zhu et al. (17) manufactured microwells from poly(dimethylsiloxane) (PDMS) with an acrylic mold. Rectangular arrays of 18×28 mm, optimized for a protein kinase assay, consisted of 10×14 wells with a volume of 300 nl; but arrays of smaller dimensions, for high-throughput screening, could be produced using the most recent molding techniques. Resulting elastomer sheets were placed on microscope slides for handling purposes. For protein immobilization, PDMS was modified with 5 M H_2SO_4, 10 M NaOH, hydrogen peroxide, or 3-glycidoxypropyltrimethoxysilane (GPTS), the latter resulting in the greatest protein adsorption, namely, up to 8×10^{-9} μg/μm^2 HRP anti-mouse Ig.

References

1. Danczyk, R., Krieder, B., North, A., Webster, T., HogenEsch H., Rundell, A. (2003) Comparison of Antibody Functionality Using different Immobilization Methods. *Biotechnology and Bioengineerin.* **84**, 216–223.
2. Blawas, A. S., Reichert, W. M. (1998) Protein Patterning. *Biomaterials* **19**, 595–609.
3. Angenendt, P., Glökler, J., Murphy, D., Lehrach, H., Cahill, D. J. (2002) Towards optimized antibody microarrays: a comparison of current microarray support materials. *Analytical Biochemistry* **309**, 253–266.
4. Guilleaume, B., Buneß, A., Schmidt, C., Klimek, F., Moldenhauer, G., Huber, W., Arlt, D., Korf, U., Wiemann, S., Pouska, A. (2005) Systematic comparison of surface coatings for protein microarrays. *Proteomics* **5**, 4705–4712.
5. Seurynck-Servoss, S. L., White, A. M., Baird, C. L., Rodland, K. D., Zangar, R. C. (2007) Evaluation of surface chemistries for antibody microarrays. *Analytical Biochemistry* **371**, 105–115.
6. Rusmini, F., Zhong, Z., Feijen, J. (2007) Protein Immobilization Strategies for Protein Biochips. *Biomacromolecules* **8**, 1775–1789.
7. Angenendt, P., Gloekler, J., Sobek, J., Lehrach, H., Cahill, D. J. (2003) Next generation of protein microarray support materials: evaluation for protein and microarray applications. J. *Chromatography A* **1009**, 97–104.

8. Schaeferling, M., Schiller, S., Paul, H., Kruschina, M., Pavlickova, P., Meerkamp, M., Giammasi, C., Kambhampati, D. (2002) Appplication of self-assembly techniques in the design of biocompatible protein microarray surfaces. *Electrophoresis* **23**, 3097–3105.
9. Kusnezow, W., Hoheisel, J. D. (2003) Solid supports for microarray immunoassays. *J Mol Recognit* **16**, 165–176.
10. Jung, Y., Jeong, J. Y., Chung, B. H. (2008) Recent advances in immobilization methods of antibodies on solid supports. *The Analyst* **133**, 697–701.
11. Butler, J. E., Spradling, J. E., Suter, M., Dierks, S. E., Heyermann, H., Petermann, J. H. (1986) The Immunochemistry of Sandwich ELISAs – I. The Binding Characteristics of Immunoglobulins to Monoclonal and Polyclonal Capture Antibodies Adsorbed on Plastic and Their Detection by Symmetrical and Assymetrical Antibody-Enzyme Conjugates. *Mollecular Immunology* **23**, 971–982.
12. Domnanich, P., Sauer, U., Pultar, J., Preininger, C. (2009) Protein microarray for the analysis of human melanoma biomarkers. *Sensors and Actuators B* **139**, 2–8.
13. Kusnezow, W., Syagailo, Y. V., Goychuk I., Hoheisel, J. D., Wild, D. G. (2006) Antibody microarrays: the crucial impact of mass transport on assay kinetics and sensitivity. *Expert Rev. Mol. Diagn.* **6**, 111–124.
14. Vijayendran, R. A., Leckband, D. E. (2001) A Quantitative Assessment of Heterogeneity for Surface-Immobilized Proteins. *Anal. Chem.* **73**, 471–480.
15. Sauer, U., Preininger, C., Hany-Schmatzberger, R. (2005) Quick and simple: quality control of microarray data. *Bioinformatics* **21**, 1572–1578.
16. Culf, A. S., Cuperlovic-Culf, M., Ouellette, R. J. (2006) Carbohydrate microarrays:survey of facbrication techniques. *OMICS* **10**, 289–310.
17. Zhu, H., Klemic, J. F., Chnag, S., Bertone, P., Casamyor, A., Klemic, K. G., Smith, D., Gerstein, M., Reed, M. A., Snyder, M. (2000) Analysis of yeast protein kinases using protein chips. *Nature Genetics* **26**, 283–289.
18. Bilitewski, U. (2006) Protein-sensing assay formats and devices. *Analytica Chimica Acta* **568**, 232–247.
19. Butler, J.E., Ni, L., Nessler, R., Joshi, K.S., Suter, M., Rosenberg, B., Chang, J., Brown, W. R., Cantarero, L.A. (1992) The physical and functional behavior of capture antibodies adsorbed on polystyrene. *Journal of Immunological Methods* **150**, 77–90.
20. Urabowska, T., Mangialaio, S., Hartmann, C., Legay, F. (2003) Development of protein microarray technology to monitor biomarkers of rheumatoid arthritis disease. *Cell Biol Toxicol* **19**, 189–202.
21. Stillman, B. A., Tonkinson, J. L. (2000) FAST slides: a novel surface for microarrays. *BioTechniques* **29**, 630–635.
22. Yin, L. T., Hu, C. Y., Chang, C. H. (2008) A single layer nitrocellulose for fabricating protein chips. *Sens. Actuat. B* **130**, 374–378.
23. Reck, M., Stahl, F., Walter, J. G., Hollas, M., Melzner, D., Scheper, T. (2007) Optimization of a Microarray Sandwich-ELISA against hINF-y on a Modified Nitrocellulose Membrane. *Biotechnol. Prog.* **23**, 1498–1505.
24. Afanassiev, V., Hanemann, V., Wölfl, S. (2000) Preparation of DNA and protein micro arrays on glass slides coated with an agarose film. *Nucleic Acids Research* **28**, 12, e66.
25. Lv, L-L., Liu, B-C., Zhang C-X., Tang Z-M., Zhang L., Lu, Z-H.(2007) Construction of an antibody microarray based on agarose-coated slides. *Electrophoresis* **28**, 406–413.
26. Rubina, A. Yu., Dyukova, V. I., Dementieva E. I., Stomakhin A. A., Nesmeyanov V. A. Grishin, E. V., Zasedatelev A. S. (2005) Quantitavie immunoassay of biotoxins on hydrogel-based protein microchips. *Anal Biochem* **340**, 317–329.
27. Derwinska, K., Gheber, L. A., Sauer, U., Schron, L., Preininger, C. (2007) Effect of Surface Parameters on the Performance of IgG-Arrayed Hydrogel Chips: A Comprehensive Study. *Langmuir* **23**, 10551–10558.
28. Derwinska, K., Sauer, U., Preininger, C. (2008) Adsorption versus covalent, statistically oriented and covalent, site-specific IgG immobilization on poly(vinyl alcohol)-based surfaces. *Talanta* **77**, 652–658.
29. Zhou, Y., Andersson, O., Lindberg, P., Liedberg, B. (2004) Protein Microarrays on Carboxymethylated Dextran Hydrogels. *Microchimica Acta*, **147**, 21–30.
30. Akkoyun, A., Bilitewski, U. (2002) Optimisation of glass surfaces for optical immmunosensors. *Biosensors and Bioelectronics* **17**, 655–664.
31. Yadavalli, V. K., Koh, W.-G., Lazur, G.L., Pishko, M. V. (2004) Microfabricated protein-containing poly(ethylene) glycol hydrogel arrays for biosensing. *Sensors & Actuators B* **97**, 290–297.
32. Moorthy, J., Burgess, R., Yethirai, A., Beebe, D. (2007) Microfluidic Based Platform for characterization of Protein Interactions in Hydorgel Nanoenvironments. *Anal. Chem.* **79**, 5322–5327.

33. Dominguez, M. M., Wathier, M., Grinstaff, S. E., Schaus, S. E. (2007) Immobilized Hydrogels for Screening of Molecular Interactions. *Anal. Chem.* **79**, 1064–1066.
34. Zubtsov, D. A., Ivanov, S. M., Rubina, A. Y., Dementieva E. I., et al. (2006) Effect of mixing on reaction-diffusion kinetics for protein hydrogel based microchips. *J. Biotechnol.* **122**, 16–27.
35. Preininger, C., Sauer, U., Kern, W., Dayteg, J. (2004) Photoactivatable Copolymers of Vinybenzyl Thiocyanate as Immobilization Matrix for Biochips. *Anal. Chem.* **76**, 6130–6136.
36. MacBeath, G., Schrieber, S. L. (2000) Printing Proteins as Microarrays for High-Throughput Function Determination. *Science* **289**, 1760–1763.
37. Lu, B., Smyth, M. R., O'Kennedy, R. (1996) Oriented Immobilization of Antibodies and Its Applications in Immunoassays and Immunosensors. *The Analyst* **121**, 29R–32R.
38. Seong, S.-Y. (2002) Microimmunoassay Using a Protein Chip: Optimizing Conditions for Protein Immobilization. *Clinical and Diagnostic Laboratory Immunology* **9**, 4, 927–930.
39. Olle, E. W., Messamore, J., Deogracias, M. P., McClintock, S. D., Anderson, T. D., Johnson, K., J. (2005) Comparison of antibody array substrates and the use of glycerol to normalize spot morphology. *Experimental and Molecular Pathology* **79**, 206–209.
40. El Khoury, G., Laurenceau, E., Chevelot, Y., Mérieux, Y., Desbos, A., Fabien, N., Rigal, D., Souteyrand, E., Cloarec, J-P. (2010) Development of miniaturized immunoassay: Influence of surface chemistry and comparison with enzyme-linked immunosorbent assay and Western blot. *Anal. Biochem.* **400**, 10–18.
41. Cao, T., Wang, A., Liang, X., Tang, H., Auner, G. W., Salley, S. O., Ng, K.Y.S. (2007) Investigation of Spacer Length Effect on Immobilized *Escherichia coli* Pili-Antibody Molecular Recognition by AFM. *Biotechnology and Bioengineering* **98**, 1109–1121.
42. Pathak, S., Singh, A. K., McElhanon, J. R., Dentinger, P. M. (2004) Dendrimer-Activated Surfaces for High Density and High Activity Protein Chip Applications. *Langmiur* **20**, 6075–6079.
43. Yam, C. M., Deluge, M., Tang, D., Kumar, A., Cai, C. (2006) Preparation, characterization, resistance to protein adsorption, and specific avidin-biotin binding of poly(amidoamine) dendrimers functionalized with oligo(ethylene glycol) on gold. *J Colloid Interface Sci* **196**, 118–130.
44. Rowe, C. A., Scruggs, S. B., Feldstein, M. J., Golden, J. P., Ligler, F. S. (1999) An Array Immunosensor for Simultaneous Detection of Clinical Analytes. *Anal. Chem.* **71**, 433–439.
45. Bathia, S. K., Shriver-Lake, L. C., Prior, K. J., Georger J.H., Calvert, J. M., Bredehorst, R., Ligler, F. S. (1989) Use of thiol-terminated silanes and heterobifunctional crosslinkers for immobilization of antibodies on silica surfaces. *Anal. Biochem.* **178**, 403–413.
46. Cho, I-H., Paek, E-H., Lee, H., Kang, J. Y., Kim, T. S., Paek, S-H. (2007) Site-directed biotinylation of antibodies for controlled immobilization on solid surfaces. *Analytical Biochemistry* **365**, 14–23.
47. Peluso, P., Wilson, D.S., Do, D., Tran, H., Venkatasubbaiah, M., Quincy, D., Heidecker, B., Poindexter, K., Tolani, N., Phelan, M., Witte, K., Jung, L. S., Wagner, P., Nock, S. (2003) Optimizing antibody immobilization strategies for the construction of protein microarrays. *Anal. Biochem.* **312**, 113–124.
48. Khan, F., He, M., Taussig, M.J. (2006) Double-Hexahistidine Tag with High-Affinity Binding for Protein Immobilization, Purification, and Detection on Ni-Nitrilotriacetic Acid Surfaces. *Anal. Chem.* **78**, 3072–3079.
49. Muir, B. W., Barden, M. B., Collett, S. P., Gorse A.-D., Monteiro, R., Yang, L., McDougall, N. A., Gould, S., Maeji, N. J. (2007) High-throughput optimization of surfaces for antibody immobilization using metal complexes. *Anal. Biochem.* **363**, 97–107.
50. Boozer, C., Ladd, J., Chen, S., Yu, Q., Homola, J., Jiang, S. (2004) DNA Directed Protein Immoblization on Mixed ssDNA/Oligo(ethylene glycol) Self-Assembled Monolayers for Sensitive Biosensors. *Anal. Chem.* **76**, 6967–6972.
51. Niemeyer C. M., Boldt, L., Ceyhan, B., Blohm, D. (1999) DNA-Directed Immobilization: Efficient, Reversible, and Site-Selective Surface Binding of Proteins by Means of Covalent DNA-Streptavidin Conjugates. *Anal. Biochem.* **268**, 54–63.
52. Wacker, R., Schröder, H., Niemeyer, C. M. (2004) Performance of antibody microarrays fabricated by either DNA-directed immobilization, direct spotting, or streptavidin-biotin attachment: a comparative study. *Anal. Biochem.* **330**, 281–287.
53. Shriver-Lake, L., Donner, B., Edelstein, R., Breslin, K., Bhatia, S. K., Ligler, F. S. (1997) Antibody immobilization using heterobifunctional crosslinkers. *Biosensors & Bioelectronics* **12**, 1101–1106.

54. Kusnezow, W., Jacob, A., Walijew, A., Diehl, F., Hoheisel, J.D. (2003) Antibody microarrays: An evaluation of production parameters. *Proteomics* **3**, 254–264.

55. Cretich, M., Di Carlo, G., Giudici, c., Pokoj, s., Lauer, I., Scheurerand, S., Chiari, M. (2009) Detection of allergen specific immunoglobulins by microarrays coupled to microfluidics. *Proteomics* **9**, 2098–2107.

56. Nijdam, J. A., Cheng, M. M. C., Geho, D. H., Fedele, R., Herrmann, P., Killian, K., Espina, V., Petricoin III, E. F., Liotta, L. A., Ferrari, M. (2007) Physicochemically modified silicon as a substrate for protein microarrays. *Biomaterials* **28**, 550–558.

57. Lee, Y., Lee, E. K., Cho Y. W., Matsui, T., Kang, I-C., Kim, T-S., Han M. H. (2003) ProteoChip: A highly sensitive protein microarray prepared by a novel method of protein immobilization for application of protein-protein interaction studies. *Proteomics* **3**, 2289–2304.

58. Oh, S. W., Moon, J. D., Lim, H. J., Park, S. Y., Kim T., Park, J. B., Han, M. H., Snyder, M., Choi, E. Y. (2005) Calixarene derivative as a tool for highly sensitive detection and oriented immobilization of proteins in a microarray format through noncovalent molecular interaction. *The FASEB Journal* **19**, 1335–1337.

Chapter 25

Contact Printing of Protein Microarrays

John Austin and Antonia H. Holway

Abstract

A review is provided of contact-printing technologies for the fabrication of planar protein microarrays. The key printing performance parameters for creating protein arrays are reviewed. Solid pin and quill pin technologies are described and their strengths and weaknesses compared.

Key words: Protein arrays, Antibody arrays, Lysate arrays, Reverse-phase arrays, Spotted arrays, Arrayers, Contact printing, Solid pins, Quill pins

1. Introduction

Microarray-based assays have brought high-throughput processing techniques to the field of molecular biology. Oligonucleotide arrays have become widely accepted as a standard means of performing multiplexed genomic studies. From this beginning in the genomic world, microarray applications have expanded rapidly and a wide range of materials can now be examined in a high-throughput, multiplexed format, including DNA, micro-RNA, small molecules, antibodies, recombinant proteins, carbohydrates, lipids, and lysates of cultured cells, tissue, plasma, sera, urine, synovial fluid, vitreous humor, and other bodily fluids. Similarly, the types of media upon which microarrays are fabricated have expanded rapidly, from the early coated glass microscope slides to semiconductor and gold surfaces, plate-glass, membranes, gels, cartridges, printed circuit boards, and microtiter plates.

Whereas oligo microarrays are now fabricated primarily using in-situ synthesis methods, this is certainly not true for protein arrays. Although some in-situ protein array fabrication techniques have been developed and show great promise (1), the great majority of

Ulrike Korf (ed.), *Protein Microarrays: Methods and Protocols*, Methods in Molecular Biology, vol. 785, DOI 10.1007/978-1-61779-286-1_25,

protein arrays at present are "spotted" arrays deposited on planar surfaces using presynthesized proteomic samples.

"Spotted" microarrays were developed at the advent of genomic microarray technology (2–6). Even in the early days of genomic microarrays, it was clear that there was no single array fabrication technology that met all requirements, since the early manufacturers of microarrayers adopted a range of technical solutions, including split pins, capillary pins or pens, capillary tubes, solid pins, and various jetting technologies. The fact that a wide range of technical arraying solutions still exists to this day is clear evidence that no single technology has been able to meet the needs of all microarray users.

Spotted microarray printing systems can be broadly divided into two categories: contact printing systems using pins, pens, or related devices that touch, or almost touch, the surface of the microarray substrate, and jetting systems that form and propel droplets toward the substrate without making physical contact. Both technologies have their relative advantages and disadvantages and both technologies can provide high printing speeds and excellent printing consistency.

Contact printing systems have the advantage of a straightforward droplet forming technology and relatively simple wash systems to clean the deposition element between samples. Jetting technologies, which can use techniques for droplet formation/propulsion, such as piezoelectric technology, syringe pumps, and acoustic focusing, are more technically complex and require more challenging wash systems, but when operating as designed, they can offer high-speed arraying and consistent spot formation for some protein sample types. However, the extremely wide range of proteomic samples and their associated buffer properties can pose a significant challenge. Compared to contact printing systems, inkjet systems are quite constrained in terms of the viscosity range that they can handle; in fact, optimization of the droplet dispensing parameters (voltage, frequency, pressure, etc.) can be required even between samples of the same type because of very small differences in fluid properties. Whereas the noncontact feature of inkjet systems is obviously advantageous when delicate surfaces are used in protein array applications, it should be noted that some modern contact printers, particularly those using solid pin printing technology, routinely print onto extremely delicate substrates with no effect on the binding surface. With regard to compatibility with the wide range of surfaces used for protein arrays, it should also be noted that inkjet systems, in particular, require the deposition surfaces to be relatively hydrophobic in order to control the final droplet shape following its fairly high-speed impact on the surface.

The various performance parameters by which protein microarraying technologies must be compared cover a wide range. The relative emphasis placed on these parameters for a specific application determines which printing technology is best suited for each individual application. These performance parameters include the ones described in Sections 1.1 to 1.11 below.

1.1. Gridding Accuracy

Microarrays are commonly arranged as spots on a square or rectangular grid, sometimes with offsets between adjacent rows or columns. The accuracy with which the spots are deposited directly affects the processing of the image that is obtained from the array. Microarray analysis software may be unable to locate a spot if its placement on the substrate is too far from the expected position. High-precision robotic motion control is, therefore, a common requirement for all spotted-array production methods, since all spotted array fabrication techniques require relative motion between the "deposition head" that supplies the biological material to the substrate and the substrate onto which it is to be deposited.

Modern arrayers use either open-loop stepper motors or, preferably, servo controlled motors (linear or stepper) with feedback control to position the deposition head (the device holding the deposition pins, pens, or inkjet dispenser) over the microarray substrate. Most arrayers provide position accuracy of the deposition head of a few tens of microns, with the more advanced instruments achieving micron or submicron positioning accuracy. However, there are additional factors that can lead to spot position errors; if these factors are outside the positioning "control loop," even a servo positioning system will be unable to correct for them. For jetting systems, these factors can include variability in the droplet's trajectory once it is ejected from the jetting hardware; such variability typically results from effects, such as surface tension between the forming droplet and dried material built-up at the jetting orifice, electrostatic attraction or repulsion of the droplet, or air currents encountered during the passage of the droplet between the orifice and the substrate. Contact arrayers are required to provide compliance of the pin or pen to allow it to move upward once the tip of the device has made contact with the surface of the substrate. For contact printing systems, "out of the loop" position errors can be generated by inconsistent seating of the deposition element (pin or pen) and/or rotation of the pin in its seat if it is not straight.

1.2. Arraying Flexibility

In addition to simple gridding precision, array layout flexibility may be of importance. There are growing numbers of applications of spotted protein microarrays which require the printing of spots in specific or unusual patterns or in a specific order. Examples include:

1. Small arrays used for diagnostic applications, printed on unusually sized or unusually shaped substrates.

2. Printing of reverse phase protein arrays, in which samples are often printed such that the spots of each dilution series are printed linearly or at least within close proximity of each other, to facilitate visual examination and analysis.
3. Repeated printing of the same material onto the same spot location to build up signal on absorbent substrates.
4. Printing of different materials on the same spot location on the array to effect chemical reactions within the spot.

1.3. Substrate Type

Because of the wide range of assay types and binding surfaces for protein arrays, consideration should be given to the ease with which different or unusual substrate types/shapes may be accommodated by the arraying system. An example is the increasing common requirement for the printing of arrays into the wells of well plates.

1.4. Printing Speed and Overall Throughput

The relative importance of printing speed depends on the number of spots to be deposited on each substrate and the number of substrates to be printed within a given time period. Protein microarrays rarely have the very large spot numbers that are common in genomic applications. Demands for high printing speeds in protein array applications are more likely to be related to production quantities (e.g., high-volume production of diagnostic protein arrays containing a small number of features), or may be related to biological material issues (e.g., minimization of sample exposure time and evaporation).

1.5. Consistency of Printing

Printing consistency relates to (a) the reliability of the printing (reflecting how often spots that should have been printed on the array are missing), (b) the consistency of spot size and signal intensity, and (c) the consistency of the spot shape. Poor spot morphology can certainly be caused by faulty deposition, but other factors, such as buffer properties, surfactants, drying conditions, substrate uniformity, and substrate hydrophobicity or hydrophilicity, can also play a significant role. Examples of spot morphology issues that can be related to fluid drying properties include the so-called doughnut spots with a strong outer band but a weak center region, or “bulls-eye” spots with a weak outer region and a bright center. When using salt-based buffers, fine-grain differences are always seen within the dried spots because of the random initiation of crystal formation after the fluid becomes super-saturated. Microenvironment effects on arrays are also common, irrespective of the arraying technology, since spots at the edge of the array can experience an asymmetric humidity environment compared to spots within the array; this effect can lead to spots that dry faster on one side and therefore show asymmetry in signal intensity.

1.6. Viscosity Range

The viscosity range of the buffer materials to be used is a major determinant in choosing an appropriate protein arraying technology because of the wide range of buffers used in the field. This can be of particular concern for inkjet systems.

1.7. Sample Volume

Unlike genomic materials that are often reproduced at will, proteomic samples are sometimes available in very limited quantities. Consideration should, therefore, be given to the minimum sample volume that must be provided to the arrayer and the amount of sample volume that is consumed or is nonrecoverable following printing. As microarraying technologies have advanced, the sample source plates have increased in density from 96 wells to 384 and even 1,536 wells. The ability to use the small well size of 1,536 plates may be of advantage in reducing required sample volumes in some protein array applications.

1.8. Sample Evaporation

For relatively volatile buffers, the change in protein concentration within the well plate due to evaporation, during the course of the printing run, should be taken into account. This may, in turn, influence the required printing speed, depending on the number of features per substrate and the number of substrates to be printed.

1.9. Surface Damage

Substrate coatings used to bind protein samples are sometimes more delicate than the traditional silanes, aldehydes, etc., used for genomic arrays. Examples of such delicate substrates include nitrocellulose-coated slides and 3-D gel-coated slides. If these coatings are to be used, the printing technology must not damage or change the binding properties of the surface. Whereas this was once a major benefit of inkjet printing, some modern contact printers can similarly print on such delicate materials without affecting their binding performance.

1.10. Sample Carryover

The ability to effectively remove sample materials from the deposition device during the wash cycle is essential to avoid sample carryover from spot to spot. Proteins can adhere to the surface of the deposition device, and the wash protocol must efficiently remove them from all surfaces, including recessed areas or internal structures.

1.11. Sample Cooling

Experience has shown that the great majority of protein samples can be arrayed successfully at room temperatures. For the relatively small number that actually require continuous low-temperature conditions, a means of sample handling and printing at low temperatures is required.

2. Contact Printing Technologies

2.1. Background and Origins

Contact printing probably originated from an extension of the function of solid pin "replicators," which are used to replicate samples between well plates. Such replicators house a bank of solid pins, the tips of which are immersed together in sample fluids and then manually touched, en masse, onto the receiving medium. These replicators are commercially available and are still in use today.

Genetic analysis using modern contact printing arose from the work of Pat Brown and Ron Davis at Stanford (2). Pat Brown and coworkers developed arraying technologies, including split deposition pins (6) which held a reservoir of sample between the arms of the pin; a small quantity of sample could, therefore, be carried with the pin, thereby avoiding the need to revisit the source well plate after every deposition. Such "split pins" typically needed to be "tapped" onto the surface of the receiving substrate to overcome surface tension and eject a small amount of sample from the reservoir onto the surface. The Brown Lab also developed an affordable design for robotically moving the pins over the source wells (microtiter plate) and a bank of substrates, thereby facilitating high-volume printing and avoiding the inconsistencies of manual droplet deposition. The Brown Lab even published detailed instructions and parts lists for other researchers to build similar arrayers of their own (7).

Commercial versions of such "gantry" arrayers soon followed (the use of the word gantry meaning that the head containing the deposition pins is moved over the substrates, in either one or two axes), including offerings by BioRobotics{R} in the UK and GeneMachines{R} in the USA. The early split pins were soon replaced by more sophisticated designs producing much more consistent spotting, the manufacturing technology for which was pioneered by TeleChem/Array-It.com{R} based on electrical discharge machining (EDM). Using this manufacturing technology, pins with various tip sizes and reservoir capacities can be fabricated repeatably. These pins are known in the literature by various names, including quill pins, capillary pens, and microspotting pins. The term quill pins is used in this chapter.

Contact printers using various capillary tube technologies have also been developed. These are briefly reviewed later in this chapter; however, the focus of the chapter is on solid pin and quill pin applications since these are the dominant technologies used in present protein array applications.

2.2. Solid Pin Printing

Among the benefits of solid pin printing is the relative simplicity of the physical structure of the pin compared to quill pins (however, it should be recognized that any manufactured part with dimensions of a few hundred microns or less is very challenging to produce and requires advanced machining technology). Solid pins tend to be somewhat more robust since there is no fine structure (the slit and reservoir of quill pins) contained within the diameter of the tip. The method (simple surface affinity and surface tension) whereby solid pins capture and carry the fluid droplet on the tip of the pin makes solid pins the most versatile in dealing with a wide range of sample viscosities – a particularly important parameter for many protein array applications. Aushon BioSystems{R}, for example, has demonstrated the ability of solid pins to successfully create high-quality arrays of materials with a wide viscosity range, including lysates in SDS, urea and beta-mercaptoethanol buffers, small molecules in DMSO (up to and including 100% DMSO), proteins in glycerol buffers (up to 80%), carbohydrates in PBS-based buffers, and various bodily fluids, such as serum and plasma.

Solid pins are typically made with a flat tip. The volume of fluid transported by a solid pin is dependent on whether fluid is carried on the tip of the pin alone, or if sample adheres to the side of the pin in addition to the tip. In either case, the fluid volume extracted from the source well of a microtiter plate is clearly related to the diameter of the pin, and so the manufacturing tolerance on the pin diameters directly correlates with pin-to-pin spot size variability. In a simplistic example, if one considers a hemisphere of fluid captured only on the tip of a flat-bottomed pin of radius R, then the fluid volume extracted is $(1/2 \times 4/3 \times \pi \times R^3)$. The extreme importance of consistently matched pin diameters should therefore be apparent, since the fluid volume extracted is related to the cube of the pin diameter. Fortunately, modern manufacturing techniques allow solid pin diameters to be matched to a few microns. Aushon BioSystems{R} (8) has developed noncontact machining technology that allows pins to be made with tolerances orders of magnitude less than a micron. Solid pins manufactured by Aushon BioSystems{R} are shown mounted in a printhead in Fig. 1.

For precious or low-abundance samples, an advantage of solid pin printing is that material wastage is minimized, since the pin only picks up the material that is to be immediately deposited. In contrast, any sample left over in the reservoir of a quill pin is irrecoverable and is lost during the wash cycle.

The wash cycle stringency for solid pin printing is often less severe than that for other technologies because the surfaces exposed to sample are all on the exterior of the pin and are readily accessible to the wash process.

A very important feature of solid pin printing is that the deposition conditions for printing the first spot can be exactly the same as for any other spot on the array, in that there is no variation in deposited volume related to the fill state of a reservoir, as occurs in quill pin and capillary tube printing. This same advantage, however, also underlies a historical limitation of solid pin printing in that there is a requirement for the pin to revisit the sample well

Fig. 1. Aushon BioSystems solid pins mounted in a 48-pin printhead at 4.5-mm spacing (reproduced by permission of Aushon BioSystems Inc.).

after each and every deposition, which can lead to excessive travel of the pin head and lower printing speeds. However, this limitation has been minimized in recent solid pin protein arrayers (9) by the use of motion architectures specifically designed around the use of solid pins and by providing for large numbers of pins to be used in parallel – potentially up to hundreds or, in some implementation, even thousands at once.

Genetic Microsystems{R} (later acquired by Affymetrix{R}) developed an interesting variation on the use of solid pins based on the pin-and-ring design. This hybrid technology, which not only used a solid deposition pin, but also carried a sample reservoir with the pin, is well-described elsewhere in the literature (10). In the Genetic Microsystems{R} design, an aliquot of sample is captured in a ring positioned under, and moving at all times with the pin. The solid pin is lowered through the sample meniscus held within the ring, to deposit the spot on the substrate. Limitations on the design, however, include relatively low-sample volume transfer to the substrate, high-sample usage and restrictions on the use of small wells because of the size of the ring.

Various solid pin tip geometries have been developed for controlling the fluid droplet shape that is carried on the pin and for maximizing the droplet volume that is carried. For example, MiraBio{R} (Hitachi Software{R}) developed a pin with a cross-shaped indentation on the tip to maximize fluid volume (11).

Pins have been fabricated from a range of different materials; however, most manufacturers use stainless steel or tungsten for strength and durability.

Evaporation of the fluid droplet captured by a solid pin prior to deposition can be an issue when working in evaporative conditions and when using volatile sample buffers. Note that the volume of a fluid droplet carried by a typical solid pin of 100-μm diameter is only in the range of a few hundred picolitres and so evaporation of the droplet can occur quickly, even within a few seconds. The gantry-based robotic architecture of many commercial contact arrayers requires that the printhead and its deposition pins travel over a large bed on which slides are mounted. Most arrayers move the pin head in either three axes (this applies to a horizontal fixed platen on which the slides are mounted), or in two axes (this applies to the case where the slide platen is moved in one axis and the pin head is moved in the other two axes). In either case, there is a significant difference in the length of printhead travel from the microtiter plate to the closest slide versus the furthest slide. Under strongly evaporative conditions, this can lead to both variation in the volume of fluid transferred and variation in the sample concentration between substrates. However, refinements to the motion architecture of recent solid pin arrayers (9) have been able to eliminate sample evaporation as a source of variability both within and between substrates.

Protein arrays tend to be fabricated on a wide range of binding surfaces. Although they have been shown to work on hydrophobic coatings, solid pins are more suited to substrates with lower contact angles (i.e., not severely hydrophobic) to aid the transfer of the materials from the pin tip onto the substrate. Unlike other deposition technologies, including inkjets and quill pins, solid pins also work well with hydrophilic surfaces.

Figure 2 shows example arrays printed using the Aushon BioSystems 2470 Microarrayer{R} using solid pins. Figure 2a shows an example of spot size and shape consistency with bovine serum albumin (BSA) and a test dye printed onto a nitrocellulose-coated slide using different pin sizes. Figure 2b shows a close-up fluorescent image of a dilution series of cell lysates in SDS buffer printed in quadruplicate onto a nitrocellulose-coated slide (Grace BioLabs{R}) in a reverse phase format. Such reverse phase lysate arrays, reviewed in other chapters of this book, show much potential for enabling high-throughput proteomic analysis (12–17). Figure 2c shows a fluorescent image of an antibody array printed in 96-well microtiter plate format. Figure 2d shows an antibody array printed onto a 3-D hydrogel surface.

2.3. Split and Quill Pins

Contact printing using split or quill pins has been the most commonly used technology for do-it-yourself genomic arraying on glass slides because of the cost and printing-speed advantages.

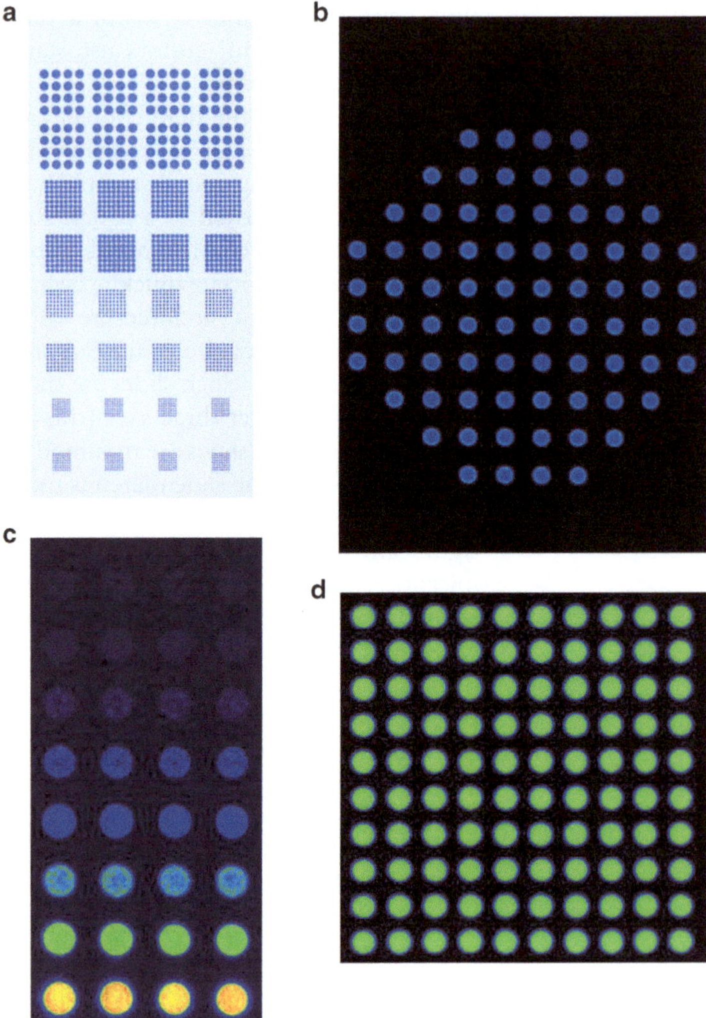

Fig. 2. Protein arrays examples printed with solid deposition pins on the Aushon BioSystems 2470 Arrayer. (**a**) BSA plus bromophenol-blue dye printed onto a Grace BioLabs{R} nitrocellulose-coated slide using, from the top, 350-, 185-, 110-, and 85-μm pin diameters (SDS buffer). (**b**) Antibody array printed in 96-well microtiter plate format (spots in one well only shown). Spot diameter 190 μm. (**c**) Dilution series of cell lysates in SDS buffer printed in duplicate onto a nitrocellulose-coated slide (Grace BioLabs{R}) in a reverse phase format. Spot diameter 330 μm. (**d**) Antibody array printed onto a Schott Slide H [R] hydrogel surface. Spot diameter 180 μm (reproduced by permission of Aushon BioSystems Inc.).

Quill-pin arrayers are also now routinely used for successfully producing a range of protein and antibody arrays.

From the early days of split pins, quill pins have made dramatic improvements as advanced pin manufacturing methods were introduced. Early split pins required a substantial impact on the surface of the substrate to eject the sample fluid, but that is not required with modern quill pins. Telechem/Arrayit.com{R} (18) developed a patented manufacturing method using EDM technology and today a range of pin tip diameters are available, as are pins with a range of different reservoir sizes (the use of a larger reservoir allows a greater number of spots to be printed between recharges of the pin). Telechem{R} pins have a square cross-section at the tip.

Parallel Synthesis{R} (19) also makes a range of quill pins, in this case using lithographic techniques to etch the pins from silicon wafers. Because of the precision of this manufacturing technique, highly reproducible pins can be produced. Example quill pins made by Telechem/Arrayit.com{R} and Parallel Synthesis{R} are shown in Fig. 3.

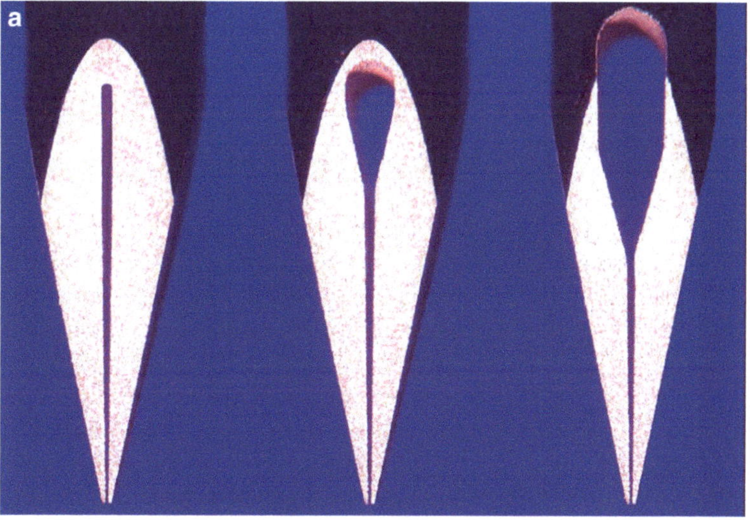

b

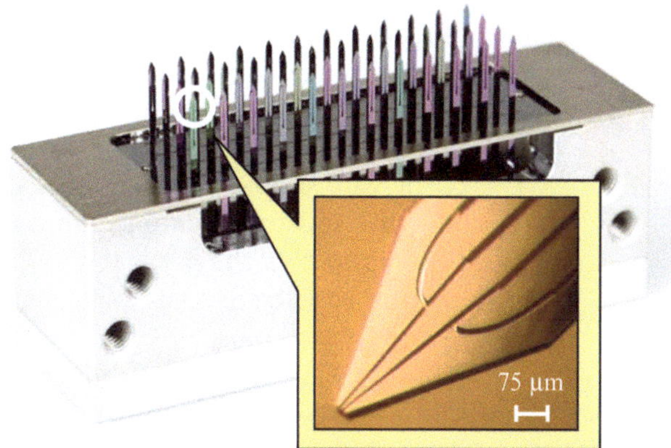

Fig. 3. Examples of Quill printing pins. (**a**) Arrayit Microspotting (Quill) Pins with three different reservoir sizes (image provided by Arrayit.com, copyright 2010 Arrayit Corporation, all rights reserved). (**b**) Parallel Synthesis Technology's Silicon Quill Pins and Printhead (reproduced by permission of Parallel Synthesis Technologies Inc.).

Quill pins are initially dipped into the sample fluid, typically contained within the well of a microtiter plate, at which time sample enters the reservoir via capillary action. It is common for there to be an excess of fluid on the tip immediately after dipping, and therefore "blotting" of the excess material onto a waste slide is sometimes required. Once the excess is removed, sample fluid flows continuously down from the reservoir to the tip as the pin moves rapidly from spot to spot across substrates. With some quill pin designs, it is possible to produce several hundred spots before a recharge of the fluid reservoir is required, although a very slight change in spot size and volume can occur as the reservoir discharges (20). It is inadvisable to print with a reservoir that is close to being fully discharged since partially formed or missing spots may result. Evaporation of the sample carried by the pin also occurs with quill pins, but the effect is less severe than that for solid pins since the sample is carried in a reservoir that is at least partially protected from the air.

Since the reservoir of quill pins is directly connected to the tip, such that continuous flow from the reservoir to the substrate can occur (wicking), the use of relatively hydrophobic substrates is preferred to minimize spot size variability.

Wash protocols for quill pins are similar to those for solid pins, except that ultrasonic baths are sometimes used to ensure sample removal from inside the fine slit in the pin (note that ultrasonic baths are not recommended for silicon pins). A drying step is also required to remove the wash fluid that is unavoidably trapped in the reservoir and slit after washing. Remarkably, silicon pins provide another option for infrequent but thorough cleaning; the pins are sufficiently inert and the melting point is sufficiently high that contamination can be removed from the tips by heating them with a propane torch!

Modern quill pins have been optimized for specific viscosity ranges (often associated with typical sample buffers used for genomic materials) and particular care should be taken to ensure that unusual or viscous protein sample buffers are compatible with these pins. Since clogging of the narrow slit of quill pins can easily occur, care should be taken to ensure that samples are well-filtered and that substrate coatings are not likely to be dislodged during printing to clog the pin.

Figure 4 shows an example protein arrays printed using Parallel Synthesis quill pins.

2.4. Other Contact Printing Technologies

Other contact printing technologies for protein arrays have been developed: they are mentioned here briefly for completeness, but they have not been widely adopted to date and so are not described in detail. More details may be found at the references cited.

Molecular Dynamics{R} (later purchased by Amersham Pharmacia{R} and subsequently by General Electric Healthcare{R})

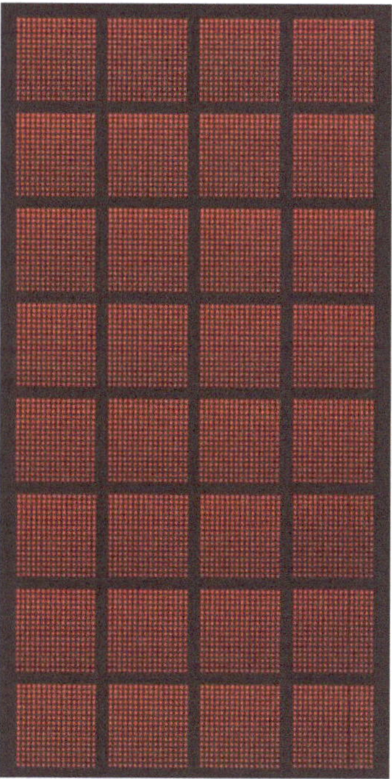

Fig. 4. Protein array example using Parallel Synthesis Quill pins. CY2 {R} labeled antibodies printed using 75 μm by 75 μm Parallel Synthesis {R}silicon pins on GE Healthcare FAST{R} nitrocellulose-coated slides. The spots are ~140 μm in diameter spaced on 250 μm centers (reproduced by permission of Parallel Synthesis Technologies Inc.).

developed a contact deposition technology for genomic arraying based on stainless steel capillary tubes, each of which had a slit up one side to act as a vent and to facilitate cleaning. This technology is described in detail in Schena (10). The technology is rarely used now since it is no longer commercially supported.

An arraying system using ceramic pins with a capillary at the center has been developed by NextLab{R} (21), based on earlier pin technology developed by Matrix{R}/Apogent Discoveries{R}. Continuous tube-based sample replenishment (from a remote source) is possible with some versions of these pins. A capillary pin system using glass capillaries has also been developed by UCSF (22). In another variant on capillaries, a very high-capacity "pin" using a spring-mounted hollow "needle" that is fed from a larger reservoir within the body of the pin has been developed by Parallel Synthesis (19).

Whereas most protein arrayers produce spots with diameters in the range of 50 μm to several millimeters, much smaller spot diameters, of the order of a few μm, can also be created using nanoscale or microelectromechanical (MEMs) technology. The utility of such miniaturized protein arrays is still to be determined; one clear obstacle to their adoption is the requirement for users to acquire scanners compatible with these spot sizes. However, some potentially interesting applications exist in the field of cell capture, since it may be possible to create arrays of cells using such technology. NanoInk{R} has arraying products based on dip-pen technology using solid tips (23). Cantilevered nanopipettes have also been described in the literature (24).

3. Considerations Common to all Contact Printing Technologies

3.1. Deposition Element Dynamics

The rate of motion of the pins/pens used in contact printing, within certain parts of their motion profiles, can affect arraying performance. The dwell time of quill pins within a source well can affect the amount of fluid captured within the reservoir. For both solid and quill pins, the rate of withdrawal of the pin from the source fluid can affect the manner in which the sample fluid is shed from the sides of the pin. In general, a slower withdrawal is preferred. The speed with which the deposition pin touches the surface can also affect performance. When used for printing onto delicate surfaces, such as nitrocellulose films and 3-D gels, it is important that the pressure exerted by the pin on the surface be minimized. When the pins lift out of their seats during deposition, the gravitational component of the pin pressure applied to the surface is related to the weight of the pins and the surface area of the tip of the pin. By dramatically slowing the pin head as it approaches the substrate, the addition of momentum-related pressure can be avoided.

3.2. Printing Environment

Control of the printing environment is crucial for achieving high-quality arraying. To make high quality, repeatable arrays, all sources of printing variability need to be understood and controlled, and this includes the environmental conditions within the arraying system.

3.2.1. Temperature

Temperature fluctuations during printing should be avoided since this may change fluid viscosities, drying rates, and reaction rates. Fortunately, most modern laboratories have climate control systems (heat and air-conditioning) that will maintain laboratory temperatures within acceptable bounds for protein microarrays. If protein samples must be chilled during arraying, however, additional provisions are needed. A common misconception is that chilled operation is required to reduce sample evaporation. While temperature is certainly a factor in evaporation rates, the relative humidity in the storage environment and the minimization of the time with the lid off the well plate are more important parameters. Some arrayer manufacturers have provided chilled plate storage and handling; however, it should be noted that the existence of chilled elements within a humidified chamber can be counterproductive.

The chilling will result in significant condensation on the sample plates and such condensation will in turn cause fluctuations in humidity levels. The best solution for printing arrays of temperature sensitive proteins is to perform the printing in a chilled room. Portable chilled enclosures are not prohibitively expensive nor complicated to install.

3.2.2. Humidity

Maintaining consistent and controllable humidity is similarly important for maintaining arraying consistency. In general, operating contact printers at relatively high-humidity levels (40–80%) is desirable (a) to maintain the spot in a liquid phase to allow time for efficient binding to the substrate, (b) to minimize evaporation from the deposition pins, and (c) to minimize sample evaporation from the well plates. The actual operating humidity that can be used in practice, however, may be determined by the allowable humidity specification of the selected substrate.

3.2.3. Air Filtering

The air in relatively clean laboratories can still contain significant quantities of microscopic airborne dust. A number of protein contact printers provide built-in HEPA filters to remove dust entering the instrument, but that obviously does not address sample and slide preparation and processing steps outside the instrument. Ideally, arraying operations should be performed within a clean room environment to avoid these problems.

4. Summary

When compared to inkjet approaches, contact printing generally provides a more robust solution for research applications because of the wide range of sample types, buffers, and surfaces used for protein microarrays. Inkjet arrayers can create excellent reproducibility and low coefficients of variation for a limited range of protein array applications; these applications are typically centered on the use of hydrophobic substrates and the use of lower viscosity buffers or buffers with very consistent fluid properties. Jetting systems may be required where noncontact deposition is imperative, or for extremely hydrophobic surfaces. Contact printing using quill pins or solid pins generally supports a wider range of sample and substrate options for creating protein arrays. Modern contact arrayer implementations can provide high printing speeds and low coefficients of variation in protein printing. Quill pin arraying generally offers high printing speeds for the same number of pins used, but is limited in the range of buffer viscosities that can be used. Solid pins are able to accommodate a wider range of sample viscosities and substrate types, and, in some arrayer implementations, can provide lower spot-to-spot and substrate-to-substrate variability.

References

1. Ramachandran, N., Hainsworth, E., Bhullar, B., Eisenstein, S., Rosen, B., Lau, AY., Walter, J. C., LaBaer, J. (2004) Self-assembling protein microarrays, Science, Jul 2, 305(5680), 86–90.
2. Schena, M., Shalon, D., Davis, R. W., Brown, P. O. (1995) Quantitative monitoring of gene expression with a complementary-cDNA microarray, Science, Vol. 270, Page 467–470.
3. Schena, M., Shalon, D., Heller, R., Chai, A., Davis, R. (1996) Parallel human genome analysis: microarray-based expression monitoring of 1000 genes, PNAS, Oct 1, Vol. 93, No. 20, 10614–10619.
4. Brown, P., Botstein, D., Exploring the new world of the genome with DNA microarrays, Nat. Genet. Suppl., 21:33–37.
5. Cheung, V., Morley, M., Aguilar, F., Massimi, A., Kucherlapati, R., Childs, G., Making and reading microarrays, Nat. Genet. (Suppl.), 21:15–19.
6. Brown, P., Shalon, D., Methods of fabricating microarrays of biological substances, (1998), U.S. Patent 5,807,522.
7. Available at the Brown Lab website at Stanford: cmgm.stanford.edu/pbrown/mguide/index.html.
8. Aushon BioSystems website: www.aushon.com
9. Aushon BioSystems website: http://www.aushon.com/2470-Arrayer.php.
10. Schena, M. (editor), Microarray Biochip technology, (2000), Eaton Publishing; ISBN 1-88299-37-6.
11. Hitachi Software Engineering America website: http://www.miraibio.com/download-document/spbio-ii-spotting-pin-evaporation-1.html.
12. Mundinger, G., Espina, V., Liotta, L., Petricoin, E., Clinical phosphoproteomic profiling for personalized targeted medicine using reverse phase protein microarrays, (2006) Targ. Oncol., DOI 10.1007/s11523-006-0025-2, Springer-Verlag.
13. Wulfkuhle, J., Edmiston, K., Liotta, L., Petricoin, E. (2006) Technology Insight: pharmacoproteomics for cancer – promises of patient-tailored medicine using protein microarrays, Nature Clinical Practice – Onocology, May, Vol 3, No 5.
14. Nishizuka, S., Spurrier, B., Honkanen, P., Austin, J., Wakabayashi, G. (2008) Application of quantitative proteomic analysis for cancer therapy using "reverse-phase" protein lysate microarrays, Cancer & Chemotherapy, 35(2):200–5.
15. Nishizuka, S., Spurrier, B. (2008) Experimental validation for quantitative protein network models, Current Opinion in Biotechnology,19(1):41–9.
16. Spurrier, B., Honkanen, P., Holway, A., Kumamoto, K., Terashima, M., Takenoshita, S., Wakabayashi, G., Austin, J., Nishizuka, S. (2008) Protein and lysate array technologies in cancer research, Biotechnology Advances, 26(4):361–9.
17. Spurrier, B., Ramalingam, S., Nishizuka, S. (2008) Reverse-phase protein lysate microarrays for cell signaling analysis, Nature Protocols 3, 1796–1808.
18. Array-It Corporation website: http://arrayit.com/
19. Parallel Synthesis website: http://www.parallel-synthesis.com/.
20. Gulati, R. (2004) Testing and establishing optimal protein array conditions to be applied towards the DARPP-32 Phosphoprotein model system, University of New Haven, MD 690, Research Project, Spring.
21. LabNext website: http://www.labnext.com/xtend_MP_microarray_pins.htm.
22. Clark, S., Hamilton, G., Nordmeyer, R., Uber, D., Cornell, E., Brown, N., Segraves, R., Davis, R., Albertson, D., Pinkel, D. (2008) High-efficiency microarray printer using fused-silica capillary tube printing pins, Analytical Chemistry, 80(19):7639–42.
23. NanoInk website: http://www.nanoink.net/.
24. Taha, H. et al. (2003) Protein printing with an atomic force sensing nanofountainpen, Applied Physics Letters, Volume 83, 1041–1043.

Index

A

B

C

D

E

Ulrike Korf (ed.), *Protein Microarrays: Methods and Protocols*, Methods in Molecular Biology, vol. 785,
DOI 10.1007/978-1-61779-286-1, © Springer Science+Business Media, LLC 2011

T

U

V

W

MIX
Papier aus verantwortungsvollen Quellen
Paper from responsible sources
FSC® C105338

If you have any concerns about our products,
you can contact us on
ProductSafety@springernature.com

In case Publisher is established outside the EU,
the EU authorized representative is:
Springer Nature Customer Service Center GmbH
Europaplatz 3, 69115 Heidelberg, Germany

Printed by Libri Plureos GmbH
in Hamburg, Germany